Generalized Analytic Functions

Generalized Analytic Functions

Theory and Applications to Mechanics

Edited by

Helmut Florian
Technical University, Graz

Klaus Hackl
Technical University, Graz

Franz Josef Schnitzer
Mining University, Leoben

and

Wolfgang Tutschke
Technical University, Graz

KLUWER ACADEMIC PUBLISHERS
DORDRECHT / BOSTON / LONDON

A C.I.P. Catalogue record for this book is available from the Library of Congress.

ISBN 0-7923-5043-X

Published by Kluwer Academic Publishers,
P.O. Box 17, 3300 AA Dordrecht, The Netherlands.

Sold and distributed in North, Central and South America
by Kluwer Academic Publishers,
101 Philip Drive, Norwell, MA 02061, U.S.A.

In all other countries, sold and distributed
by Kluwer Academic Publishers,
P.O. Box 322, 3300 AH Dordrecht, The Netherlands.

Printed on acid-free paper

Printed in the Netherlands

Contents

Preface

From its very beginning, Complex Analysis was closely connected with applications, mainly to Mechanics. This is particularly true for the theory of generalized analytic functions which, at least partly, was initiated by problems from Mechanics. Although theories generalizing the concept of holomorphy have a long history, an intensive investigation of generalized analytic functions started only half a century ago when I. N. Vekua applied the concept of distributional solutions of Partial Differential Equations to Complex Analysis. That way various important problems such as boundary value problems have been solved very effectively. Nevertheless, these ideas are far from being exhausted. This is also true when applying complex methods to higher dimensions.

In order to promote the development of the theory of generalized analytic functions and its applications to Mechanics, the Department of Mathematics and the Strength of Materials Division of the Technical University of Graz have jointly organized a Conference on

"Generalized Analytic Functions
—Theory and Applications"

which took place from January 6 to January 10, 1997. In accordance with the goals of the conference, the present Proceedings have two parts dealing with

Generalizations of Complex Analysis

and

Applications of Partial Differential Equations to Mechanics,

respectively. We would like to acknowledge the valuable assistance of Dr. Annemarie Maitz who took over the tedious task of preparing the manuscript.

During this Conference also the 65th birthday of R. P. Gilbert was celebrated, a man who has made numerous contributions to both subjects mentioned above. This is the reason why most of the papers published in these Proceedings are dedicated to him.

The Editors

Contributing Authors

M. B. Balk
Smolensk Pedagogical Insitute
214000 Smolensk Przhevalski Str. 4
Russia
e-mail: fiz@smolensk.rospac.ru

H. Begehr
Freie Universität Berlin
Institut für Mathematik I
Arnimalle 2-6
D-14195 Berlin
e-mail: begehr@math.fu-berlin.de

P. Berglez
Technical University of Graz
Department of Mathematics A
A-8010 Graz, Steyrergasse 30/2, Austria
e-mail: berglez@weyl.math.tu-graz.ac.at

J. L. Buchanan
Mathematics Department
United States Naval Academy
Annapolis, Maryland 21402 USA
e-mail: jlb@sma.usna.navy.mil

M. Čanak
Landwirtschaftliche Fakultät Beograd-Zemun
Brzakova 4
YU-11000 Beograd

R. Carroll
University of Illinois
Urbana, IL 61801
e-mail: rcarroll@symcom.math.uiuc.edu

A. O. Çelebi
Middle East Technical University,
Department of Mathematics
06531 Ankara, Turkey

A. Cialdea
Dipartimento di Matematica,
Università della Basilicata
Via N. Sauro 85, Potenza, 85100, Italy
e-mail: cialdes@unibas.it

A. Dzhuraev
Tajik Academy of Sciences
Mathematical Institute
Prosp. Rudkai 33
734025 Dushanbe
Tajikistan
e-mail: ajuraev@td.silk.org

R. P. Gilbert
Department of Mathematical Sciences
University of Delaware
Newark, Delaware 19716 USA
e-mail: gilbert@mesalina.math.udel.edu

S. Graubner
Ostwald-Gymnasium Leipzig
e-mail: sonnmath@aol.com

K. Habetha
Rektor der RWTH Aachen
RWTH Aachen
D-52056 Aachen
e-mail: habetha@rektorat.rwth-aachen.de

K. Hackl
Strength of Materials Division,
Dept. of Mechanical Engineering
Technical University Graz, A-8010 Graz, Austria
e-mail: hackl@euler.tu-graz.ac.at

G. C. Hsiao
Department of Mathematical Sciences
University of Delaware
Newark, Delaware 19716-2553
USA
e-mail: hsiao@math.udel.edu

F. Hübner
Faculty of Mathematics and Computer Science
TU Bergakademie Freiberg
Bernhard von Cotta Str. 2
09596 Freiberg
e-mail: huebner@mathe.tu-freiberg.de

A. Jeffrey
University of Newcastle
Department of Engineering Mathematics
Room 1.16, Bruce Building
Newcastle upon Tyne, NE1 7RU, England
e-mail: alan.jeffrey@newcastle.ac.uk

Z. J. Lin
Department of Mathematical Sciences
University of Delaware
Newark, Delaware 19716 USA

M. D. Marcozzi
Department of Mathematical Sciences
United States Military Academy
West Point, New York 10996
USA

Z. Naniewicz
Warsaw University,
Institute of Applied Mathematics and Mechanics,
Banach Str. 2, 02-097 Warsaw, Poland,
e-mail: znaniew@mimuw.edu.pl

E. Obolashvili
Georgian Academy of Sciences
Institue of Mathematics
Z. Rukhadse Street 1
Tbilisi 93
Georgia
e-mail: helen@imath.acnet.ge

P. D. Panagiotopoulos
Aristotle University of Thessaloniki,
Department of Civil Engineering,
GR-54006 Thessaloniki, Greece and
Faculty of Mathematics and Physics,
RWTH, Aachen, Germany,
e-mail: pdpana@heron.civil.auth.gr

J. Püngel
Technical University of Graz
Department of Mathematics D
A-8010 Graz, Steyrergasse 30/3, Austria
e-mail: f501dmat@mbox.tu-graz.ac.at

M. Reissig
Faculty of Mathematics and Computer Science
TU Bergakademie Freiberg
Bernhard von Cotta Str. 2
09596 Freiberg
e-mail: reissig@mathe.tu-freiberg.de

K. Taira
Department of Mathematics,
Hiroshima University,
Higashi-Hiroshima 739, JAPAN
e-mail: taira@math.sci.hiroshima-u.ac.jp

A. Ju. Timofeev
University Syktyvkar
Department of Mathematics
Russia
e-mail: root@sucnit.komi.su

W. Tutschke
Technical University of Graz
Department of Mathematics D
A-8010 Graz, Steyrergasse 30/3, Austria
e-mail: tutschke@matd.tu-graz.ac.at

Z. D. Usmanov
Institute of Mathematics
Tajik Academy of Sciences
prosp. Rudaki 33,
Dushanbe 734 025 TAJIKISTAN
e-mail: usmanov@td.silk.glas.apc.org

V. Vassilenkov
Smolensk Pedagogical Insitute
214000 Smolensk Przhevalski Str. 4
Russia
e-mail: fiz@smolensk.rospac.ru

Y. Xu
Department of Mathematics
University of Tennessee at Chattanooga
Chattanooga, TN 37403
e-mail: yxu@cecasun.utc.edu

U. Yüksel
Ankara University,
Department of Mathematics
06100 Tandoğan-Ankara, Turkey

Laudatio for Robert Pertsch Gilbert
on the occasion of his 65th anniversary

HIS LIFE

Ladies and Gentlemen,
dear Mrs. Gilbert, dear Bob!

It is a special pleasure for me to say some words on the occasion of the special day we have just today. You, dear Bob, have been born 65 years ago, this is an important event and it took place in an important town, in New York.

I was asked to say some words about your 65 years and I do this with great pleasure. You started your carreer in the town of your birth, in 1952 you got the Bachelor of Science in the Brooklyn College of the City University of New York. But then you moved to Pittsburgh, where three years later the Masters of Science in Mathematics as well as in Physics followed and another three years later you passed your Ph.D., all this in Carnegie-Mellon-University.

But you began to work at the University of Pittsburgh, only several hundred meters from Carnegie-Mellon, first as an instructor, then as an assistant professor. Your interests in applications showed up very early - or did you simply need money? - you have been a consultant of the U.S. Bureau of Mines from 58 to 60.

A year at Michigan State followed and then four years as an assistant and research assistant professor in the University of Maryland, where you also have been a consultant at the Naval Ordnance Laboratory.

After a year at Georgetown University in Washington you settled down for nine years from 66 to 75 at Indiana University in Bloomington. During this period we met for the first time. I remember with great pleasure your visit in Dortmund during October 72. We, the Dortmunders, had to learn very much about operators that shift harmonic or holomorphic functions to solutions of partial differential equations, what was new for me and the Dortmund group. I remember very well the terrible slides containing one side of an equation for the kernel function and the next slide the other side. But I at least lost my fears of such huge differential equations and I got familiar with these things. I think the colleagues will tell us something about your work in that direction. I remember that visit in Dortmund also as the starting point for a cooperation and friendship lasting now nearly a quarter of a century.

In 1975 you took the Unidel Chair of Mathematics at the University of Delaware in Newark and there you are also today. You worked with many scholars and collaborators on a broad range of mathematical topics mostly from applications, I assume that we shall hear something about this. But you are also very much interested in teaching problems which is expressed in your lectures but also in your books and in the several conferences you organized. Your method of teaching calculus can also be found in a book and I think this to be a very interesting method not only but especially for your students.

13 doctoral students worked out their thesis under your supervision and most of them are now professors in different universities.

From your rich production of books and over 170 articles in scientific journals I really cannot report, I can only mention it. You edited several conference proceedings and similar books but let me especially mention that you are in the editorial board of 12 journals and scientific series. This was not enough, you founded two journals of your own with the publisher Gordon & Breach, one is Applicable Analysis and the other is Complex Variables. Here we met again very narrowly as you invited me to be also an editor of Complex Variables and we cooperated very well for the last 15 years in publishing this journal. I am very happy about this and also that I could manage to stay this task during my rectorate.

It is not necessary to mention that you are always travelling around in the world, this is a bit mirrored by the participants of this conference. So if I am looking over all that you have done during the last 40 years I am wondering how you managed to do all the things you really did. And I think this shows that you are an extraordinary worker not to say a workaholic, but that is a question you, Mrs. Gilbert, could tell us more about. I think also that your energy and enthusiasm for mathematics cannot be measured with normal physical formulas, it is simply impressive.

At last I did some research about your name and I found the following: Gilbert is also a German first name as well as a German last name, it is a short form of Giselbert. Giselbert has two parts, the first one is Gisel with unknown meaning, related to several German first names, very familiar is the female Gisela. And the second part is bert, this is an "old high German" (althochdeutsches) word "beraht" and means brilliant(glänzend) and so your name already shows what is about with you as a mathematician.

And I think to be allowed to say in the name of all attendants of this conference, in the name of all of your students, collaborators, and friends: We wish you a long series of healthy and fruitful years, we are glad to cooperate with you. It may be wise if you slow down a bit with 65, but stay with us doing mathematics.

K. Habetha

Laudatio for Robert Pertsch Gilbert on the occasion of his 65th anniversary

His contributions to Complex Analysis

Exactly 20 years ago I spent six weeks of my first subbatical at the university of Delaware in Newark visiting Bob Gilbert and starting my collaboration with him. I first had met him at the 1972 conference in Bonn. In 1974/75 he had spent more than one year in Berlin at first visiting the Hahn–Meitner–institute and then the FU for one term even with a senior Alexander von Humboldt award. Since then we became close friends and we have jointly written several papers and this two volumes monograph.
It is always exciting staying with Bob because of his fascinating personality. He is including every one close to him in his daily life. Many colleagues visiting him in Newark have been invited to his nice home in Briar Lane 112 where he himself often is cooking for them. And you can be sure he will take you to the beach in St. Louis for a few days where he has a beautiful beach house just next to the ocean and some sailing boats and a flag of your home country.

Collaborating with Bob is very challanging. Every day he will confront you with another idea. Because you cannot keep up you just stick to what you had started to work and forget about of his suggestions. But when you get stuck with your problem you can be sure he will find a way out and get the work continued. Forgetting some suggestion of Bob does not work always. Sometimes he is insisting. One example is Bob's idea of founding a scientific society. When he first mentioned it I thought: "Well, forget it." But he insisted and again and again got excited about this idea and this summer ISAAC will officially be founded at the conference in Newark. Two other such ideas are his journals "Applicable Analysis" and "Complex Variables". The second one co–edited with Klaus Habetha proves his strong relations with German mathematicians. Some years ago he succeeded in winning his wife Nancy to help him editing the journals and I am sure after a few more years Jennifer will be involved in this family business too.

Bob's mathematical work is closely related to the field of complex methods in partial differential equations which has developed rapidly within the last five decades. In extending classical results for Laplace and Poisson equations St. Bergman and I.N. Vekua had treated more general linear elliptic second order equations in the plane. They introduced complex integral transforms which relate holomorphic functions of one complex variable to complex solutions to these equations. Their first papers were published in 1937 already but only after Bergman's monograph has appeared in 1961 an explosive development of this field has started. Shortly before Bob had finished his thesis which fitted very well to what Bergman and Vekua had done. Bob was a student of Zeev Nehari and I am sure Nehari has had contact to Bergman. In his thesis Bob developed his "Envelope Method". It describes the set of possible singularities of analytic functions of several complex variables given as a contour integral in the plane over functions of $n+1$ variables. This is a generalization of some result of Hadamard on singularities of convolutions of analytic functions. The envelope method applied to the Bergman–Vekua operator or for three variables to the Bergman–Whittaker operator leads to the set of singularities of harmonic functions. Bob right away was in the middle of this new field of research and he became one of the main figures in this area. Recently two books have appeared about this subject

> Kracht und Kreyszig: Methods of complex analysis in partial differential equations with applications, 1988

and

> Lanckau: Complex integral operators in mathematical physics, 1993.

In the first book 30 papers of Bob are listed in the second more than 50. I am not able to review about what Bob all has achieved in this area. You can get a good insight in his results from his classical Academic Press Monograph, his Springer Lecture Notes and the more recent excellent monograph with James Buchanan which is a beautiful report on new development in complex and Clifford analysis methods for partial differential equations. Let me just mention another well–known method of his, the method of ascent. He had realized some function – now called Gilbert's G–function – as independent on the dimension and thus he could generalize Bergman's integral operator representing solutions to certain second–order equations in the plane by means of harmonic functions to higher dimensions. This operator is now called Bergman–Gilbert operator and later became an important example for a transmutation operator. One field of application is the generalized axically symmetric potential (GASP) theory originated by A. Weinstein. Recently Bob has applied this operator to underwater acoustic.

Some 20 years ago Bob together with his student Jerry Hile has created the theory of generalized hyperanalytic functions as a combination of Vekua's theory of generalized analytic functions and Douglis' hyperanalytic function theory in order to treat elliptic systems of more than two first order equations. A few

years before Bojarski had explored his theory of generalized analytic vectors, which he just reported on yesterday. With all his activities Bob quickly became a central figure in the theory of complex methods and consequently and very early he got contact to the dominant schools around Vekua and Muskhelishvili. Even today where we celebrate his 65th anniversary the main former scholars from Vekua Bogdan Bojarski, Abduhamid Dzhuraev and Elena Obolashvili are participating in this conference.
Bob not only is working on complex methods in pde but also in applications e.g. to mechanics especially elasticity theory and more generally to other areas of applied mathematics in physics. Here he takes advantage of his scientific education where he in the same year has graduated in mathematics and in physics as well. His achievements in mathematical physics will be eluminated by his friend, colleague and collaborator for a long time George Hsiao. Another newer field of interest is symbolic computation with computer algebra where the book with W. Koepf and A. Ben Israel has appeared. I have mentioned Bob's close contact to mathematicians from the former SU and from Germany. His contacts go much beyond this. He has close relations to mathematicians in many countries as you also can see from the participants here. I just mention Great Britain, France, Denmark, Italy, Poland, Greece, Turkey, South Africa, China, Japan, Mexico and of course Austria. Two third of his publications (books as well as papers) are jointly written with colleagues. There are mathematicians who are constantly doing research and being just satisfied from the beautiness of their results. Not so Bob! Although he is very interested in his mathematics he would stop working if there would not be colleagues to share his success with. He seems to need colleagues like flowers need the sun. Mathematic is just some beautiful mean for Bob to interfear with other human beings. Sailing, photographing, ballet, music, arts, reading and last but not least his nice family are important facets in satisfying his desire for contact with other human beings. From this desire we all participate in his friendship and therefore we are here at this conference to thank him for this friendship and to congratulate him for his birthday.

Bob, we all wish you Bob a great deal of health, strength and good luck for the coming years.

H. Begehr

Laudatio for Robert Pertsch Gilbert on the occasion of his 65th anniversary

His contributions to Mechanics

Professor W. Tutschke asked me to report on the contributions of Bob Gilbert to the applications of partial differential equations in mechanics. Needless to say, I am delighted and honored to do so. I have known Bob for more than 25 years as a colleague, a long-time collaborator and, most of all, a close friend. I have always been amazed with Bob's energy and admired with his talent and scientific ability.

Gilbert is an applied analyst and is primarily known for his treatises on function theoretic methods applied to partial differential equations. He has written four such monographs. These are

- *Function Theoretic Methods in Partial Differential Equations*, Academic Press, New York, 1969.
- *Constructive Methods for Elliptic Partial Differential Equations*, Springer-Verlag, Lecture Note Series No. 365, Heidelberg, 1974.
- *First Order Elliptic Systems: A Function Theoretic Approach*, Academic Press, New York, 1993 (with J. Buchanan).
- *Transmissions, Transformations and Reproducing Kernels*, Vols. I, II, Pitman, London, 1993 (with H. Begehr).

As a research mathematician, Gilbert has published over 297 articles in professional journals and conference proceedings. Looking through Gilbert's publication lists, I believe that it will take me too long to list all his research achievements and his contributions to the applications of pdes in mechanics and mathematical physics, since Gilbert has done so much and has made significant contributions to so many differnet fields. In order not to be too repetitious, I shall only comment that most of Gilbert's early work was based on the function theoretic methods. In particular, he has applied the methods to problems in ***axially symmetric potential theory in scattering, elastic plates, shell theory*** and problems in ***Hele Shaw Flows*** (see the Laudatio by Begehr). Gilbert's recent work in connection with the fundamental problems in mechanics can be summarized according to four areas: (I) **Linear and non-linear elasticity**, (II)

Wave propagation, (III) **Free and moving boundary-value problems**, (IV) **Singular perturbation and homogenization**. In each of these areas, Gilbert has made fundamental contributions.

In (I), Gilbert has worked problems concerning nonlinear orthotropic plates (with H. Begehr, C. Y. Lo), theory of elastic-plastic and thermo-elastic plates (with K. Hackl) and obtained explicit solutions of elastic shell equations (with J. Buchanan, Y.S. Xu). In (II), He has investigated problems in underwater sound propagation (with D. H. Wood), acoustic waves and farfield patterns in oceans with porous elastic-seabeds (with Y. S. Xu), as well as direct and inverse problems in ocean acoustics (with J. Buchanan, J. Z. Lin, Y. S. Xu). In (III), this covers problems in non-Newtonian Hele-Shaw flows (with H. Begehr, P. Shi), contact problems in linear thermoelasticity (with p. Shillor, P. Shi), moving bvps for flow in porous medium (with R. Magnanini, G. Wen), and moving bvps in injection molding (with Z. Xu). Finally in (IV), Gilbert and his colleagues have studied problems in connection with boundary-layer theory for orthotropic plates (with G. Hsiao, M. Schneider), and asymptotics of reverse in situ combustion (with A. Jeffrey), to name a few.

Gilbert's current research efforts are in the areas of inverse problems, theory of pasto-elasticity, and the flow of viscous fluids. The work on inverse problems concentrates on acoustic inverse problems in the ocean, in particular shallow oceans, where there is an interactive poro- elastic seabed. The work on elasto-plasticity is involved with asymptotic theories for elastic-plastic shells and plates. His interest in the flow of viscous fluids originates out of the polymer injection molding process. Here there is a coupling of the Hele-Shaw, or as the case may be Darcy, flow of non-Newtonian fluid with an energy balance equation. The viscosity of the fluid depends heavily on the temperature of the process. Moreover, the rate of shear causes flow. Current work in this area concerns flow between curvilinear shells, that is, flow on manifolds.

At the age of 65, Gilbert is still full of energy, productive and as enthusiastic as ever working with young people. We all express our sincere gratitude to him for his continuous and faithful guidance in applied analysis.

Bob, many happy returns !

G. Hsiao

I Present Trends in the Theory of Generalized Analytic Functions and Generalizations of Complex Analysis

1 GENERALIZED ANALYTIC FUNCTIONS IN MATHEMATICAL ANALYSIS

W. Tutschke

Abstract: Discussing some recent trends in the theory of Generalized Analytic Functions, the paper is concentrated on the position of Complex Analysis within the framework of Mathematical Analysis as a whole.

1.1 MATHEMATICAL ANALYSIS IN THE AGE OF ELECTRONIC DATA PROCESSING

Considering the huge possibilities created by Electronic Data Processing, Mathematical Analysis is faced by new tasks. This concerns not only numerical methods and methods of Symbolic Computation but also the structure of the theory and interactions between different branches. In order to remain applicable for a broad circle of users, Mathematical Analysis has to formulate basic principles.

H. Florian et al. (eds.), Generalized Analytic Functions, 3–15.

1.2 COMPLEX ANALYSIS AS AN ESSENTIAL PART OF MATHEMATICAL ANALYSIS

Complex Analysis is an irreplaceable part of Mathematics. For instance, a general theory of power-series representations and the concept of holomorphy are necessary contributions of Complex Analysis to Mathematical Analysis. Moreover, essential contributions of Complex Analysis to Mathematics are given in the field of Fundamental Theorem of Algebra and in the theory of manifolds in higher dimensions as well.

1.3 INTERACTIONS OF COMPLEX ANALYSIS WITH PARTIAL DIFFERENTIAL EQUATIONS

A most stimulating influence to Complex Analysis has been given by partial differential equations. The following points of view are decisive for the interactions between partial differential equations and Complex Analysis:

1.3.1 Partial complex differentiations

An essential idea for enlarging the applicability of Complex Analysis is the definition of partial complex differentiations

$$\begin{aligned} \partial_z &= \frac{1}{2}(\partial_x - i\partial_y) \\ \partial_{\bar{z}} &= \frac{1}{2}(\partial_x + i\partial_y). \end{aligned}$$

Then the GREEN-GAUSS Integral Formulae can be rewritten in the form

$$\begin{aligned} \iint_\Omega \partial_{\bar{z}} w dx dy &= \frac{1}{2i}\int_{\partial\Omega} w dz \\ \iint_\Omega \partial_z w dx dy &= -\frac{1}{2i}\int_{\partial\Omega} w d\bar{z} \end{aligned}$$

where $z = x + iy$ and $\partial\Omega$ is the boundary of Ω, as usual.

1.3.2 Solutions in distributional sense

Applying the above complex versions of the GREEN-GAUSS Integral Formulae to $w = w_1 w_2$, one gets

$$\iint_\Omega \partial_{\bar{z}}(w_1 \cdot \partial_{\bar{z}} w_2 + \partial_{\bar{z}} w_1 \cdot w_2)\, dx dy = \frac{1}{2i}\int_{\partial\Omega} w_1 w_2 dz.$$

and an analogous formula for the derivative with respect to z. Choosing $w_2 = \varphi$ where φ is a test function (i.e., φ is continuously differentiable and vanishes identically in a neighbourhood of $\partial\Omega$), and denoting w_1 and $\partial_{\bar{z}} w_1$ by f and g resp., one obtains the relation

$$\iint_{\Omega} (\partial_{\bar{z}} f \cdot \partial_{\bar{z}} \varphi + g \cdot \varphi)\, dx dy = 0 \tag{1.1}$$

being satisfied for each choose of the test function φ. This relation can be used in order to define partial complex derivatives in the distributional sense:

> Let f be an integrable function. In case there exists an integrable function g such that the relation (1.1) is satisfied for every choice of the test function φ, the function g is called the *distributional* derivative of f with respect to $\bar{z}$.

An analogous definition can be given for the distributional derivative with respect to z.

The famous WEYL Lemma states that a function whose distributional derivative with respect to $\bar{z}$ vanishes identically is necessarily a holomorphic function in the classical sense.

1.3.3 Domain integral operators of Complex Analysis

Another essential idea for enlarging the scope of Complex Analysis is to invert the partial complex differentiations by integral operators. A far-reaching result is the following theorem:

Suppose h is integrable in the (bounded) domain Ω. Then the function T_Ω defined by

$$-\frac{1}{\pi} \iint_{\Omega} \frac{h(\zeta)}{\zeta - z} d\xi d\eta$$

(where $\zeta = \xi + i\eta$ is the integration variable) has h as distributional derivative with respect to $\bar{z}$ in Ω, i.e., one has

$$\partial_{\bar{z}} T_\Omega h = h$$

in Ω (while T_Ω is holomorphic outside $\overline{\Omega}$).

Notice, further, that the distributional derivative of T_Ω with respect to z is given by the strongly singular Π-operator

$$\partial_z T_\Omega h = \Pi_\Omega h = -\frac{1}{\pi} \iint_{\Omega} \frac{h(\zeta)}{(\zeta - z)^2} d\xi d\eta.$$

1.3.4 Inversion of differential operators by complex line integrals

Suppose $w = w(z)$ is holomorphic in a simply connected domain Ω. Let γ be a curve in Ω connecting a fixedly chosen point z_0 with an arbitrary point z. Then

$$W(z) = \int_\gamma w(\zeta)d\zeta \tag{1.2}$$

depends only on the ending point z of γ and defines a holomorphic function $W = W(z)$ whose (ordinary) complex derivative equals the given holomorphic function $w = w(z)$, i.e., one has

$$\frac{dW}{dz} = w.$$

Since holomorphic functions can be characterized by the *Cauchy-Riemann* equation $\partial_{\overline{z}} w = 0$ and since

$$\frac{d}{dz} = \partial_z$$

for holomorphic functions, the last result can be formulated as follows:

> In the space of holomorphic functions the partial complex differentiation ∂_z can be inverted by the complex line integral (1.2).

Analogously, in the space of anti-holomorphic functions (defined by $\partial_z w = 0$) the partial complex differentiation $\partial_{\overline{z}}$ can be inverted by the (path-independent) complex line integral

$$\int_\gamma w(\zeta)\overline{d\zeta}. \tag{1.3}$$

Generalizing the line integrals (1.2) and (1.3), we are going to invert more general first order differential operators (containing the differentiations ∂_z and $\partial_{\overline{z}}$) by complex line integrals of type

$$\int_\gamma \left(p d\zeta + q\overline{d\zeta}\right). \tag{1.4}$$

While the complex line integrals (1.2) and (1.3) have been considered in the spaces of holomorphic and anti-homomorphic functions resp., we consider integrals of type (1.4) in a space of generalized analytic functions defined by the equation

$$\mathcal{G}w \equiv \partial_{\overline{z}} w - a(z)w - b(z)\overline{w} = 0 \tag{1.5}$$

(which is the canonical normal form of a uniformly elliptic system for two desired real valued functions in the plane). Denote by $\mathcal{W}_{\mathcal{G}}(\Omega)$ the set of all solutions of $\mathcal{G}w = 0$ in Ω. Since in this space $\mathcal{W}_{\mathcal{G}}(\Omega)$ the partial complex derivative of w with respect to $\overline{z}$ can be expressed by w and $\overline{w}$, the most general (linear and homogeneous) first order differential operator $\mathcal{F}$ has the form

$$\mathcal{F} \equiv C(z)\partial_z w + A(z)w + B(z)\overline{w} \tag{1.6}$$

(provided it does not contain $\overline{\partial_z w}$). To invert $\mathcal{F}$ in the space $\mathcal{W}_{\mathcal{G}}(\Omega)$ means, consequently, to find an element $W \in \mathcal{W}_{\mathcal{G}}(\Omega)$ such that

$$\mathcal{F}W = w \tag{1.7}$$

$$\mathcal{G}W = 0 \tag{1.8}$$

where $w \in \mathcal{W}_{\mathcal{G}}(\Omega)$ is given. Solving the system (1.7), (1.8) for $\partial_z W$ and $\partial_{\overline{z}} W$, one obtains

$$\partial_z W = \frac{1}{C}\left(w - AW - B\overline{W}\right) \tag{1.9}$$

$$\partial_{\overline{z}} W = aW + bB\overline{W}. \tag{1.10}$$

In order to represent W in the form (1.4), the derivatives

$$p = \partial_z W \quad \text{and} \quad q = \partial_{\overline{z}} W$$

have to satisfy the compatibility condition

$$\partial_{\overline{z}} p - \partial_z q = 0. \tag{1.11}$$

Substituting (1.9), (1.10) into (1.11), and taking into account the differential equation $\mathcal{G}w = 0$, one gets the following sufficient conditions for compatibility:

$$\begin{aligned}
\partial_{\overline{z}} C &= 0 \\
B - b \cdot \overline{C} &= 0 \\
\partial_z a \cdot C + \partial_{\overline{z}} A &= 0 \\
\partial_z C + \partial_{\overline{z}} b \cdot \overline{C} + b \cdot \overline{\partial_z C} + b\left(\overline{a}C - a\overline{C}\right) + b\left(A - \overline{A}\right) &= 0
\end{aligned}$$

(see [18], cf. also [14]).

In Subsection 1.4.4 we shall see that these relations are identical with those for so-called associated pairs of first order operators $\mathcal{F}$, $\mathcal{G}$.

1.3.5 Non-linear partial differential equations. Fixed-point methods

The T_Ω-operator introduced in Subsection 1.3.3 can be used in order to reduce non-linear differential equations of from

$$\partial_{\bar{z}} = f(z, w, \partial_z w) \tag{1.12}$$

into the integro-differential equation

$$w = \Phi + T_\Omega f(\cdot, w, \partial_z w) \tag{1.13}$$

where Φ is holomorphic. Suitably choosing Φ, boundary value problems for (1.12) can be reduced to fixed-point problems for the integro-differential operator on the right-hand side of (1.13) (see [11]).

A special case of equation (1.12) is the linear equation (1.5) whose solutions are called *generalized analytic functions.*

1.3.6 Vector differential equations

Note that $w = w(z)$ in (1.12) can be a vector $w = (w_1, ..., w_m)$ having m complex components. That way complex methods can also be applied to systems for $2m$ desired real-valued functions in the plane.

1.4 SOME RECENT TRENDS IN COMPLEX ANALYSIS

1.4.1 Boundary value problems

Using integral representations with holomorphic functions, boundary value problems for solutions of complex equations (of elliptic type) can be reduced to analogous boundary value problems for holomorphic functions. Since the real part of a holomorphic function is a solution of the LAPLACE equation, methods for solving the latter equation are applicable, too.

Using RITZ-GALERKIN approximations for the LAPLACE equation, in the paper [16] (jointly written with H. L. VASUDEVA) the DIRICHLET boundary value problem is solved for generalized analytic functions. Provided the boundary values are HÖLDER continuous with exponent $1/2$, the constructed solution belongs to the SOBOLEV space $\mathcal{W}_2^1(\Omega)$. The proof makes use of SOBOLEV's Embedding Theorem.

1.4.2 Initial value problems

Complex methods can be used not only for solving boundary value problems but also for solving initial value problems. Consider an initial value problem of type

$$\partial_t u = \mathcal{F}(t, x, u) \tag{1.14}$$
$$u(0, x) = u_0(x) \tag{1.15}$$

where $\mathcal{F}$ is a first order differential operator acting with respect to the spacelike variable $x = (x_1, ..., x_n)$. Notice that the initial value problem (1.14), (1.15) can be rewritten as integro-differential equation

$$u(t, x) = u_0(x) + \int_0^t \mathcal{F}(\tau, x, u(\tau, x))d\tau. \tag{1.16}$$

A special case of the initial value problem (1.14), (1.15) is the classical CAUCHY-KOVALEVSKAYA problem for which both the initial function u_0 in (1.15) and the right-hand side of (1.14) are holomorphic functions (in their variables). M. NAGUMO [5] was the first who solved the classical CAUCHY-KOVALEVSKAYA problem by using its integral rewriting (1.16). His proof makes use of the fact that the ordinary complex derivative of a holomorphic function is holomorphic, again. The necessary estimates of the integro-differential operator in (1.16) are based on an interior estimate of the derivative of a holomorphic function (an interior estimate holds uniformly at points having a positive minimal distance from the boundary).

In the case of the classical CAUCHY-KOVALEVSKAYA problem W. WALTER [20] solved the integro-differential equation (1.16) by the contraction-mapping principle in a suitably defined BANACH space equipped with a weighted supremum norm. In the paper [15] the compactness of the integro-differential operator in (1.16) is investigated. Result: While the operator is compact in certain FRÉCHET spaces, it is not compact, in general, in the weighted BANACH space mentioned above.

The integro-differential equation (1.16) and, consequently, the initial value problem (1.14), (1.15) can also be solved in spaces of generalized analytic functions provided the operator $\mathcal{F}$ transforms (for each t) this space into itself and, further, the initial function u_0 belongs to the same space. In other words, if the space of generalized analytic functions is defined by $\mathcal{G}w = 0$ (see (1.5), then $\mathcal{F}$ and $\mathcal{G}$ have to satisfy the following condition:

$$\mathcal{G}w = 0 \quad \text{implies} \quad \mathcal{G}(\mathcal{F}w) = 0 \tag{1.17}$$

(see [10, 13] and A. CRODEL's Thesis [4] in the case of generalized analytic vectors).

1.4.3 *Associated differential operators*

Pairs of operators satisfying the condition (1.17) are said to be *associated.*

An axiomatic system for solving initial value problems of type (1.14), (1.15) in more general associated spaces is given in [19].

1.4.4 *Inversion of differential operators in solution spaces of associated differential operators*

The paper [8] (see also [10, 13]) contains (sufficient) conditions under which operators $\mathcal{F}$, $\mathcal{G}$ of type (1.5), (1.6) are associated. These conditions turn out to be identical with the compatibility conditions of Subsection 1.3.4. Thus the following statement is true:

Provided $\mathcal{F}$, $\mathcal{G}$ are associated, the differential operator $\mathcal{F}$ can be inverted by a path-independent complex line integral in $\mathcal{W}_{\mathcal{G}}(\Omega)$.

Cf. also the paper [14] containing a Volterra integral equation leading to generalizations of CAUCHY's Integral Theorem and MORERA's Theorem as well.

1.4.5 *Problems in several complex variables*

The domain integral operators introduced in Subsection 1.3.3 can also be used when investigating non-holomorphic functions in several complex variables $z_1, ..., z_n$. In the polycylinder $\Omega = \Omega_1 \times \cdots \times \Omega_n$ a special solution of an (integrable) inhomogeneous CAUCHY-RIEMANN system

$$\partial_{\bar{z}_j} w = h_j, \quad j = 1, 2, ..., n,$$

is given by

$$\sum_{\lambda=1}^{n} (-1)^{\lambda+1} \sum\nolimits^{*} T_{\Omega_{j_1}} \cdots T_{\Omega_{j_\lambda}} h_{j_1 \cdots j_\lambda} \tag{1.18}$$

where

$$h_{j_1 \cdots j_\lambda} = \partial_{\bar{z}_{j_\lambda}} \cdots \partial_{\bar{z}_{j_2}} h_{j_1}$$

and $\sum^{*}$ means summation over combinations of indices being different from each other. Moreover, HARTOGS's Continuity Theorem implies that the general solution of the above inhomogeneous CAUCHY-RIEMANN system can be obtained by adding an arbitrary holomorphic function in $z_1, ..., z_n$ to the special solution (1.18), see [9].

Taking into account theorems on the boundary behaviour of holomorphic functions in several complex variables, this representation formula for the general solution of the inhomogeneous CAUCHY-RIEMANN system implies also theorems on the boundary behaviour of non-holomorphic functions (see [1]).

Finally, solutions $w = w(z_1, ..., z_n)$ of systems of type

$$\partial_{\overline{z}_j} w = f_j(z, w, \partial_{z_i} w), \quad j = 1, ..., n,$$

can be factorized in the form

$$w = \Phi \cdot \exp \omega$$

provided the right-hand sides f_j (and some of their derivatives) satisfy suitable LIPSCHITZ conditions (details are contained in [9]).

1.4.6 Complex methods in higher-dimensional real Euclidian spaces

In the real Euclidian space $\mathbb{R}^{n+1}$ (with the coordinates $x_0, x_1, ..., x_n$ and the basis vectors $e_0 = 1, e_1, ..., e_n$) the operator

$$D = \sum_{j=0}^{n} e_j \partial_{x_j}$$

generalizes the CAUCHY-RIEMANN operator of the complex plane. This operator D is called the CAUCHY-RIEMANN operator of CLIFFORD Analysis, it leads to complex methods in $\mathbb{R}^{n+1}$. For instance, a CAUCHY Integral Formula is true for (left)-monogenic (CLIFFORD-algebra-valued) functions u which are defined by $Du = 0$ (see the monograph [3] written by F. BRACKX, R. DELANGHE and F. SOMMEN).

Another possibility to apply complex methods in higher-dimensional real Euclidian spaces can be given within the framework of associated differential operators.

An associated pair $\mathcal{F}$, $\mathcal{G}$ is called ***dimension-preserving*** (see [14]) if for any given $u \in \mathcal{W}_{\mathcal{G}}(\Omega)$ the system

$$\begin{aligned} \mathcal{F}U &= u \\ \mathcal{G}U &= 0 \end{aligned}$$

can be solved for all $\partial_{x_j} U$, $j = 0, 1, ..., n$. Similarly as in $\mathbb{C}$ one gets again a VOLTERRA integral equation (cf. Subsection 1.3.4) yielding a CAUCHY Integral Theorem and a MORERA Type Theorem in $\mathbb{R}^{n+1}$ within the framework of associated differential operators.

1.4.7 *Some applications of Complex Analysis in the theory of functional equations*

A function $h = h(x, y)$ depending on two real variables x and y can be represented in the form

$$h(x,y) = \sum_{k=1}^{n} f_k(x) g_k(y) \tag{1.19}$$

if and only if

$$D_{xy}^n h = \begin{vmatrix} h & \partial_x h & \partial_x^2 h & \dots & \partial_x^n h \\ \partial_y h & \partial_x \partial_y h & \partial_x^2 \partial_y h & \dots & \partial_x^n \partial_y h \\ \vdots & \vdots & \vdots & \ddots & \vdots \\ \partial_y^n h & \partial_x \partial_y^n h & \partial_x^2 \partial_y^n h & \dots & \partial_x^n \partial_y^n h \end{vmatrix} \neq 0$$

(see [6, 7]). In other words, the condition $D_{xy}^n h \neq 0$ is necessary and sufficient for the solvability of the functional equation (1.19) where $h(x, y)$ is given.

Using methods of Complex Analysis, the following result on functional equation can be proved (see [17]):
A function $h = h(x, y)$ can be represented in the form

$$h(z) = \sum_{k=1}^{n} g_k(z) \overline{f_k(z)}$$

where the g_k and the f_k are holomorphic in $z = x+iy$ if and only if an analogous condition $D_{\bar{z}z}^n h \neq 0$ is satisfied.

1.4.8 *Complex methods for partial differential equations of second and higher order*

The present survey is concentrated on partial complex differential equations of first order. Comprehensive surveys of complex methods for partial differential equations of second and higher order are given in the book [21] of WEN GUO-CHUN and H. BEGEHR and in the two volumes [2] of H. BEGEHR and R. P. GILBERT. In these books one can find further references.

1.5 NEW APPROACHES TO COMPLEX ANALYSIS

The theory of generalized analytic functions has lasting effects on the structure of Mathematical Analysis, too. Using the concept of partial complex derivatives (in the classical and in the distributional sense), Mathematical Analysis can be simplified and unified as the following approach shows:

One starts from the concept of a differential operator of *divergence type*. This is a differential operator $\mathcal{L}$ (of order k) for which a GREEN Integral Formula of type

$$\int_{\Omega} \left(v\mathcal{L}u + (-1)^{k+1}u\mathcal{L}^*v\right) dx = \int_{\partial\Omega} P[u,v]d\mu$$

holds ($d\,m$ is the measure element of $\partial\Omega$). Then a fundamental solution $\mathcal{E}^*(x,\xi)$ of the adjoint equation $\mathcal{L}^*v = 0$ with singularity at ξ leads to the integral representation

$$u(\xi) = (-1)^{k+1}\int_{\partial\Omega} P[u(x),\mathcal{E}^*(x,\xi)]d\mu + (-1)^k\int_{\Omega} \mathcal{E}^*(x,\xi)\mathcal{L}u(x)dx$$

which for $\mathcal{L} = \partial_{\overline{z}}$ passes over into

$$w(\zeta) = \frac{1}{2\pi i}\int_{\partial\Omega} \frac{w(z)}{z-\zeta}dz - \frac{1}{\pi}\iint_{\Omega} \frac{1}{z-\zeta}\partial_{\overline{z}}w(z)dxdy$$

because in this case a fundamental solution is given by

$$\mathcal{E}(z,\zeta) = \frac{1}{\pi}\frac{1}{z-\zeta}.$$

The last representation formula leads to CAUCHY's Integral Formula for holomorphic functions immediately if one defines holomorphic functions by the differential equation

$$\partial_{\overline{z}}w = 0.$$

Starting from $\mathcal{L} = \Delta$, one comes to Potential Theory and so on.

Starting from GREEN's Integral Formula for differential operators of divergence type, one can also introduce the concept of ***distributional solutions*** for both real and complex partial differential equations.

Finally, using (real or complex) fundamental solutions and applying GREEN's Integral Formula, one can construct solutions of non-linear differential equations by solving fixed-point problems for related integral operators.

References

[1] M. B. BALK, W. TUTSCHKE and V. P. VASILENKOV, FATOU and LUSIN-PRIVALOV type theorems for polyanalytic functions in two complex variables in Bicylinders (in Russian). Contained in the booklet "Polynanalytic functions. Boundary behaviour and the boundary value problems". Smolensk 1997, pp. 13-19

[2] H. BEGEHR and R. P. GILBERT, Transformations, transmutations and reproducing kernels, Vol. I and Vol. II, Pitman-Longman 1993 and 1994.

[3] F. BRACKX, R. DELANGHE and F. SOMMEN, CLIFFORD Analysis. *Pitman Research Notes in Mathematics*, vol. **76**, 1982.

[4] A. CRODEL, Sätze vom CAUCHY-KOWALEWSKAJA-Typ für partielle komplexe Differentialgleichungssysteme in Klassen verallgemeinerter analytischer Vektoren. *Thesis (Dissertation A)*, MARTIN LUTHER *University*, 1986.

[5] M. NAGUMO, Über das Anfangswertproblem partieller Differentialgleichungen. *Japan. Journ. Math.*, vol. **18**, 41-47, 1941.

[6] F. NEUMAN, Functions of two variables and matrices involving factorizations. *C. R. Math. Rep. Acad. Sci. Canada*, vol. 3, 7-11, 1981.

[7] - , Factorizations of matrices and functions of two variables. *Czechoslovak Math. Journ.*, vol. 32, 582-588, 1982.

[8] W. TUTSCHKE, Associated operators in Complex Analysis, *Soobch. Akad. Nauk Grus. SSR, vol.* **107**, *Nr. 3* (1982), 481-484 (in Russian).

[9] - , Partielle komplexe Differentialgleichungen in einer und in mehreren komplexen Variablen. Berlin 1977.

[10] - , problem with initial values for generalized analytic functions depending on time (generalizations of the Cauchy-Kowalewski and Holmgren theorems) *Doklady Akad. Nauk*, vol. **262**, Nr. 5, 1081-1085, 1982 (in Russian). Engl. transl.: *Soviet Math. Dokl.*, vol. **25**, Nr. 1, 201-205, 1982.

[11] - , Partial differential equations. Classical, functional-analytic, and complex methods. Teubner-Text Leipzig 1983 (in German).

[12] - , Contraction-mapping principles for initial value problems with generalized analytic initial functions. *Proceedings Conf. Mod. Probl. in Math. Physics*, vol. **1**, 101-113, Tbilissi 1987.

[13] - , Solution of initial value problems in classes of generalized analytic functions. Teubner Leipzig and Springer Verlag 1989.

[14] W. TUTSCHKE and C. WITHALM, On integration and differentiation of generalized analytic functions. Complex Variables, vol.29, 319-332, 1996.

[15] W. TUTSCHKE and H. L. VASUDEVA, Compactness of an Integro-Differential Operator of CAUCHY-KOVALEVSKAYA Theory. Zeitschr. Anal. Anwend., vol. 15, 559-564, 1996.

[16] - , A variational principle for generalized analytic functions. Complex Variables (in print).

[17] W. TUTSCHKE, Lower-dimensional decompositions using complex variables, Archivum Mathematicum, vol. 34, 1998 (in print).

[18] - , Another approach to complex integration (in preparation).

[19] - , Interior estimates in the theory of partial differential equations and their application to initial value problems. Memoirs on Differential Equations and Mathematical Physics, vol. 12 (in print).

[20] W. WALTER, An elementary proof of the CAUCHY-KOWALEVSKY theorem. *Amer. Math. Monthly*, vol. **92**, 115-125, 1985.

[21] WEN GUO-CHUN and H. BEGEHR, Boundary value problems for elliptic equations and systems. *Longman Scientific & Technical* 1990.

2 ON SINGULAR INTEGRAL EQUATIONS APPROACH TO GENERALIZED ANALYTIC FUNCTIONS

A. Dzhuraev

Abstract: Many exhaustive results in boundary value problems for elliptic partial differential equations and systems in the plane were obtained by means of the technic of complex variables through singular integral equations over the boundary of the considered bounded domains-the boundary integral equations [2, 3, 4]. By Vekua the method of two dimentional singular integral equations over unit disc for investigation of boundary value problems for generalized analytical functions was proposed and it was prolonged by B.Bojarski and V.Vinogradov. For an arbitrary multiply-connected domain the method of two dimentional singular integral equations was pro-posed by A.Dzhuraev in his book *Methods of singular integral equations.* Here we outline some results from this book.

2.1

Let G be open set in the complex plane of the variable $z = x + iy$. The complex valued function $w(z)$ called to be a generalized analytical function in G if w is a solution of the following first order elliptic partial differential equation

$$\bar{\partial} w + A(z)w + B(z)w = f(z)$$

H. Florian et al. (eds.), Generalized Analytic Functions, 17–25.

in G, where $A(z), B(z), f(z)$ are given in G functions belonging to the space $L_p(G)$ with $p > 2$. The Riemann-Hilbert problem is to find a generalized analytical function in G satisfying the following condition on the boundary of G:

$$Re(\lambda w) = h,$$

where $\lambda(t), h(t)$ are functions given on the boundary of the domain G. We consider this problem for more general first order elliptic system

$$\bar{\partial} w + q\partial w + q^* \bar{\partial} \bar{w} + A(z)w + B(z)w = f(z) \tag{2.1}$$

with coefficients q bounded in G. Although this problem was investigated for any Hölder continuous function λ nonvanishing on the boundary here for simplicity we demonstrate this method for the case $\lambda = 1$, i.e. we consider the Schwarz problem:

$$Re\ w(t) = h(t) \tag{2.2}$$

on the boundary of G for system (2.1). If we denote z derivative of the unknown function through ρ, then in case the conditions

$$Re \iint \rho\varphi_j dxdy = 0\ , \quad j = 1, ..., m$$

are fulfilled for $\varphi_j = 2\partial u_j$, where u_j is the harmonic measure of the curve C_j with respect to the domain G. In this case all solutions of the problem are given by the formula (see A.Dzh):

$$w(z) = T\rho + iC,$$

where C is an arbitrary real constant and

$$T\rho = \iint \left[\left(L^*(z,\zeta) - \frac{1}{\zeta - z} \right) \rho(\zeta) - K^*(z,\zeta)\bar{\rho}(\zeta) \right] dG \tag{2.3}$$

where $L(z,\zeta)$ is a holomorphic function which is equal to mixed derivative of the regular part of the Green function with respect to z and ζ and $L^*(z,\zeta)$ is a primitive of $L(z,\zeta)$ with respect to z. $K^*(z,\zeta)$ is the primitive with respect to the variable z of the Bergmann kernel function $K(z,\zeta)$ of the domain G. Substituting (2.3) into (2.1) we obtain the following integral equation:

$$\rho - qS\rho - q^*\bar{S}\bar{\rho} + \bar{A}T\rho + \bar{B}\bar{T}\bar{\rho} = g \tag{2.4}$$

where S is the following singular integral operator

$$S\rho = \iint \left[\left(L(z,\zeta) - \frac{1}{(\zeta - z)^2} \right) \rho(\zeta) - K(z,\bar{\zeta}) \right] dG.$$

According to the Calderon-Zygmund theorem the operator S is linear and continuous in $L_p(G), p > 1$ with norm continuous in p. Moreover it can be shown that the operator S has an $L_2(G)$-norm equal to one. This means that singular integral equation with ellipticity condition

$$|q| + |q^*| < 1$$

can be regularized to the Fredholm's integral equaion, which is equivalent to equation (2.4). By means of the similarity principle it is proved that the last Fredholm's equation is uniquely solvable, which is summarized to the following result:

Theorem 1 *The Schwarz Problem (2.2) for elliptic system (2.1) is solvable if and only if either m or $m+1$-orthogonality conditions hold, where $m+1$ is the number of the connectivity of the domain G and corresponding homogeneous problem has one nontrivial solution or it has no one nontrivial solution.*

2.2

The same approach can be applied for investigation of the Dirichlet and Neumann problems for linear elliptic systems of equations in the plane domain if this system is strongly elliptic. Let us consider for example the following equation:

$$a(z)U_{z\bar{z}} + b(z)\bar{U}_{\bar{z}\bar{z}} + A_o(z)U_z + B_o(z)\bar{U}_z = F(z) \tag{2.5}$$

The ellipticity of this equation means that

$$|a(z)| \neq |b(z)|$$

for z in G and strong ellipticity means

$$|a(z)| > |b(z)|$$

for z in G. Let us look for a solution of the equation (2.5), satisfying the homogeneous Dirichlet condition

$$U(t) = 0$$

on the boundary of the domain G. Obviously such a solution may be represented in view

$$U(z) = -\frac{2}{\pi}\iint g(z,\zeta)F_1(\zeta)dG, \tag{2.6}$$

where $g(z,\zeta)$ is the Green function of the domain G and

$$F_1(z) = \frac{1}{a(z)}\left[F(z) - b(z)\bar{U}_{\bar{z}\bar{z}} - A_o(z)U_z - B_o(z)\bar{U}_{\bar{z}}\right]$$

provided if $a(z) \neq 0$ in G. Integrating by parts on the right-hand side of (2.6) and taking into account that $g(z,\zeta)$ for z in G and for ζ in the boundary of G, we obtain the following integral equation with respect to unknown $\omega(z) = U_z(z)$:

$$\omega(z) + \frac{b(z)}{a(z)}\bar{\omega}(z) - \iint \frac{b(\zeta)}{a(\zeta)}K(z,\xi)\bar{\omega}(\zeta)dG + K^*\omega = F(z) \qquad (2.7)$$

K^* is an integral operator compact in L_p and $K(z,\zeta)$ is the Bergman kernel function of the domain G, which is expressed in terms of the derivatives of Greens functin

$$K(z,\zeta) = -\frac{2}{\pi}\partial_{z\bar{\zeta}}g(z,\zeta)$$

The integral equation (2.7) is singular, because of the strong singularity of the kernel $K(z,\bar{\zeta})$ at the boundary points. Nevertheless it can be shown that this equation has the Fredholm' s property Here we do all calculations for the case when a, b are constant and the domain G is the unit disc $|z| < 1$, i.e. let us consider the equation

$$\omega(z) + \lambda\bar{\omega} - \lambda K\bar{\omega} = g(z) \qquad (2.8)$$

in disc $|z| < 1$, where

$$K\bar{\omega} = \iint \frac{\bar{\omega}(\zeta)}{(1-\bar{\zeta}z)^2}d\xi d\eta, \qquad \lambda = \frac{a}{b}.$$

From (2.8) it follows that

$$\omega(z) = \frac{1}{1-|\lambda|^2}(\varphi(z) - \lambda\bar{\varphi}(z) + g(z) - \lambda\bar{g}(z), \qquad (2.9)$$

because the function

$$\varphi(z) = K\bar{\omega} = \iint \frac{\bar{\omega}(\zeta)}{(1-\bar{\zeta}z)^2}d\xi d\eta$$

is holomorphic in the disc $|z| < 1$ for any ω from $L_p, p > 1$. Substituting (2.9) into (2.8) we obtain the integral equation for the holomorphic function $\varphi(z)$:

$$\varphi - \lambda K\bar{\varphi} = \lambda(Kg - \lambda K\bar{g}), \qquad (2.10)$$

because of the reproducing property

$$K\varphi = \varphi.$$

Hence by Green' s formula and the residue theorem for any holomorphic function $\varphi(z), K\bar{\varphi} = \bar{\varphi}(0)$ and (2.10) becomes

$$\varphi(z) = \lambda\bar{\varphi}(0) + \lambda(Kg - \lambda K\bar{g}).$$

Since $|\lambda| \neq 1$ and

$$(K\bar{g} - \bar{\lambda}Kg)z = 0 = \frac{1}{\pi}\iint \bar{g}(\zeta)d\xi d\eta - \frac{\bar{\lambda}}{\pi}\iint g(\zeta)d\xi d\eta,$$

then

$$\varphi(0) = \frac{1}{\pi}\iint g(\zeta)d\xi d\eta.$$

Thus the solution of (2.8) is

$$\begin{aligned} \omega(z) &= \frac{\lambda}{\pi}(1-|\lambda|^2)\iint(\bar{g}(\zeta) - \bar{\lambda}g(\zeta))d\xi d\eta = \frac{g(z) - \lambda\bar{g}(z)}{1-|\lambda|^2} \\ &= \frac{\lambda}{1-|\lambda|^2}(Kg - \lambda K\bar{g}) = \frac{|\lambda|^2}{1-|\lambda|^2}(\bar{\lambda}K^*g - K^*\bar{g}), \end{aligned}$$

where

$$K^*\bar{g} = \frac{1}{\pi}\iint \frac{g(\zeta)}{(1-\zeta\bar{z})^2}d\xi d\eta,$$

i.e. (2.8) is uniquely solvable and a general equation reduces to Fredholm' s equation as we desired. As we see above approach is not available if the function $a(z)$ vanishes in G: it is natural because in this case the equation is not strongly elliptic and the Dirichlet problem is not a good problem for such an elliptic equations. The following problem

$$Re\ U(t) = 0, \ \ Re\ t'(s)\partial\bar{z}U(t) = 0$$

turns out to be more natural problem for above second order elliptic system independently it is strongly elliptic or not: this can be shown also by means of our singular integral equations approach.

2.3

There is close link between above kind singular integral equations and the conjugation boundary problem. Here we demonstrate it for the case of the following singular integral equation:

$$a(z)\omega(z) - \frac{b(z)}{\pi}\iint \frac{\bar{\omega}(\zeta)}{(\bar{\zeta}-\bar{z})^2}d\xi d\eta = g(z_0) \tag{2.11}$$

We assume that the inequality

$$|a(z)| \neq |b(z)|$$

in closure of the multiply connected domain G holds. If equation (2.11) is solvable and $\omega(z)$ is its solution from L_p, with $p > 2$, then the function

$$W(z) = T\omega = -\frac{1}{\pi}\iint \frac{\omega(\zeta)}{\zeta - z} dG$$

is continuous in the whole complex plane, holomorphic outside of the closure of G, vanishes at infinity, but in domain G it has generalized first order derivatives such that

$$\begin{aligned} \phi(z) &= a(z)W(z) + b(z)\bar{W}(z) \quad \text{in } G \text{ and} \\ \phi(z) &= W(z) \qquad\qquad\qquad \text{outside of the closure of } G \end{aligned}$$

to conclude that it is a generalized analytical function:

$$\partial_{\bar{z}}\phi - A(z)\phi - B(z)\bar{\phi} = g(z)$$

with

$$\begin{aligned} A(z) &= \frac{\bar{a}(z)\partial_{\bar{z}}a(z) - \bar{b}(z)\partial_{\bar{z}}b(z)}{|a(z)|^2 - |b(z)|^2}, \\ B(z) &= \frac{a(z)\partial_{\bar{z}}b(z) - b(z)\partial_{\bar{z}}a(z)}{|a(z)|^2 - |b(z)|^2} \end{aligned}$$

in the domain. Outside the domain G the function ϕ is holomorphic and vanishes at infinity. On the boundary of the domain G this function satisfies the following jump condition:

$$\phi^+(t) = a(t)\phi^-(t) + b(t)\bar{\phi}^-(t).$$

Conversely if we can solve this last problem for generalized analytical functions, then we solve our integral equation. This last problem for holomorphic functions was considered by B.Bojarski. Using this link between integral equation and the jump boundary problem for generalized analytical function we come in particular to the following result:

Theorem 2 *Let $|a(z)| \neq |b(z)|$ in the closure of G and let $a(t) \neq 0$ on the boundary of G. Let $\kappa = Ind\ a(t)$ be the winding number of the function $a(t)$. Then singular integral equation has a finite index which equal exactly 2κ. In* *particular if $a(z) \neq 0$ in the closure of G, then it has zero index and in case $|a(z)| > |b(z)|$ in closure of G it is uniquely solvable.*

Let us consider more general characteristic singular integral equation

$$a(z)\omega + b(z)\bar{\omega} - \frac{c(z)}{\pi}\iint \frac{\omega(\zeta)}{(\zeta - z)^2}dG - \frac{d(z)}{\pi}\iint \frac{\bar{\omega}(\zeta)}{(\bar{\zeta} - \bar{z})^2}dG = g(z) \tag{2.12}$$

Theorem 3 *If the coefficients of equation (2.12) satisfy the condition*

$$|a(z)\bar{c}(z) - \bar{b}(z)d(z)| + |a(z)\bar{d}(z) - \bar{b}(z)c(z)| < ||a(z)|^2 - |b(z)|^2|$$

in the closure of G, then this equation is uniquely solvable in L_p, with $p > 2$ for any right-hand side $g(z)$, belonging to L_p with $p > 2$, provided the coefficients of (2.12) are continuously differentiable in G.

Corollary 1 *The equations*

$$\omega(z) + \frac{q(z)}{\pi}\iint \frac{\omega(\zeta)}{(\zeta - z)^2}dG = g(z),$$

$$\omega(z) + \frac{q(z)}{\pi}\iint \frac{\bar{\omega}(\zeta)}{(\bar{\zeta} - \bar{z})^2}dG = g(z)$$

are solvable in L_p for any right-hand side $g(z)$ from L_p with $p > 2$, provided $q(z)$ is bounded in the closure of G function satisfying the inequality $|q(z)| < 1$.

Note that this corollary first was proved by I. Vekua using contractive mapping principle.

2.4

Here we give one more application of above singular integral equation. Let Ω be the bi-disk $\{w = (z, \zeta), |z| < 1, |\zeta| < 1\}$. The problem is to find a function $U(z, \zeta)$ in Ω, which is piecewise continuous, bounded and satisfies the following equation

$$U_{\bar{z}\bar{z}} + U_{\zeta\zeta} = 0 \tag{2.13}$$

in Ω, the initial conditions

$$U(0, \zeta) = \varphi(\zeta), \qquad U_z(0, \zeta) = \psi(\zeta) \tag{2.14}$$

on the plane $z = 0$ and the boundary condition

$$Re\ U_{\bar{z}}(0, \zeta) = h(\zeta) \tag{2.15}$$

on the circle $|\zeta| = 1$, where $\varphi, \psi \in L_2(C)$ are given functions on the plane C and h is given Hölder continuous function on the circle $|\zeta| = 1$. To solve this problem it is an easy to see that a general solution of (2.13) is represented through two arbitrary piecewise continuous bounded functions of one variable F and G on complex plane by

$$U(z,\zeta) = F(\zeta + i\bar{z}) + G(\zeta - i\bar{z}) \tag{2.16}$$

Now second condition in (2.14) yields

$$(F - G)_{\bar{\zeta}}(\zeta) = \frac{1}{2}\psi(\zeta), \qquad |\zeta| < 1,$$

so that

$$F(\zeta) - G(\zeta) = -\Phi(\zeta) - \frac{1}{2\pi}\iint \frac{\psi(\eta)}{\eta - \zeta} dD,$$

where Φ is an arbitrary function holomorphic in the disc $|\zeta| < 1$. By means of the first condition (2.14) we have

$$\begin{aligned} F(\zeta) &= -\tfrac{1}{2}\Phi(\zeta) + \tfrac{1}{2}\varphi(\zeta) - \tfrac{1}{4\pi}\iint \tfrac{\psi(\eta)}{\eta-\zeta} dD, \\ G(\zeta) &= \tfrac{1}{2}\Phi(\zeta) + \tfrac{1}{2}\varphi(\zeta) + \tfrac{1}{4\pi}\iint \tfrac{\psi(\eta)}{\eta-\zeta} dD \end{aligned}$$

and by (2.16)

$$\begin{aligned} U(z,\zeta) &= -\tfrac{1}{2}\{\Phi(\zeta + i\bar{z}) - \Phi(\zeta - i\bar{z})\} + \tfrac{1}{2}\{\varphi(\zeta + i\bar{z}) - \varphi(\zeta - i\bar{z})\} \\ &\quad - \frac{1}{4\pi}\iint \psi(\eta)\left(\frac{1}{\eta - \zeta - i\bar{z}} - \frac{1}{\eta - \zeta + i\bar{z}}\right) dD \end{aligned}$$

Now assuming $\alpha = \zeta + i\bar{z}$, $\beta = \zeta - i\bar{z}$ from (2.16) we have

$$U_{\bar{z}}(0,\zeta) = -i\Phi'(\zeta) - \frac{i}{2\pi}\iint \frac{\psi(\eta)}{(\eta - \zeta)^2} dD \tag{2.17}$$

Therefore condition (2.15) yields the Riemann-Hilbert boundary condition

$$Re\ i\Phi'(\zeta) = \bar{h}(\zeta).$$

Let us replace the c second initial condition (2.15) into the following

$$a(\zeta)U_z(0,\zeta) + b(\zeta)U_{\bar{z}}(0,\zeta) = \psi(\zeta) \tag{2.18}$$

on the plane $z = 0$. In this case we denote for a moment the value $Uz(0,\zeta)$ of unknown derivative of U with respect to z on the plane by $\chi(\zeta)$ and represent the unknown holomorphic function Φ' as

$$\Phi'(\zeta) = \frac{\zeta^2}{i\pi}\iint \frac{\bar{\chi}(\zeta)}{(1 - \zeta\bar{\eta})^2} dD + \theta(\zeta)$$

with unknow holomorphic function $\theta(\zeta)$ in $|\zeta| < 1$. Substituting (2.17) into the boundary condition (2.15) we obtain

$$Re\ \theta(\zeta) = h(\zeta)$$

on $|\zeta| = 1$. From this condition we determine the unknown holomorphic function through Schwarz integral up to an imaginary constant C and then substituting it into (2.17) we find $U_{\bar{z}}(0,\zeta)$, so that the condition is reduced to the following singular integral equation

$$a(\zeta)\theta(\zeta) + \frac{b(\zeta)}{\pi}\iint\left[\frac{\theta(\zeta)}{(\eta-\zeta)^2} - \frac{\zeta^2\bar{\theta}(\eta)}{(1-\bar{\eta}\zeta)^2}\right]dD = \bar{g}$$

which we investigated under the condition $|a| \neq |b|$.

References

[1] H Begehr & A.Dzhuraev, An introduction to several complex variables and partial differential equations, Addison Wesley ,Longman,,Harlow ,1997.

[2] F.D. Gachov, Boundary value problems, Fizmatgiz, Moscow, 1963(russian); Pergamon Press, Oxford, 1966.

[3] N.Mushelishvili, Singular integral equations, Nordhoff, Groningen, 1953.

[4] I.N.Vekua, Generalized analitic functions, Moscow, 1959(russian); Pergamon Press, London, 1962.

3 ON SOME CLASSES OF PSEUDO-ANALYTIC FUNCTIONS

P. Berglez

Dedicated to Prof. Robert P. Gilbert on the occasion of his 65th birthday

3.1

The pseudo–analytic functions were introduced by L. Bers [8] generalizing the analytic functions of one complex variable. Independently I.N. Vekua [13] considered generalized analytic functions which are in close connection to the pseudo–analytic functions. Both authors developed several methods for the representation of these classes of functions and there we can find deep investigations of the function theoretic properties of these functions.

To apply the representations given by these authors to special problems one needs on the one hand the generating pair due to L. Bers or on the other hand the kernels of certain integral operators introduced by I.N. Vekua. Unfortunately these functions are known explicitly only in few cases.

The class of pseudo–analytic functions is related to particular elliptic differential equations of second order. For the latter there exist several different representations among which one is best suited for the explicit calculation of the solutions. It is the method introduced by K.W. Bauer [4] who made use of certain differential operators acting on holomorphic functions producing solutions of these elliptic equations. Because of the particular relation to the pseudo–analytic functions one can use these differential operators for the representation of pseudo–analytic functions also. The disadvantage of this method lies in the

H. Florian et al. (eds.), Generalized Analytic Functions, 27–32.

fact that there exist such operators only for certain classes of pseudo–analytic functions.

In the present paper we give some results concerning the existence and shape of such differential operators and further investigations on this topic.

The concepts described here for the representation of pseudo–analytic functions can be extended to certain systems of partial differential equations which leads to the so called generalized analytic vectors which themselves may be considered in the context of the theory of first order elliptic system as it has been developed e.g. by R.P. Gilbert and J.L. Buchanan in [9]. Particular investigations on this topic especially on the application of differential operators of suitable type for solving these systems may be found in [7].

3.2

The set of pseudo–analytic functions may be described by the class of solutions of the differential equation

$$w_{\bar{z}} = c\,\overline{w} \tag{3.1}$$

Let $\mathcal{D}$ be a fundamental domain for this equation in Vekua's sense. In $\mathcal{D}$ it's solutions can be represented by an integral operator as

$$w(z,\bar{z}) = \varphi(z) + \int_{z_1}^{z} \Gamma_1(z,\bar{z},t,\overline{z_0})\,\varphi(t)\,dt + \int_{\overline{z_1}}^{\bar{z}} \Gamma_2(z,\bar{z},z_0,\tau)\,\varphi^*(\tau)\,d\tau$$

where Γ_1, Γ_2 denote the Vekua–resolvents and φ is an arbitrary function holomorphic in $\mathcal{D}$ (the conjugate of which is φ^*) and z_0, z_1 are arbitrary but fixed points in $\mathcal{D}$.

With this operator it is possible to solve boundary value problems for pseudo-analytic functions provided the resolvents are known.

The resolvents are determined uniquely by an initial value problem (cf. [12], p. 70) which can be converted into the following one: Let u be a twoparametric solution of (3.1) which satisfies the conditions

$$u(z,\tau;t,\tau) = c(z,\tau)\,, \quad u(t,\bar{z};t,\tau) = c(t,\tau)$$

Then the resolvents are given by

$$\Gamma_2(z,\bar{z};t,\tau) = u(z,\bar{z};t,\tau)\,, \quad \Gamma_1(z,\bar{z};t,\tau) = \frac{1}{c(z,\bar{z})}\overline{\left[\frac{\partial}{\partial \bar{z}}\Gamma_2(z,\bar{z};t,\tau)\right]}$$

With the pseudo–analytic functions so called complex potentials are related (cf. [13], p. 141) which satisfy a certain differential equation of second order.

Let $R(z, \bar{z}; t, \tau)$ be the Riemann function for this equation then the second Vekua resolvent for (3.1) is given by

$$\Gamma_2(z, \bar{z}; t, \tau) = c(t, \tau)\, R(t, \tau; z, \bar{z})$$

Another possibility for the representation of pseudo-analytic functions is given by differential operators of the form

$$K_n := \sum_{k=0}^{n} a_k(z, \bar{z}) \frac{\partial^k}{\partial z^k}\,, \quad \tilde{K}_m := \sum_{k=0}^{m} b_k(z, \bar{z}) \frac{\partial^k}{\partial \bar{z}^k}\,, \quad n, m \in \mathbb{N}$$

They are acting on holomorphic and antiholomorphic functions and give solutions of (3.1) provided $c \neq 0$ in $\mathcal{D}$. In [6] we proved that it is necessary and sufficient for the existence of such an operator K_n that there is a natural number n such that with

$$B_0 = -c\bar{c}\,, \quad A_0 = c^{-1}$$

$$A_k = A_{k-1}\, B_{k-1}\,, \quad B_k = B_{k-1} + (\log A_k)_{z\bar{z}}\,, k = 1, \ldots, n$$

the condition $B_n \equiv 0$ in $\mathcal{D}$ is satisfied. This condition is a nonlinear partial differential equation of order $2n$ for the coefficient c in equation (3.1). Provided the operator K_n exists it is given by

$$K_n = F_1 \cdots F_n\,, \quad F_k = \frac{\partial}{\partial z} + (\log A_k)_z$$

Then the function

$$w = K_n f + \frac{1}{\bar{c}} \frac{\partial}{\partial z} \overline{K_n f} = K_n f + c\, \overline{K_{n-1} f} \tag{3.2}$$

is a solution of (3.1) in $\mathcal{D}$ for each f holomorphic in $\mathcal{D}$.

Conversely with a theorem proved in [5] we can state that every solution w of (3.1) may be represented in the form (3.2) using a function f holomorphic in $\mathcal{D}$.

Under the assumption that such a differential operator K_n exists the Vekua resolvents Γ_1 and Γ_2 can be calculated immediately from the coefficient c in (3.1) as

$$\Gamma_2(z, \bar{z}, t, \tau) = c(t, \tau)\, K_n \left(E_0 \cdots E_{n-1} \left(\frac{A_n(t, \tau)}{A_n(z, \tau)} \right) \right)$$

with

$$E_k\, X := \frac{1}{B_{n-k}(t, \tau)} \left(\frac{\partial}{\partial \tau} X \right), \quad k = 0, \ldots, n-1$$

The first Vekua resolvent is then given by

$$\Gamma_1(z,\bar{z};t,\tau) = \left[\int_{\eta=\tau}^{\zeta} \Gamma_2(z,\zeta;t,\eta)\, c^*(\eta,t)\, d\eta\right]_{\zeta=\bar{z}}$$

In [11] the condition for the existence of such a differential operator K_n has been investigated. For $n = 1$ all factors have been given explicitly such that there exists an operator K_1 for (3.1). In this report relations between a pseudo–analytic function w and its generating holomorphic function f are investigated, especially the generator of the zero–solution is given. Moreover we find representations of pseudo–analytic functions in the neighbourhood of isolated singularities.

3.3

Finally we consider some particular classes of coefficients c in equation (3.1) such that the solutions may be represented by differential operators.

In [3] the equation

$$w_{\bar{z}} = \frac{n\alpha'}{\alpha+\bar{\alpha}}\,\overline{w}\,,\quad \alpha(z)\,\text{holomorphic in}\ \mathcal{D},\ \alpha'(\alpha+\bar{\alpha}) \neq 0\,,\quad n \in \mathbb{N}$$

was investigated. K.W. Bauer considered in [1] equation (3.1) with the coefficient $c = \gamma_{\bar{z}}/\gamma$ where $\gamma \neq 0$ is a real valued solution of the differential equation

$$u_{z\bar{z}} - \frac{n(n+1)}{(z+\bar{z})^2}\,u = 0\,,\ n \in \mathbb{N} \tag{3.3}$$

whereas in [2] the coefficient has the form

$$c = \frac{p_z q_{zz} - q_z p_{zz}}{p_{\bar{z}} q_z - p_z q_{\bar{z}}}$$

Here p and q are real valued solutions of

$$u_{z\bar{z}} - \frac{1}{z+\bar{z}}(\lambda u_z + \mu u_{\bar{z}}) = 0\,,\ \lambda,\mu \in \mathbf{Z}$$

Recently in [10] the following differential equations were discussed

$$w_{\bar{z}} = -(\log(Y_z - Y_{\bar{z}}))_z\,\overline{w}$$

$$w_{\bar{z}} = \left[\log\left(Y_z + Y_{\bar{z}} - \frac{2(n+1)}{z+\bar{z}}Y\right)\right]_z \overline{w}$$

$$w_{\bar{z}} = \left[\log\left(Y_z + Y_{\bar{z}} + \frac{2n}{z+\bar{z}}Y\right)\right]_z \overline{w}$$

where Y is an arbitrary particular solution of (3.3).

References

[1] K.W. BAUER, *Zur Darstellung pseudoanalytischer Funktionen*, in: Function Theoretic Methods for Partial Differential Equations, (Proc. Internat. Sympos., Darmstadt 1976), 101–111, Lecture Notes in Math. **561**, Springer, Berlin, 1976.

[2] K.W. BAUER, *Differentialoperatoren bei verallgemeinerten Euler–Gleichungen*, Ber. Math.-Statist. Sekt. Forschungszentrum Graz **121** (1979), 1–17.

[3] K.W. BAUER - ST. RUSCHEWEYH, *Ein Darstellungssatz für eine Klasse pseudoanalytischer Funktionen*, Ber. Gesellsch. Math. Datenv. Bonn **75** (1973), 3 – 15.

[4] K.W. BAUER - ST. RUSCHEWEYH, *Differential Operators for Partial Differential Equations and Function Theoretic Applications*, Lecture Notes in Math., **791** (1980).

[5] P. BERGLEZ, *Characterization of certain differential operators in the representation of pseudo–analytic functions,*

Rend. Ist. Mat. Univ. Trieste **14** (1982), 27 – 31.

[6] P. BERGLEZ, *Differentialoperatoren bei partiellen Differentialgleichungen*, Ann. Mat. Pura Appl., **143** (1986), 155–185.

[7] P. BERGLEZ, *On the representation of generalized analytic vectors,* Z. Anal. Anwendungen **14**, (1995), 421–430.

[8] L. BERS, *Theory of Pseudo–Analytic Functions,* New York University, 1953.

[9] R.P. GILBERT, J.L. BUCHANAN, *First Order Elliptic Systems: A Function Theoretic Approach,* Academic Press, New York, 1983.

[10] K. NUSSER, *Über die Darstellung pseudoanalytischer Funktionen,* Diplomarbeit, Graz, 1994.

[11] S. PERUCI, *Allgemeine Darstellungssätze für pseudoanalytische Funktionen,* Diplomarbeit, Graz, 1997.

[12] I. N. VEKUA, *New Methods for Solving Elliptic Equations,* North Holland Publ. Co, Amsterdam, 1967.

[13] I. N. VEKUA, *Verallgemeinerte analytische Funktionen*, Akademie Verlag, Berlin, 1963.

4 SOME YUGOSLAVIAN CONTRIBUTIONS TO THE THEORY OF GENERALIZED ANALYTIC FUNCTIONS

M. Čanak

Dedicated to Professor Robert P. Gilbert on the occasion of his 65th birthday

After World War II a "Yugoslavian School" developed contributing to the theory of generalized analytic functions. Choosing only a few of a large number of papers, we hint to some methods making easier to solve the VEKUA equation in some special cases.

4.1

Let

$$D = (u'_x - v'_y) + i(u'_y + v'_x) = 2w'_{\overline{z}}$$

be the well-known differential operator of KOLOSSOV, and let $A = A(z)$ and $B = B(z)$ be given analytic functions. Then in his paper [1] M. ČANAK investigates systems of partial differential equations which can be written in the form

$$Dw = A\overline{w} + B \tag{4.1}$$

Using the method of so-called modified areolair series, he shows that the general solution of (4.1) has the form

$$w(z, \overline{z}) = w_0 + A\sum_{k=1}^{\infty} \varphi_{k-1}\overline{\varphi}_k + \frac{B}{2}\overline{z} + \frac{A}{2}\sum_{k=1}^{\infty} \beta_{k-1}\overline{\beta}_k \tag{4.2}$$

H. Florian et al. (eds.), Generalized Analytic Functions, 33–39.

where $w_0 = w_0(z)$ is an arbitrary analytic function, while the φ_j and the β_j are defined by the recursions

$$\begin{aligned} \varphi_0 &= 1 \\ 2\varphi_1 &= \int_0^z w_0(z)dz \\ &\vdots \\ 2\varphi_n &= \int_0^z A(z)\varphi_{n-2}(z)dz \end{aligned}$$

and

$$\begin{aligned} \beta_0 &= z \\ 2\beta_1 &= \int_0^z B(z)dz \\ 2\beta_2 &= \int_0^z zA(z)dz \\ &\vdots \\ 2\beta_n &= \int_0^z A(z)\beta_{n-2}(z)dz \end{aligned}$$

resp. It is also proved that the two areolair series

$$\sum_{k=1}^{\infty} \varphi_{k-1}\overline{\varphi}_k \quad \text{and} \quad \sum_{k=1}^{\infty} \beta_{k-1}\overline{\beta}_k$$

are uniformly convergent in a finite simply connected domain. Further, the solution (4.2) satisfies an initial condition on a given curve (cf. also the next section 4.2.).

Independently on [1], the method of areolair series of the form

$$w(z,\overline{z}) = \sum_{p,q=0}^{\infty} c_{pq} z^p \overline{z}^q$$

is also used by D. DIMITROVSKI, N. RALEVIĆ and B. ILIEVSKI, see [2, 3, 4].

4.2

Suppose a complex equation of the form

$$\overline{z} = g(z) \tag{4.3}$$

is given where $g(z)$ is an arbitrary analytic function. This equation (4.3) determines a closed or a non-closed contour, or it determines a set of isolated points. In the sequel the set of all solutions of (4.3) will be denoted as K-contour. Notice that many important contours in the complex plane such as straight lines, circles and so on can be represented in the form (4.3).

In his paper [5] and in some earlier papers as well, M. Čanak defined a mapping $\alpha_{g(z)}$ of the set of complex-valued continuous functions $w = w(z, \overline{z})$ on the set of analytic functions in the following way:

Suppose $g(z)$ is analytic and $w(z, \overline{z})$ is a power-series in z and $\overline{z}$. Then the composed function $w(z, g(z))$ is analytic and will be denoted by $\alpha_{g(z)}w$. The geometrical meaning of this operator $\alpha_{g(z)}$ is the following: Provided $\overline{z} = g(z)$ defines a closed contour, the functions $w(z, \overline{z})$ and $\alpha_{g(z)}w$ have the same boundary values on this contour.

The representation of contours in the form $\overline{z} = g(z)$ and applications of the mapping $\alpha_{g(z)}$ are the main ideas of the so-called CRC-method where CRC means complex representations of contours. The author used this method in the theory of non-analytic functions and for solving boundary value problems, see [5] where one can find also an example for the CRC-method.

4.3

In his paper [6] S. Fempl shows that the general solution of the equation

$$Dw + A(z, \overline{z})w + B(z, \overline{z}) = 0 \tag{4.4}$$

with (continuous) coefficients A, B can be represented in the form

$$w = \left[Q(z) - \frac{1}{2}\int B \exp\left(\frac{1}{2}\int A d\overline{z}\right) d\overline{z}\right] \exp\left(-\frac{1}{2}\int A d\overline{z}\right)$$

where $Q(z)$ is an arbitrary analytic function.

The solutions of (4.4) are also called generalized analytic functions in the sense of Fempl or F-analytic functions. Of course, the differential equation (4.4) is only a special case of the differential equation which are under consideration in Vekua's theory. On the other hand, F-analytic functions can be represented by areolair series of the form

$$\sum_{k=1}^{\infty} \overline{z}^k \varphi_k(z),$$

whereas such a convenient representation is not possible for the solutions of more general equations. Moreover, following A. BILIMOVIĆ [7], geometric aspects of F-analytic functions can be investigated. In [8] various boundary value problems are solved for this class of functions. The left-hand side of the differential equation (4.4), finally, leads to the definition of the FEMPL differential operator

$$F_{A,B}w = Dw + A(z,\overline{z})w + B(z,\overline{z}) = \Phi(z,\overline{z}) \tag{4.5}$$

where the characteristics $A(z,\overline{z})$ and $B(z,\overline{z})$ are given. This operator assigns a (continuous) function $\Phi(z,\overline{z})$ to a given differentiable function $w(z,\overline{z})$. The operator $F_{A,B}$ plays an important role in the theory of complex differential equations.

Consider, for instance, the linear complex differential equation

$$D^2w + a(z,\overline{z})Dw + b(z,\overline{z})w = c(z,\overline{z}). \tag{4.6}$$

We ask under which conditions the solutions of this equation (4.6) are mapped into themselves when applying the differential transformation (4.5). In the case of real differential equations of second order these question was posed by I. ŠAPKAREV [9]. After a longer calculation one gets the condition

$$2Da + a^2 - 4b = 0.$$

If this condition is satisfied, then the general solution of (4.6) can be written in the form

$$w(z,\overline{z}) = P(z)R(z,\overline{z}) + Q(z)S(z,\overline{z}) + T(z,\overline{z})$$

where $P(z)$ and $Q(z)$ are arbitrary analytic functions, while $R(z,\overline{z})$, $S(z,\overline{z})$ and $T(z,\overline{z})$ are known complex-valued functions which can be expressed by $a(z,\overline{z}), b(z,\overline{z})$ and $c(z,\overline{z})$.

Various boundary value problems can be solved for the partial differential equation (4.6). Consider, for instance, the following CARLEMAN type boundary value problem:

Boundary value problem C. Find the solution of the partial complex second order equation

$$D^2w + \frac{2}{\overline{z}}Dw - \frac{1}{\overline{z}^2}w = 0 \tag{4.7}$$

satisfying the CARLEMAN type boundary condition

$$w\left(\frac{1}{t}\right) = tw(t) + 1 - t \tag{4.8}$$

$$Dw\left(\frac{1}{t}\right) = \frac{1}{t}Dw(t) + \frac{1}{t} - 1 \tag{4.9}$$

on the unit circle K which is defined by

$$\overline{z} = \frac{1}{z}.$$

Using the differential transformation

$$\begin{aligned} Dw + \frac{w}{\overline{z}} &= \Phi, \quad \text{i.e.,} \\ D^2 w + \frac{\overline{z} Dw - 2w}{\overline{z}^2} &= D\Phi, \end{aligned}$$

equation (4.7) passes over into

$$D\Phi + \frac{\Phi}{\overline{z}} = 0.$$

A longer calculation shows, finally, that the general solution of (4.7) is given by

$$w(z, \overline{z}) = P(z)\overline{z}^{\frac{1}{2}} + Q(z)\overline{z}^{-\frac{1}{2}} \tag{4.10}$$

where $P(z)$ and $Q(z)$ are arbitrary analytic functions. Substituting the value (4.10) and its areolair derivative into (4.8) and (4.9), we get

$$P\left(\frac{1}{t}\right) t^{\frac{1}{2}} + Q\left(\frac{1}{t}\right) t^{-\frac{1}{2}} = P(t)t^{\frac{1}{2}} + Q(t)t^{\frac{3}{2}} + 1 - t \tag{4.11}$$

$$P\left(\frac{1}{t}\right) t^{-\frac{1}{2}} - Q\left(\frac{1}{t}\right) t^{-\frac{3}{2}} = P(t)t^{-\frac{1}{2}} - Q(t)t^{\frac{1}{2}} + \frac{1}{t} - 1. \tag{4.12}$$

Eliminating the function P, these equations (4.11) and (4.12) lead to

$$Q\left(\frac{1}{t}\right) = t^2 Q(t).$$

This, however, is nothing else than the homogeneous CARLEMAN boundary value problem for analytic functions. A solution of this boundary value problem is

$$Q(z) = \frac{1}{z}$$

(see [10]). Substituting this value into (4.11), (4.12), one obtains

$$P\left(\frac{1}{t}\right) - P(t) = \frac{1 - t}{t^{\frac{1}{2}}}.$$

This is a so-called jump problem. Its solution is

$$P(z) = z^{\frac{1}{2}},$$

cf. [10]. Finally, using this result, one gets the solution of the boundary value problem (4.7), (4.8), (4.9) in the form

$$w(z,\overline{z}) = z^{\frac{1}{2}}\overline{z}^{\frac{1}{2}} + \frac{1}{z+\overline{z}^{\frac{1}{2}}}.$$

4.4

The Yugoslavian mathematicians investigated various generalizations of the VEKUA system. Here we shall quote only two of them.

Using the pseudo-linear transformation $V = \alpha W + \beta\overline{W} + \gamma$, D. DIMITROVSKI, M. RAJOVIĆ and D. STOJILKOVIĆ [11] proved that the generalized VEKUA equation

$$aV'_{\overline{z}} + bV'_{z} = AV + B\overline{V} + F$$

can be reduced to

$$W'_{\overline{z}} = AW + B\overline{W} + F$$

provided the condition $a\beta + b\overline{\alpha} = 0$ holds. Moreover, the latter equation is solved by iterations. Similarly,the same authors solve the equation

$$aV'_{\overline{z}} + bV'_{z} + c\overline{V}'_{\overline{z}} + d\overline{V}'_{z} = AV + B\overline{V} + F,$$

too.

N. RALEVIĆ [3] and M. ČANAK [12] defined the so-called p-polyanalytic functions which are an important generalization of the ordinary polyanalytic function by

$$P_n(z,\overline{z}) = \sum_{k=0}^{n}(z+\overline{z})^k f_k(z,\overline{z})$$

or by

$$P_n(z,\overline{z}) = \sum_{k=0}^{n}\overline{z}^k f_k(z,\overline{z})$$

where the $f_k(z,\overline{z})$ are arbitrary p-analytic functions which are defined by the well-known POLOŽIJ system

$$u'_x = \frac{1}{p}v'_y, \quad u'_y = -\frac{1}{p}v'_x.$$

where $f_k = u + iv$. M. ČANAK solved also some boundary value problems for this class of functions.

References

[1] M. ČANAK, Systeme von Differentialgleichungen erster Ordnung vom elliptischen Typus mit analytischen Koeffizienten und Methode der verallgemeinerten areolären Reihen. Publications de L'Inst. Math., Beograd, vol. 33 (47),1983, pp. 35-39.

[2] D. DIMITROVSKI and B. ILIEVSKI, L'équation différentielle aréolaire analytique. PRILOZI, Macedonian academy of sciences, vol. 1-2, Skopje, 1984, pp. 25-39.

[3] N. RALEVIĆ, Analitičke i numeričke metode za rešavanje graničnih problema za eliptičke parcijalne diferencijalne jednačine. Thesis, Beograd 1986.

[4] B. ILIEVSKI, Linearni areolarni ravenki. Thesis, Skopje 1992.

[5] M. ČANAK, Anwendung der KRK-Methode auf eine inverse Randwertaufgabe. Zeitschr. angew. Math. Mech. (=ZAMM), vol. 70, 1990, pp. 573-574.

[6] S. FEMPL, Reguläre Lösungen eines Systems partieller Differentialgleichungen. Publications de L'Inst. Math., Beograd, vol. 4 (18), 1964, pp. 115-120.

[7] A. BILIMOVIĆ, Sur la géometrie differentielle d'une fonction non analytique. Glas de L'Académie Serbe des Sciences, Beograd vol. CCXLII, N. 19, 1960, pp. 1-81.

[8] N. RALEVIĆ and M. ČANAK, Some boundary value problems for an elliptic system of partial equations of Fempl's type. Zeitschr. angew. Math. Mech. (=ZAMM), vol. 66, 1986, pp. 390-392.

[9] I. ŠAPKAREV, Sur des équations différentielles linéaires transformables en elles-mémes par un changement de fonctions. Glasnik Mat. Fiz. Astr. 19, 1964, pp. 211-215.

[10] G. S. LITVINČUK, Kraevie zadači i singuljarnie integraljnie uravnenija so sdvigom. Moskva, Nauka, 1977.

[11] D. DIMITROVSKI, M. RAJOVIĆ and D. STOJILKOVIĆ, The generalization of the VEKUA equation with analytic coefficients in $z, \bar{z}$. Filomat, Niš, 1997.

[12] M. ČANAK, Randwertaufgabe von Riemanntypus für die p–polyanalytischen Funktionen auf der spiralförmigen Kontur. Matematički Vesnik, Beograd, vol. 40, 1988, pp. 197-203.

5 APPLICATIONS OF REPRODUCING KERNELS TO PARTIAL DIFFERENTIAL SYSTEMS

R. P. Gilbert

Abstract: In an earlier work Gilbert and Hile [12] developed the idea of a reproducing kernel on a Hilbert Function module as these modules seem to have the appropriate structure for systems of partial differential equations. Their work was based on the pioneering researches of Goldstein–Horwitz [13] and Saworotnow [16] on Hilbert modules. In the present paper we collect several examples of Hilbert function modules which arise naturally in the theory of differential equations, and also in the theory of elastic shells and show the existence of a reproducing kernel.

For further details concerning reproducing kernels see the above cited works and also the monograph by Begehr–Gilbert [2]. [1]

5.1 HILBERT MODULES

A Hilbert module is a linear space H whose *scalars* are elements from a finite dimensional H^*–algebra $\mathcal{A}$. Furthermore there exists *addition of vectors* and *right–multiplication by scalars*. These operations are governed by the following

(a) $f + g = g + f$

(b) $(f + g) + h = f + (g + h)$

H. Florian et al. (eds.), Generalized Analytic Functions, 41–54.

(c) $f(a+b) = fa + fb$

(d) $(f+g)\,a = fa + ga$

(e) $(fa)\,b = f(ab)$

(f) $f \cdot 0 = 0\,,\; f \cdot 1 = 1$

where $f, g, h \in H, a, b \in \mathcal{A}$.

The algebra $\mathcal{A}$ may be over either the real or complex field, and as a finite H^*–algebra it has the following properties:

(i) $\mathcal{A}$ is a Banach algebra with unity I (or I), whose underlying Banach space is a finite dimensional Hilbert space with inner product $\langle , \rangle$, and norm given by $\|a\| = (\langle a, a \rangle)^{\frac{1}{2}}$.

(ii) $\mathcal{A}$ has an involution $a \to a^*$ satisfying

$$\begin{aligned}(a^*)^* &= a \\ (ab)^* &= b^* a^* \\ (\lambda a)^* &= \bar{\lambda} a^* (\lambda \text{ real or complex})\end{aligned}$$

(iii) m $\langle ab, c \rangle = \langle b, a^* c \rangle = \langle a, cb^* \rangle$

We say an element a in $\mathcal{A}$ is *symmetric* if $a = a^*$, and *positive* if both a is symmetric and $\langle ab, b \rangle \geq 0$ for all b in $\mathcal{A}$. An element e in $\mathcal{A}$ is a *projection* if $e = e^2 = e^*$. Another property of a finite dimensional H^*–algebra is:

(iv) If a is positive, then a has the representation

$$a = \sum_{i=1}^{n} \lambda_i^2 e_i$$

where the e_i's are projections in $\mathcal{A}$, $e_i e_j = 0$ if $i \neq j$, and each λ_i is a positive real number. Furthermore, a has a unique positive square root given by

$$a^{\frac{1}{2}} = \sum_{i=1}^{n} \lambda_i e_i\,.$$

As further postulates on our Hilbert module, we assume that H itself has an *inner product* (f, g), defined for f, g in H and with values in $\mathcal{A}$, and governed by the rules

(g) $(f+g, h) = (f, h) + (g, h)$

(h) $(f, g)^* = (g, f)$

(i) (f, f) is positive, and $(f, f) = 0$ if and only if $f = 0$

(j) $(f, ga) = (f, g)a$ for a in $\mathcal{A}$.

We assume that the norm in $\mathcal{A}$ is chosen such that $\|I\| = 1$. It is convenient for us to introduce *trace function* by

(1) $tr\,a = \langle a, I\rangle$

we then have

(2) $tr\,I = 1$
$tr(ba) = \langle a, b^*\rangle = tr(ab)$
$\langle a, b\rangle = tr\,b^* a.$

We define the modulus of an element a in $\mathcal{A}$ by

(3) $|a| = tr\,[(aa^*)^{\frac{1}{2}}]$

where $(aa^*)^{\frac{1}{2}}$ is the unique positive square root of the positive element aa^*. The two norms on $\mathcal{A}$ are equivalent, and in fact there exists a positive constant k such that

(4) $|a| \le \|a\| \le k\,|a|$.

To obtain the right–hand inequality, we observe that since $\mathcal{A}$ is a Banach algebra, there exists a positive constant k such that

$$\|ab\| \le k\,\|a\|\,\|b\|\,.$$

Let $aa^* = p^2$ where p is positive. Then

$$|a| = tr\,p = \langle p, I\rangle = \langle\sqrt{p}, \sqrt{p}\rangle = \|\sqrt{p}\|^2$$

and

$$\|a\|^2 = \langle a, a\rangle = \langle p^2, I\rangle = \langle p, p\rangle = \|p\|^2\,.$$

Hence

$$\|a\| = \|p\| \le k\,\|\sqrt{p}\|^2 = k\,|a|\,.$$

For the left–hand inequality, we have

$$|a| = \langle p, I\rangle \le \|p\|\,\|I\| = \|a\|\,.$$

Next we define a norm on elements of H by

(5) $\|f\| = +|(f, f)|^{\frac{1}{2}}$

which is equivalent to

(6) $\|f\|^2 = tr(f, f)$.

(We use the same notation for norms on $\mathcal{A}$ and H, but no confusion should develop.)

(k) H is *complete* with respect to the norm (5).

We are particularly interested in the case where H is a family of functions defined on some topological space Ω, with values in $\mathcal{A}$. Right–multiplication by elements in $\mathcal{A}$ is then naturally defined. We shall refer to a family $H = H(\Omega)$ of functions, defined pointwise on a topological space Ω with values in $\mathcal{A}$, as a Hilbert *function module* if it is closed under addition and under right–multiplication by elements of $\mathcal{A}$, if it is endowed with an inner product with values in $\mathcal{A}$, and if postulates (a)–(k) above hold.

5.2 REPRODUCING KERNELS FOR HILBERT MODULES

We say that a Hilbert function module $H(\Omega)$ has a *reproducing kernel* $K = K(x,y)$ if

(i) K is defined point–wise in $\Omega \times \Omega$ with values in $\mathcal{A}$,

(ii) for each fixed x in Ω the function K_x, given by $K_x(y) = K(x,y)$, in $H(\Omega)$,

(iii) $f(x) = (K_x, f)$ for $x \in \Omega$, $f \in H(\Omega)$.

We refer to a Hilbert function module $H(\Omega)$ as having *property P* if for each x in Ω there exists a non–negative constant $M(x)$ such that

$$|f(x)| \leq M(x)\,\|f\|$$

for all f in H.

The following theorem may be proved as in the Hilbert space situation, see for example, Fichera, [8] Meschkowski, [15] or Gilbert–Hile, [12].

Theorem 1 *$H(\Omega)$ has a reproducing kernel if and only if $H(\Omega)$ has property P. Moreover, the reproducing kernel, if it exists, is unique.*

Likewise, it is easy to show the

Theorem 2 *If $H(\Omega)$ has a reproducing kernel, then a sequence that converges in the norm of $H(\Omega)$ converges pointwise. If $H(\Omega)$ has property P' , convergence is uniform on relatively compact subsets of Ω.*

Theorem 3 *([1] pp. 296–298). If H is separable (i.e. there exists a countable, dense subset of H), then there exists an orthonormal basis $\{\phi_n\}_{n=1}^{\infty}$ for H, in the sense that each g in H has the representation*

$$g = \sum_n \phi_n(\phi_n, g)$$

where the series converges in the norm of H, and moreover,

$$\|g\|^2 = \sum_n |(\phi_n), g|^2 .$$

As in the Hilbert space case we have [1].

Theorem 4 *If $H(\Omega)$ has reproducing kernel K, and $\{\phi_n\}_{n=1}^{\infty}$ is an orthonormal basis for $H(\Omega)$, then K has the representation*

$$K(x,y)\Phi = \sum_n (\phi_n(y), \phi_n(x)^*) .$$

For fixed x the series, as a function of y, converges both in the norm of H and pointwise in Ω. If $H(\Omega)$ has property P', convergence is uniform on relatively compact subsets of Ω.

Moreover, the reproducing kernel for a Hilbert function module can be characterized as the solution to certain extremal problems.

With this end in mind we consider two extremal problems:

Problem A. For x in Ω, b in $\mathcal{A}$, minimize $\|f\|$ subject to $f \in H(\Omega)$, $f(x) = b$

Problem B. For x in Ω, minimize $\|f\|$ subject to

$$f \in H(\Omega)\,, \ |f(x)| = 1\,.$$

Gilbert and Hile [12] established the following:

Theorem 5 *If b has the representation $b = K(x,x)\,a$, then Problem A has the unique solution $g = K_x a$.*

and

Theorem 6 *If $K(x,x) \neq 0$, the Problem B has a solution, uniquely determined in the sense that g in $H(\Omega)$ is a solution if and only if it has the form $g = K_x a$, where $a \in \mathcal{A}$ and satisfies $a^* a = |K(x,x)|^2 I$.*

Corollary 1 *Assume $K(x,x) \neq 0$. Then (i) functions of the form $K_x a$, where $a^* a = \mu I$, $\mu > 0$, minimize $\|f\|\,|f(x)|^{-1}$ subject to $f \in H(\Omega)$, $f \neq 0$, (ii) functions of the form $K_x a$ where $z^* a = |K(x,x)|^{-1} I$, maximize $|f(x)|$ subject to $f \in H(\Omega)$, $\|f\| = 1$.*

Theorem 7 *The space $H(\Omega)$ is separable if and only if K has the representation*

$$K(x,y) = \sum_n \phi_n(y)\,\phi_n(x)^* \tag{5.1}$$

or some countable orthonormal set $\{\phi_n\}_{n=1}^{\infty}$ in $H(\Omega)$, where the series for fixed x converges as a function of y in the norm of $H(\Omega)$.

5.3 EXAMPLES OF HILBERT FUNCTION MODULES

The structure of the Clifford algebra suggests we study the algebraic systems generated by the $m \times m$ matrices $A_2, A_3, \ldots, A_n$ with complex constant entries, where the matrices satisfy

$$A_i A_j + A_j A_i = 0\,, \quad i \neq j\,, \quad 2 \leq i,j \leq n$$

$$A_i^2 = -I\,, \quad 2 \leq i \leq n\,. \tag{5.2}$$

Here I is the identity matrix. The complex matrix solutions in $\mathbf{R}^n$ of the partial differential equation

$$DF \equiv I\,\frac{\partial F}{\partial x_1} + A_2\,\frac{\partial F}{\partial x_2} + \cdots + A_n\,\frac{\partial F}{\partial x_n} = 0 \tag{5.3}$$

are referred to as hyperanalytic in the Clifford sense. Let $x := (x_1, x_2, \dots, x_n) \in \Omega \subset \mathbf{R}^n$ where Ω is a domain, and define

$$\begin{aligned} |x| &:= (x_1^2 + \cdots + x_n^2)^{\frac{1}{2}}, \\ x^* &:= x_1 I + x_2 A_2 + \cdots + x_n A_n, \\ \bar{x} &:= x_1 I - x_2 A_2 - \cdots - x_n A_n. \end{aligned} \tag{5.4}$$

It was shown in [12] that the $C^1(\Omega)$ solutions of (5.3) have a Cauchy integral representation, namely for $x \in \Omega$

$$F(x) = \omega_n^{-1} \int_{\dot{\Omega}} |y-x|^{-n}\, \overline{(y-x)\,\nu^*}(y)\, F(y)\, d\sigma(y), \tag{5.5}$$

where ν is the outwards pointing unit normal vector on $\dot{\Omega}$. ω_n is the surface area of the unit sphere on $\mathbf{R}^n$.

Using this integral representation we may show that there exists a bounded linear functional on the class of hyperanalytic functions. We do this for $\Omega \subset \mathbf{R}^n$ a fixed domain, and $\mathcal{A}$ be some subalgebra with unity of the algebra of $m \times m$ constant matrices, such that $\mathcal{A}$ is closed under conjugate–transposition.

For some fixed $x \in \Omega$ let $B(x,\rho) \subset \Omega$ to be a sphere with center x and radius ρ. Using the Cauchy representation (5.5) above on the surface $B(x,\rho)$ we have

$$F(x) = \omega_n^{-1} \int_{|y-x|=\rho} \rho^{-n}(y-x)((y-x)^*/\rho)\, F(y)\, d\sigma(y), \tag{5.6}$$

but as

$$\overline{(y-x)}(y-x)^* = |y-x|^2\, I = \rho^2 I,$$

we obtain

$$F(x) = \omega_n^{-1}\rho^{1-n} \int_{|y-x|=\rho} F(y)\, d\sigma(y), \tag{5.7}$$

which yields

$$F(x) = nR^{-n}\,\omega_n^{-1} \int_{|y-x|\le R} F(y)\, dy.$$

Since for elements $a \in \mathcal{A}$, $\|a\| = (tr(aa^*))^{\frac{1}{2}}$, $|a| = tr(aa^*)^{\frac{1}{2}}$ and that $|a| \le \|a\|$, the Cauchy–Schwarz inequality implies

$$\begin{aligned} |F(x)| &\le \|F(x)\| \le nR^{-n}\,\omega_n^{-1} \int_{|x-y\le R} \|F(y)\|\, dy \\ &\le nR^{-n}\,\omega_n^{-1} \left(\int_{|y-x|\le R} dy\right)^{\frac{1}{2}} \left(\int_{|y-x|\le R} \|F\|^2\, dy\right)^{\frac{1}{2}} \\ &\le nr^{-n}\,\omega_n^{-1}(\omega_n R^n/n)^{\frac{1}{2}}\, \|F\|_{L_2,\Omega}. \end{aligned} \tag{5.8}$$

Hence, there exists a bounded linear functional on the class of solutions,

$$|F(x)| \leq C R_n^{-n/2} \, \|F\|_{L_n,\Omega}\,, \tag{5.9}$$

where R_x is the distance from x to Ω.

Finally (5.9) implies that $H(\Omega)$ has a reproducing kernel, and axiom (k) then implies it has an orthogonal expansion (5.1).

Another example is provided by the generalized analytic vectors of Bojarski [3], which are the solutions of the system

$$\frac{\partial w}{\partial \bar{z}} - Q(z)\,\frac{\partial w}{\partial z} = Aw + B\bar{w}\,, \tag{5.10}$$

where A, B are square matrices, and Q is an $m \times m$ "quasi–diagonal" matrix of the form $Q = \{Q_j\}$, each diagonal block being a triangular matrix normalized as

$$Q_j = \begin{pmatrix} q_j & 0 & 0 \\ (1+q_j)\,B_j\,, & q_j & \\ \vdots & \vdots & \\ (1+q_j)\,B_j^{r_j+1}\,, & \ldots\ldots\ldots, & (1+q_j)\,B_j, q_j \end{pmatrix}. \tag{5.11}$$

The q_j, and B_j are complex valued and $|q_j| < 1$, and $|B_j|$ may be made arbitrarily small. The solutions of the principal part (5.10) are referred to as Q–holomorphic functions when the coefficient matrix $Q = Q(z) \in W_p^{-1}(C)$. There exists a Cauchy kernel $V = V(t,z)$ with the following properties [3]

(a) $V(t,z)$ is a matrix of quasidiagonal type.

(b) For fixed $t, V(t,z)$ is Q–holomorphic, i.e.

$$L\Phi : \Phi_{\bar{z}} - Q\Phi_z = 0\,,\ z \neq t\,.$$

(c) For fixed $z, V^*(t,z)$ is a solution of

$$L^*\Psi := -\Psi_{\bar{t}} + (Q^*\Psi)_t = 0\,,\ t \neq z\,.$$

(d) $\operatorname{Lim}_{t\to\infty} tV(t,z) = I$ for fixed z.

(e) $\operatorname{Lim}_{z\to\infty} zV(t,z) = \tilde{V}(t)$, where $\det \tilde{V}(t) \neq 0$ for each $t \in C$.

(f) If w is continuous in a neighbourhood of z, then

$$\operatorname*{Lim}_{\delta=0} \frac{1}{2\pi i} \int_{|t-z|=\delta} V(t,z)\, d_Q t w(t) = w(z)\,,$$

where

$$d_Q t \quad I dt + Q(t)\, d\bar{t}\,.$$

(g) $|V(t,z)| \leq \dfrac{C}{|t-z|}$, where C is a constant independent of t and z.

For the system $\frac{\partial w}{\partial \bar{z}} - Q(z)\,\frac{\partial w}{\partial z} = 0$ we have a Cauchy representation, $z \in \Omega$

$$\frac{1}{2\pi i}\int_{\Omega} V(t,z)\, d_Q t w(t) = \begin{cases} w(z), & z \in \Omega \\ 0, & z \notin \Omega \end{cases}.$$

The $m \times m$ matrix Q–holomorphic solutions for which $\|w\|^2 := (w,w) < \infty$ is a Hilbert function module with reproducing kernel over the algebra $\mathcal{A}$ of all $m \times m$ constant matrices, where involution is conjugate transposition. Moreover, and one is able to also develop generalized Cauchy kernels for equation (5.10), certainly in the case of quasidiagonal A, B, it is possible to show there is a unique kernel also.

5.4 REPRODUCING KERNELS FOR THE HYPERMEMBRANE

In Vekua [18] and Dikmen [6] a hierarchial system of equations is obtained by writing down the equations of equilibrium for a thin shell, namely

$$\underset{k}{M}{}^{\alpha\beta}_{|\alpha} - b^{\beta}_{\alpha}\,\underset{k}{M}{}^{\alpha 3} + \rho\,\underset{k}{l}{}^{\beta} - \underset{k}{m}{}^{\beta} = 0\,, \tag{5.12}$$

$$\underset{k}{M}{}^{\alpha 3}_{|\alpha} + b_{\alpha\beta}\,\underset{k}{M}{}^{\alpha\beta} + \rho\,\underset{k}{l}{}^{3} - \underset{k}{m}{}^{3} = 0\,, \tag{5.13}$$

$$\underset{k}{M}{}^{21} - \underset{k}{M}{}^{12} = b^{2}_{\alpha}\,\underset{k+1}{M}{}^{\alpha 1} - -b^{1}_{\alpha}\,\underset{k+1}{M}{}^{\alpha 2}\,, \tag{5.14}$$

$$\underset{k}{M}{}^{\alpha 3} = b^{\alpha}_{\beta}\,\underset{k+1}{M}{}^{\beta 3} + \frac{1}{k+1}\,\underset{k+1}{m}{}^{\alpha}\,. \tag{5.15}$$

Here the vector quantities $\underset{\bar{k}}{M}{}^{\alpha}$, $\underset{\bar{k}}{m}{}^{\alpha}$ are moment vectors ($k \in I\!N o$) given by integrals over the shell thickness [2]

$$\begin{aligned}
\underset{\bar{k}}{M}{}^{\alpha}(\theta^{\beta}) :&= \frac{1}{\sqrt{a}}\int_{h^-}^{h^+} T^{\alpha}\xi^{k}\, d\xi\,, \\
\underset{k}{m}(\theta^{\beta}) :&= \frac{1}{\sqrt{a}}\int_{h^-}^{h^+} T^{3}\xi^{k-1}\, d\xi \\
\rho\sqrt{a}\,\underset{\bar{k}}{l}(\theta^{\beta}) :&= \int_{h^-}^{h^+} \rho^{*}\sqrt{g^{*}}\, f^{*}\xi^{k}\, d\xi \\
&+ \{\xi^{k}(-h^{+}, \alpha\mathbf{T}^{\alpha} + \mathbf{T}^{3})\}_{\xi=h^+} \\
&- \{\xi^{k}(-h^{-}, \alpha\mathbf{T}^{\alpha} + \mathbf{T}^{3})\}_{\xi=h^-}\,,
\end{aligned}$$

a is the determinant of the metric tensor for the first quadratic form associated with the shell, $\mathbf{T} := (T^{ij})$ is the Cauchy stress tensor, ρ the shell density, and $b := \det(b_{\alpha\beta})$ the determinant of the second quadratic form.

Buchanan and Gilbert [4] defined a hypermembrane of order r to be a shell such that in the equations (5.12–5.15) there exists an r such that

$$\underset{k}{M}{}^{\alpha} = 0 \quad \underset{k}{m} = 0\,, \quad \underset{k}{l} = 0 \quad \text{for all} \quad k > r\,.$$

For $r = 0$ we obtain Vekua's membrane theory of shells, which can be solved by means of the theory of generalized analytic functions. The assumption that a shell is a hypermembrane permits much more generality. It is, however, a stronger statement about the moment vectors than is sometimes assumed in higher–order bending theories in that the consequences of (5.12–5.15) for $k = r$

$$\underset{r}{M}^{21} - \underset{r}{M}^{12} = 0\,, \quad \underset{r}{M}^{\alpha 3} = 0\,,$$

are not always assumed in these theories. In terms of the hypercomplex Douglis algebra, generated by $1, i$, and the idempotent e ($e^r = 0$), Buchanan and Gilbert [4] introduce the hypercomplex moments

$$\mu^{\alpha i} : \;=\; \sum_{k=0}^{r} \underset{r-k}{M}^{\alpha i}\, e^k\,, \tag{5.16}$$

$$l^{i} : \;=\; \sum_{k=0}^{r} \underset{r-k}{l}^{i}\, e^k\,, \tag{5.17}$$

$$m^{i} : \;=\; \sum_{k=0}^{r} \underset{r-k}{m}^{i}\, e^k\,, \tag{5.18}$$

$$\widehat{m}^{\alpha} : \;=\; \sum_{k=0}^{r-1} \frac{1}{r-k}\, \underset{r-k}{m}^{\alpha}\, e^k\,. \tag{5.19}$$

Equation (5.15) of the momentum balance equations becomes in hypercomplex form

$$\mu^{\alpha 3} = e b^{\alpha}_{\beta}\, \mu^{\beta 3} + e\widehat{m}^{\alpha}\,,$$

which may be solved to obtain

$$\mu^{13} = \frac{\widehat{m}^1\, e[1 - b^2_2 e] + \widehat{m}^2 b^1_2\, e^2}{D}\,,\; \mu^{23} = \frac{\widehat{m}^2\, e[1 - b^1_1 e] + \widehat{m}^1 b^2_1\, e^2}{D}\,. \tag{5.20}$$

Here

$$D := [1 - b^1_1 e][1 - b^2_2 e] - b^1_2 b^2_1\, e^2 = 1 - 2He + Ke^2\,,$$

where H is the mean curvature and K is the Gaussian curvature. It is usually assumed that the quantities $\widehat{m}^{\alpha}$ are known. Hence, it follows from (5.20) that the hypercomplex moments $\mu^{\alpha 3}$ are known. The remaining hypercomplex balance equations are

$$\begin{aligned} \mu^{\alpha\beta}_{|\alpha} &= b^{\beta} := b^{\beta}_{\alpha}\, \mu^{\alpha 3} - \rho\, l^{\beta} + m^{\beta}\,, \\ b_{\alpha\beta}\mu^{\alpha\beta} &= b^3 := \mu^{\alpha 3}_{|\alpha} - \rho\, l^3 + m^3\,, \\ \mu^{21} &- \mu^{12} = e\, [b^2_{\alpha}\mu^{\alpha 1} - b^1_{\alpha}\mu^{\alpha 2}]\,, \end{aligned}$$

where the functions b^i are known. Using covariant differentiation one gets

$$\mu^{11}_{,1} + \mu^{21}_{,2} + [2\Gamma^1_{11} + \Gamma^2_{12}]\, \mu^{11} + \Gamma^1_{12}\, \mu^{12} + [2\Gamma^1_{12} + \Gamma^2_{22}]\, \mu^{21} + \Gamma^1_{22}\, \mu^{22} = b_1\,,$$

$$\mu^{12}_{,1} + \mu^{22}_{,2} + \Gamma^2_{11}\,\mu^{11} + [2\Gamma^2_{12} + \Gamma^1_{11}]\,\mu^{12} + \Gamma^2_{12}\,\mu^{21} + [2\Gamma^2_{22} + \Gamma^1_{12}]\,\mu^{22} = b_2\,,$$

where the $\Gamma^{\gamma}_{\alpha\beta}$ are the usual Christoffel symbols. If the coordinate system is isometric–conjugate, i.e. $b_{11} = b_{22} = \hat{b}$, $b_{12} = b_{21} = 0$, this system takes on a simple form,

$$\begin{aligned} \mu^{11}_{,1} &+ \mu^{12}_{,2} + p_1\mu^{11}_{,2} + c_{11}\,\mu^{11} + c_{12}\,\mu^{12} = g_1\,, \\ \mu^{12}_{,1} &+ \mu^{11}_{,2} + c_{21}\,\mu^{11} + c_{22}\,\mu^{12} = g_2\,, \end{aligned}$$

where the coefficients are given in terms of the Christoffel symbols $\Gamma_{\alpha}\nu\mu$ as

$$\begin{aligned} c_{11} &:= 2\Gamma^1_{11} + \Gamma^1_{12} - \Gamma^1_{22} + p_{1,2} + [2\Gamma^1_{12} + \Gamma^2_{22}]\,p_1\,, \\ c_{12} &:= 3\Gamma^1_{12} + \Gamma^2_{22} + p_{2,2} + [2\Gamma^1_{12} + \Gamma^2_{22}]\,p_2\,, \\ c_{21} &:= \Gamma^2_{11} + 2\Gamma^2_{22} - \Gamma^1_{12} + \Gamma^2_{12}\,p_1\,, \\ c_{22} &:= 3\Gamma^2_{12} + \Gamma^1_{11} + \Gamma^2_{12}\,p_2\,, \\ g_1 &:= b^1 - [p_3\hat{b}^3]_{,2} - \Gamma^1_{22}\,b^3 - -p_3\,[2\Gamma^1_{12} + \Gamma^2_{22}]\,\hat{b}^3\,, \\ g_2 &:= b^2 - \hat{b}^3_{,2} - p_3\Gamma^2_{12}\,\hat{b}^3 - [2\Gamma^2_{22} + \Gamma^1_{12}]\,\hat{b}^3\,. \end{aligned} \tag{5.21}$$

This system may be put into the elegant form

$$\partial_{\bar{z}}\omega + \mathrm{Re}\,[2q\partial_z\omega] + a\omega + b\bar{\omega} = c\,,$$

where the coefficients are defined by

$$2q := \frac{1/2ip}{1 + 1/2p_2}\,,$$

$$a := \frac{1}{2}(1+q)\,\hat{a} + \frac{1}{2}\,\bar{q}\bar{\hat{b}}\,, \quad b := \frac{1}{2}(1+q)\,\hat{b} + \frac{1}{2}\,\bar{q}\bar{\hat{a}}\,,$$

$$c := \frac{1}{2}(1+q)\,q + \frac{1}{2}\,\overline{g\bar{g}}\,,$$

and where

$$\begin{aligned} \omega &:= \mu^{11} - i\mu^{12}\,, \quad z := \theta^1 + i\theta^2\,, \\ P &:= p_1 + ip_2\,, \quad h := g - \hat{a}\omega - \hat{b}\bar{\omega}\,, \\ \hat{a} &:= \frac{1}{2}\{[c_{11} + c_{22}] - i[c_{21} - c_{12}]\}\,, \\ \hat{b} &:= \frac{1}{2}\,\{[c_{11}0c_{22}] - i[c_{21} + c_{12}]\}\,, \end{aligned}$$

and

$$g := g_1 - ig_2\,.$$

This suggests that we study the hypercomplex equations of the type

$$\partial_{\bar{z}}w + q_1\partial_z w + q_2\overline{\partial_z w} + aw + b\bar{w} = f\,, \tag{5.22}$$

where q_1 and q_2 are nilpotent and hypercomplex valued. The coefficients a, b, and f, moreover, are also hypercomplex. The system (5.10) is a generalization of the equation

$$\partial_{\bar{z}} w + q \partial_z w + aw + b\bar{w} = f \tag{5.23}$$

which was investigated by Gilbert and Hile [12] along with the equation [2]

$$\partial_{\bar{z}} w + q \partial_z w + \sum_{k=0}^{r-1} e^k \sum_{\ell=0}^{k} (a_{k\ell} w_\ell + b_{k\ell} \bar{w}_\ell) = f \tag{5.24}$$

Buchanan–Gilbert [4] developed a generalized Cauchy–Pompieu representation associated with the operator

$$\bar{D}w := \partial_{\bar{z}} w + q_1 \partial_z w + q_1 \overline{\partial_z w}\,.$$

Theorem 8 *Let $q_1, q_2 \in C^m_\alpha(\widehat{G}_0)$ where $G_0 \in C^{m+1}_\alpha$ and $m \geq 1$. Let $\psi_j(\zeta, z) = \zeta - z + \Psi_j(\zeta, z)$, $j = 1, 2$, be the functions constructed in Lemma 1 below for the domain G_0. Then for $G \in C$, $G \subseteq G_0$ and $w \in C^0_\gamma(\widehat{G})$, $0 < \gamma < 1$, $\partial_z w$, $\partial_{\bar{z}} w \in L^p(\widehat{G})$, $p > 2$,*

$$\begin{aligned} w(a) &= \frac{1}{2\pi i} \int_{\partial G} \left\{ w(\zeta) \left[\frac{d\zeta}{\zeta - z} + d\tau_1(\zeta, z) \right] - \overline{w(\zeta)}\, \overline{d\tau_2(\zeta, z)} \right\} \\ &\quad - \frac{1}{\pi} \int_G \left\{ \left[\frac{1}{\zeta - z} + \partial_{\zeta \tau_1}(\zeta, z) \right] \bar{D}w(\zeta) + \overline{\partial_{\zeta \tau_2}(\zeta, z)}\, \overline{\bar{D}w(\zeta)} \right\} d\sigma_\zeta\,, \end{aligned} \tag{5.25}$$

where

$$\tau_j(\zeta, z) := \sum_{k=1}^{r} \frac{(-1)^{k-1}}{k} \left[\frac{\Psi_j(\zeta, z)}{\zeta - a} \right]^k .$$

See [4] for details of the proof.

Remark If w is a Hölder continuous solution to $\bar{D}w = 0$, then

$$w(z) = \frac{1}{2\pi i} \int_{\partial G} \left\{ w(\zeta) \left[\frac{d\zeta}{\zeta - z} + d\tau_1(\zeta, z) \right] - \overline{w(\zeta\, d) \tau_2(\zeta, z)} \right\},$$

for any domain $G \in C$ contained in G_0.

The problem where the stresses are prescribed on the boundary thereby can be expressed as a Riemann–Hilbert problem, and the existence of the generalized Cauchy–Pompieu representation (5.25) indicates that this Riemann–Hilbert problem has a unique solution.

The Cauchy–Pompieu representation provides the existence of a bounded linear functional existing on the family of solutions of equation (5.24), and, hence, by Theorem 1 the existence of a reproducing kernel, which may be used to prove the existence of a complete orthonormal set of solutions which may be used to approximate solutions uniformly on compact subsets of Ω.

Lemma 1 *Let $q_1, q_2 \in C_\alpha^m(\widehat{G})$, $G \in C_\alpha^{m+1}$ for some $m \geq 1$. Then there exist hypercomplex functions*

$$\psi_j(\zeta, z) = \zeta - z + \Psi_j(\zeta, z)\,, \quad j = 1, 2$$

with $\psi_j \in DS_\alpha^{m+1}(\widehat{G}, z)$, $\partial_z \psi_j$, $\partial_{\bar{z}} \psi_j \in LS(z) \oplus DS_\alpha^m(\widehat{G}, z)$ such that

$$V_1(\zeta, z) := \log\, \psi_1(\zeta, z)\,, \quad V_2(\zeta, z) := \log\left[\frac{\psi_2(\zeta, z)}{\zeta - z}\right]$$

are solutions to (5.22) for $\zeta \neq z$, [4].

5.5 ELLIPTIC SYSTEMS

We consider in this section the example of self–adjoint system

$$\Delta \vec{U} = C(x, y)\, \vec{U} \tag{5.26}$$

where U and C are complex valued $n \times n$ matrices and Δ is the two–dimensional Laplacian, and $(x, y) \in D$. In $\bar{D}$, $c > 0$ (positive definite) and Hermitian: $C = C^*$, where C^* is the conjugate–transpose of C; also, C will be assumed to belong to $C'(\bar{D})$. For our purposes it is useful to introduce the Dirichlet "inner product"

$$E\{V, W\} := \iint\limits_D [V_x^* W_x + V_y^* W_y + V^* C W]\, dx\, dy\,.$$

It follows that $E^*\{V, W\} = E\{W, V\}$, and that E^* is sesquilinear. The quadratic mapping $E\{V, V\}$, abbreviated $E\{V\}$, is a positive semidefinite Hermitian matrix. Gilbert and Weinacht [10] [11] showed that there exists a set of eigenmatrices $\{U_\nu\}$ such that

$$E\{U_\nu, U_\mu\} = 0\,, \quad \mu \neq \nu$$

and

$$E\{U_\nu, U_\nu\} = I_\nu$$

where I_ν is a nonzero diagonal idempotent matrix. The $\{U_\nu\}$ are complete in the Sobolev space associated with $E\{\cdot, \cdot\}$. Every $u \in \hat{\Sigma}$ has a representation with constant matrix coefficients

$$U = \sum_{\nu=1}^{\infty} U_\nu A_\nu\,, \quad I_\nu A_\nu = E\{U_\nu, U\}$$

convergent with respect to $\|E\{\}\|^{1/2}$ and in particular, there exists a reproducing kernel with the development

$$K(P, Q) = \sum_{\nu=1}^{\infty} U_\nu(P)\, A_\nu(Q)\,. \tag{5.27}$$

In particular, Gilbert–Weinacht obtain the estimates

$$\|K(P, Q) - K_N(P, Q)\| \leq (1/\lambda_1^{2N})\, \|M(P, P)\|^{1/2}\, \|M(Q, Q)\|^{1/2}$$

where K_N is the nth partial sum for k in (5.27):

$$K_N(P,Q) \equiv \sum_{\rho=0}^{N} \sum_{\nu=0}^{\rho} (-1)^{\nu} \binom{\rho}{\nu} i_{\nu+1}(P,Q),$$

and where

$$M(P,Q) = \sum_{\nu=1}^{\infty} (U_{\nu}(P)\, U_{\nu}^{*}(Q)/\lambda_{\nu}^{2}).$$

These estimates permit effective estimates for the error in the approximation of solutions of boundary value problems for (5.26). The estimates in this paper were constructive; however, the existence of a kernel function and its expansion are derivable from the Hilbert Module theory given earlier.

Notes

1. **Key Words:** kernel functions, function modules, hypermembrane, elliptic systems

2. Equations (5.23) and (5.24) form a subset of the class of equations studied by Bojarski [3] which gave rise to the generalized analytic vectors. Bojarski's system which was based on Fredholm theory did not lead to the concise formulas as ours did, as there was always the null space of an adjoint operator to be added to the representation. Our theory, while more specialized and dependent on the Douglis representation [7], allowed most of the Vekua theory to be reproduced line by line. In this regard, see the books by Buchanan–Gilbert [4] and Begehr–Gilbert [3] where this research is reported on in detail.

References

[1] **ARONSZAJN, N.**, Theory of reproducing kernels, *Trans. Am. Math. Soc.* **68**, 337–404 (1950).

[2] **BEGEHR, H.** and **R. P. GILBERT**, *Transformations, Transmutations, and Kernel Functions*; Vols. I, II, (1993/94), Longman, Essex.

[3] **BOJARSKI, B. B.**, Die Theorie des verallgemeinerten analytischen Vektors (Aufsatz in russicher Sprache, Ubersetz aus dem Russischem Bibl. der Tech. Univ. Hannover), *Anns. Pol. Math.* **17**, 281–320 (1966).

[4] **BUCHANAN, J. L.** and **R. P. GILBERT**, A solution to the equations of shell theory, *Math. Meth. Appl. Scis.* **12**, (1990), 341–363.

[5] **DELANGHE, R.**, On regular–analytic functions with values in a Clifford algebra, *Math. Annln* **185**, 91–111 (1970).

[6] **DIKMEN, M.**, *The Theory of Thin Elastic Shells*, Surveys and Reference Works in Mathematics #8, Pitman, London (1982).

[7] **DOUGLIS, A.**, A function theoretic approach to elliptic systems of equations in two variables, *Communs. Pure Appl. Math.* **6**, 259–289 (1953).

[8] **FICHERA, G.**, *Lectures on Linear Transformations* (Translated by J.B. Diaz), Liberia Eredi Virgilio Veschi, Viale Dell'–Universita Roma (1954).

[9] **GILBERT, R. P.** and **R. J. WEINACHT**, Reproducing kernels for elliptic systems in divergence form, ISNM25, Birkhäuser, Basel (1974).

[10] **GILBERT, R. P.** and **R. J. WEINACHT**, Iterative schemes for elliptic systems, in "Constructive and Computational Methods for Differential and Integral Equations," Lecture Notes in Mathematics, Vol. 430, pp. 253–260, Springer–Verlag, Heidelberg, 1974.

[11] **GILBERT, R. P.** and **R. J. WEINACHT**, Reproducing kernels for elliptic systems, *J. Approx, Theory* **15**, 243–255 (1975).

[12] **GILBERT, R. P.** and **G. N. HILE**, Hilbert Function Modules with Reproducing Kernels, *Nonlinear Analysis* **1** (2), 135–150 (1977).

[13] **GOLDSTEIN, H. H.** and **L. P. HORWITZ**, Hilbert space with nonassociative scalars II, *Math. Annln.* **164**, 291–316 (1966).

[14] **HILE, G. N.**, Representations of solutions of a special class of first order systems (to appear).

[15] **MESCHKOWSKI, A.**, *Hilbertsche Raume mit Kernfunktion*, Springer–Verlag, Berlin (1962).

[16] **SAWOROTNOW, P. P.**, A generalized Hilbert space, *Duke Mathematical Journal* **35**, 191–197 (1968).

[17] **SAWOROTNOW, P. P.** and **J. C. FRIEDELL**, Trace–class for an arbitrary H^*–algebra, *Proc. Am. Math. Soc.* **26**, 95–100 (1970).

[18] **VEKUA, I. N.** *Generalized Analytic Functions*, Pergamon, Oxford (1962).

6 TRANSMUTATIONS IN DIFFERENTIAL RINGS

J. Püngel

Dedicated to Professor Robert P. Gilbert on his 65th birthday

Abstract: In this paper we study certain generalizations of scalar differential transformations for solutions of partial differential equations. Matrix or vector valued differential transformations lead to explicit solutions of special systems of nonlinear partial differential equations. A more abstract description of this method can be given within the framework of noncommutative differential rings. Realizations of commutative differential subrings are constructed using special classes of matrices. That way we get various vector products which induce useful principles for transformations of (scalar) holomorphic functions into vector valued holomorphic functions of several complex variables. This is basic to generalize scalar differential transformations into transformations for systems. Moreover, we point out connections of operator relations used in transformation techniques with a generalization of the idea of transmutation of operators in the sense of R.W. CARROLL, R.P. GILBERT and H. BEGEHR. [1]

6.1 TRANSFORMATIONS AND OPERATOR RELATIONS

In this paper we shall deal primarily with transformations like

$$\forall u \in \mathbf{H} : \forall v \in \mathbf{H}^* : (\mathbf{L}u = o \wedge v = \mathbf{T}u \Rightarrow \mathbf{L}^* v = o) \qquad (6.1)$$

where $\mathbf{H}$ and $\mathbf{H}^*$ denotes suitable spaces of functions of two independent variables z, ζ (e.g., sets of functions holomorphic or meromorphic in some complex domains). In case of complex or real single-valued functions u, v we will refer to (6.1) as a scalar transformation. However our interest lies mainly in spaces of vector or matrix valued functions. Here $\mathbf{L}$ and $\mathbf{L}^*$ denote (linear or nonlinear) differential operators. (6.1) is called a differential transformation if $\mathbf{T}$ is also a differential operator.

H. Florian et al. (eds.), Generalized Analytic Functions, 55–63.

An operator relation

$$\exists \mathbf{T}^* : \ (\mathbf{L}^* \circ \mathbf{T} = \mathbf{T}^* \circ \mathbf{L}) \tag{6.2}$$

together with $\mathbf{T}^* o = o$ imply (6.1). $\mathbf{T}^*$ in general is an integral operator. $\mathbf{T}^*$ exists if $\mathbf{L}$ is right invertible.

6.2 EXAMPLES

Let us first give some examples of scalar differential transformations.

EXAMPLE 1: A linear differential transformation. Let

$$\mathbf{L}u := u_{z\zeta} - au_\zeta - bu_z \tag{6.3}$$

for $u(z, \zeta)$ with coefficients $a(z, \zeta), b(z, \zeta), a, b, u$ holomorphic in a domain $D \subset \mathbb{C}^2$,

$$\mathbf{T}u := (\alpha_z u_\zeta - \alpha_\zeta u_z)/d \tag{6.4}$$

where $d := \alpha_z \beta_\zeta - \alpha_\zeta \beta_z$, $\alpha_z \beta_\zeta d \neq 0$ in D and

$$\mathbf{L}^* v := v_{z\zeta} - a^* v_\zeta - b^* v_z \tag{6.5}$$

where $a^* = a + (\ln(\alpha_z/d))_z$, $b^* = b + (\ln(\alpha_\zeta/d))_\zeta$. Then (6.1) holds for every pair α, β of solutions of $\mathbf{L}\alpha = \mathbf{L}\beta = o$. We emphasize that in this case (6.2) holds too with the differential operator

$$\mathbf{T}^* = \mathbf{T} + \gamma \mathbf{I} \tag{6.6}$$

where $\mathbf{I}$ denotes the identity operator and $\gamma = (\alpha_{zz}\alpha_\zeta/\alpha_z - \alpha_{\zeta\zeta}\alpha_z/\alpha_\zeta)/d$.

EXAMPLE 2: A nonlinear differential transformation. Let $\mathbf{L}$ be as in (6.3) above but

$$\mathbf{T}u := \ln(u_\zeta/u_z) \tag{6.7}$$

$$\mathbf{L}^* v := v_{z\zeta} - (a + be^{-v})_\zeta + (b + ae^v)_z \tag{6.8}$$

Then (6.1) describes a nonlinear transformation $u \to v = \mathbf{T}u$ of solutions u of a linear equation $\mathbf{L}u = o$ into solutions v of a nonlinear equation $\mathbf{L}^* v = o$. (6.2) has to be replaced by a suitable generalization. We will discuss this problem later in a more general context.

EXAMPLE 3: A very interesting nonlinear transformation (6.1) is realized by

$$\mathbf{L}u := u_{z\zeta} \tag{6.9}$$

$$\mathbf{T}u := \ln(2u_z u_\zeta/u^2) \tag{6.10}$$

$$\mathbf{L}^* v := v_{z\zeta} - e^v \tag{6.11}$$

In case of ζ being conjugate to z the equations $\mathbf{L}u = o$ and $\mathbf{L}^*v = o$ are the complex form of the LAPLACE equation and the LIOUVILLE equation respectively.

EXAMPLE 4: The transformation (6.1) in case of

$$\mathbf{L}u := u_{z\zeta} - e^u \tag{6.12}$$

$$\mathbf{T}u := u_z \tag{6.13}$$

$$\mathbf{L}^*v := (2v_z - v^2)_\zeta \tag{6.14}$$

reduces the construction of solutions of the LIOUVILLE equation to the solution of an ordinary differential equation

$$2v_z - v^2 = \varphi(z)$$

(with respect to z) of RICCATI type.

We conclude the list of examples for scalar differential transformations with

EXAMPLE 5: Let

$$\mathbf{L}u := u_{zz} - u_\zeta \tag{6.15}$$

$$\mathbf{T}u := u_z/u \tag{6.16}$$

$$\mathbf{L}^*v := v_{zz} + 2vv_z - v_\zeta \tag{6.17}$$

In case of real variables $z = x$ and $\zeta = t$ equation $\mathbf{L}u = o$ is the heat equation, $v = \mathbf{T}u$ the COLE-HOPF transformation and $\mathbf{L}^*v = o$ the BURGERS equation.

6.3 TRANSFORMATIONS IN DIFFERENTIAL RINGS

Some of the methods for the construction of differential transformations of the type cited above can be found in [3], [4], [7] and [8].

The main goal of this paper is a generalization of scalar differential transformations in an essentially algebraic way. One of the results in this area is a direct interpretation of transformations (6.1) as differential transformations for vector or matrix valued functions u and v. So we can read (6.1) as differential transformation between systems of nonlinear partial differential equations.

Linear differential transformations for systems are useful in the construction of generalizations of analytic functions (see eg.[1]).

In each of the above mentioned examples it is very easy to prove the transformation property (6.1). Apart from the fact that functions like "ln" and "exp" in the definitions (6.10) and (6.11) can be avoided, we only make use of the well known rules for the differentiation of sums and products. All this can also be done in an adequate algebraic structure: a *differential ring* $\mathbf{R}$ with two (or more) commuting derivations ∂_1, ∂_2.

$\mathbf{R}$ need not be commutative as we can see e.g., from a realization of example 5: For all invertible solutions $u \in \mathbf{R}$ of equation

$$\partial_1^2 u = \partial_2 u \tag{6.18}$$

the elements $v \in \mathbf{R}$ defined by

$$v = u^{-1} \partial_1 u \tag{6.19}$$

are solutions of the abstract BURGERS equation

$$\partial_1^2 v + 2v\partial_1 v = \partial_2 v \tag{6.20}$$

This holds in an arbitrary noncommutative differential ring $\mathbf{R}$ with derivations ∂_1, ∂_2 such that $\partial_1 \circ \partial_2 = \partial_2 \circ \partial_1$. On the other hand the transformation (6.1) for the operators defined in Example 3 can be interpreted directly only in commutative differential rings (chose $w = e^v$ instead of v).

6.4 RINGS OF COMMUTING MATRICES AND VECTORS

Every concrete realization of a commutative differential ring, as ring of commuting matrices, gives a generalized interpretation of (6.1), as systems of partial differential-equations and transformations. We sketch one of the many ways to do this (see also [5]).

Let $\mathbf{A}$ be any $n \times n$ matrix whose entries belong to a field $\mathbf{K}$. The minimal polynomial

$$\mu(\mathbf{A}) = c_0\mathbf{E} + c_1\mathbf{A} + \cdots + c_m\mathbf{A}^m \tag{6.21}$$

where $c_0, c_1, ..c_m \in \mathbf{K}$, exists and is uniquely defined by $c_m = 1, \mu(\mathbf{A}) = \mathbf{0}$ and the assumption that the degree m is minimal. One can prove that $1 \le m \le n$. Thus any polynomial $p(\mathbf{A})$ of $\mathbf{A}$ can be reduced to a polynomial

$$p(\mathbf{A}) = p_0(\mathbf{A}) = \alpha_0\mathbf{E} + \alpha_1\mathbf{A} + \cdots + \alpha_l\mathbf{A}^l \tag{6.22}$$

of degree $l \le m-1$ using $\mathbf{A}^m = -c_0\mathbf{E} - c_1\mathbf{A} - \cdots - c_{m-1}\mathbf{A}^{m-1}$ successively. This defines a map from the commutative ring of polynomials of $\mathbf{A}$ over $\mathbf{K}$ into the vector space $\mathbf{K}^m$ by

$$p(\mathbf{A}) \to (\alpha_0, \alpha_1, \ldots, \alpha_{m-1})$$

where $\alpha_0, \alpha_1, \ldots, \alpha_{m-1}$ denote the coefficients of the polynomial (6.22) (define $\alpha_k = 0$ for $l < k \le m-1$). Now the conventional matrix product induces a commutative product $\bullet : \mathbf{K}^m \times \mathbf{K}^m \to \mathbf{K}^m$ by the following natural procedure.

Take two vectors $(\alpha_0, \ldots, \alpha_{m-1}), (\beta_0, \ldots, \beta_{m-1})$ of dimension m, identify them with the corresponding polynomials

$$p(\mathbf{A}) = \alpha_0\mathbf{E} + \cdots + \alpha_{m-1}\mathbf{A}^{m-1}\ ,\ q(\mathbf{A}) = \beta_0\mathbf{E} + \cdots + \beta_{m-1}\mathbf{A}^{m-1}$$

reduce the matrix product $s(\mathbf{A}) = p(\mathbf{A})q(\mathbf{A})$ to a polynomial

$$s(\mathbf{A}) = s_0(\mathbf{A}) = \gamma_0\mathbf{E} + \cdots + \gamma_{m-1}\mathbf{A}^{m-1}$$

of degree $l \le m-1$. Then the vector product $\bullet$ is given by

$$(\alpha_0, \ldots, \alpha_{m-1}) \bullet (\beta_0, \ldots, \beta_{m-1}) := (\gamma_0, \ldots, \gamma_{m-1}).$$

So $\bullet$ depends only on the choice of the matrix $\mathbf{A}$.

In order to illustrate this we consider the class of all matrices of type

$$\mathbf{A} = \begin{pmatrix} 0 & 1 \\ -pq & p+q \end{pmatrix} \tag{6.23}$$

with arbitrary $p, q \in \mathbf{C}$. The minimal polynomial of $\mathbf{A}$ is given by

$$(\mathbf{A} - p\mathbf{E})(\mathbf{A} - q\mathbf{E}) = 0 \tag{6.24}$$

Consequently any polynomial p of $\mathbf{A}$ with complex coefficients can be written as

$$p(\mathbf{A}) = \alpha\mathbf{E} + \beta\mathbf{A}, \qquad \alpha, \beta \in \mathbf{C} \tag{6.25}$$

Thus this matrices $\mathbf{A}$ define the vector products

$$(\alpha, \beta) \bullet (\lambda, \mu) := (\alpha\lambda - pq\beta\mu, \alpha\mu + \beta\lambda + (p+q)\beta\mu) \tag{6.26}$$

in $\mathbf{C}^2$. In view of the consideration of power series of a vector variable (α, β) with respect to this products we are interested in explicit representations of powers $(\alpha, \beta)^n$, that is to say, reduced powers of polynomials (6.25). By (6.24) in particular we have

$$\mathbf{A}^n = c_n\mathbf{E} + d_n\mathbf{A}, n \in \mathbf{N}, \tag{6.27}$$

Obviously from $\mathbf{A}^0 = \mathbf{E}$ and $\mathbf{A}^{n+1} = \mathbf{A}^n\mathbf{A}$, using $\mathbf{A}^2 = (p+q)\mathbf{A} - pq\mathbf{E}$, we obtain

$$c_0 = 1, d_0 = 0, c_{n+1} = -pqd_n, d_{n+1} = c_n + (p+q)d_n. \tag{6.28}$$

Now let

$$x_1(t) = \sum c_n t^n/n!, \quad x_2(t) = \sum d_n t^n/n! \tag{6.29}$$

be generating functions of the coefficients $c_n, d_n \in \mathbf{C}$. Due to (6.28) $x(t) = (x_1(t), x_2(t))$ is the solution of the initial value problem

$$x'(t) = \mathbf{A}^T x(t), \qquad x(0) = (1, 0). \tag{6.30}$$

Thus (6.27) with coefficients given by the power series of the unique solutions

$$x_1(t) = \frac{pe^{qt} - qe^{pt}}{p-q}, \quad x_2(t) = \frac{e^{pt} - e^{qt}}{p-q} \tag{6.31a}$$

of (6.30) in case of $p \neq q$ and

$$x_1(t) = (1 - pt)e^{pt}, \quad x_2(t) = te^{pt} \tag{6.31b}$$

if $p = q$ leads respectively to

$$(\alpha, \beta)^n = \frac{1}{p-q}\Big(p(\alpha + q\beta)^n - q(\alpha + p\beta)^n, (\alpha + p\beta)^n - (\alpha + q\beta)^n\Big) \tag{6.32a}$$

in case of $p \neq q$ and

$$(\alpha, \beta)^n = \Big((\alpha + p\beta)^n - np\beta(\alpha + p\beta)^{n-1}, n\beta(\alpha + p\beta)^{n-1}\Big) \tag{6.32b}$$

if $p = q$.

6.5 TRANSMISSION PRINCIPLES

To every $n \times n$ matrix $\mathbf{A}$ one can associate an inner composition

$$\bullet_{\mathbf{A}} : \mathbb{C}^m \times \mathbb{C}^m \to \mathbb{C}^m$$

as explained above and additionally a "transmission principle"

$$\tau_{\mathbf{A}} : \mathbf{H}(G, \mathbb{C}) \to \mathbf{H}(B, \mathbb{C}^m) \tag{6.33}$$

(generated by $\bullet_{\mathbf{A}}$) from the set of functions holomorphic in a domain $G \subset \mathbb{C}$ into the set of vector valued functions holomorphic in a domain $B \subset \mathbb{C}^m$ in such a way that $(\mathbb{C}^m; +, \bullet_{\mathbf{A}})$ is a commutative ring and $\tau_{\mathbf{A}}$ ensures that a wide class of scalar differential transformations can be interpreted as transformations between systems of differential equations.

To this end consider e.g., the special 2×2 matrices $\mathbf{A}$ (6.23). If one defines the sum of power series of (α, β) as limits of the partial sums as usual, then every $f \in \mathbf{H}(G, \mathbb{C})$, $G \subset \mathbb{C}$, (with local power series representation) leads to $f_{\mathbf{A}} := \tau_{\mathbf{A}} f \in \mathbf{H}(B, \mathbb{C}^2)$, $B \subset \mathbb{C}^2$, where

$$f(t) \to f_{\mathbf{A}}(\alpha, \beta)$$

with

$$f_{\mathbf{A}}(\alpha, \beta) := \frac{1}{p-q}\Big(pf(\alpha + q\beta) - qf(\alpha + p\beta), f(\alpha + p\beta) - f(\alpha + q\beta)\Big) \tag{6.34a}$$

and

$$f_{\mathbf{A}}(\alpha, \beta) := \Big(f(\alpha + p\beta) - p\beta f'(\alpha + p\beta), \beta f'(\alpha + p\beta)\Big) \tag{6.34b}$$

in case of $p \neq q$ and $p = q$, respectively, where we made use of (6.32a) and (6.32b). In that way for example one obtain

$$\begin{aligned}
\exp_{\mathbf{A}}(\alpha, \beta) &= \frac{e^{\alpha}}{p-q}\left(pe^{q\beta} - qe^{p\beta}, e^{p\beta} - e^{q\beta}\right), \quad p \neq q, \\
\exp_{\mathbf{A}}(\alpha, \beta) &= e^{\alpha + p\beta}(1 - p\beta, \beta), \quad p = q, \\
\ln_{\mathbf{A}}(\alpha, \beta) &= \frac{1}{p-q}\left(\ln \frac{(\alpha + q\beta)^p}{(\alpha + p\beta)^q}, \ln \frac{\alpha + p\beta}{\alpha + q\beta}\right), \quad p \neq q, \\
\ln_{\mathbf{A}}(\alpha, \beta) &= \left(\ln(\alpha + p\beta) - \frac{p\beta}{\alpha + p\beta}, \frac{\beta}{\alpha + p\beta}\right), \quad p = q,
\end{aligned}$$

and (6.1) with (6.9), (6.10) are explicit representations of the solutions $v(z, \zeta)$, $w(z, \zeta)$ of the systems

$$\begin{aligned}
(p - q)v_{z\zeta} &= e^{v}(pe^{qw} - qe^{pw}) \\
(p - q)w_{z\zeta} &= e^{v}(e^{pw} - e^{qw}),
\end{aligned} \tag{6.35a}$$

with $p, q \in \mathbb{C}$, $p \neq q$, and

$$\begin{aligned} v_{z\zeta} &= (1-pw)e^{v+pw} \\ w_{z\zeta} &= we^{v+pw}, \end{aligned} \tag{6.35b}$$

with $p \in \mathbb{C}$, respectively. (6.35a,6.35b) can be interpreted as generalizations of the scalar "LIOUVILLE equation" $v_{z\zeta} = e^v$. Corresponding results one can obtain for all transformations mentioned in section 6.2.

Certain transmission principles can be obtained if one consider derivations defined on *composition* rings $(\mathbf{R}; +, \cdot, \circ)$ postulating the chain rule $\partial(f \circ g) = ((\partial f) \circ g) \cdot \partial g$ for all derivations ∂.

Suppose (6.1) can be interpreted as a differential transformation in a differential ring $\mathbf{R}$ and a transmission principle

$$f(t) \quad \to \quad F : \mathbf{R} \to \mathbf{R}$$

is at one' s disposal. Then we have the possibility of an application of scalar differential transformations to abstract systems of partial differential equations without much difficulty.

6.6 GENERALIZED TRANSMUTATIONS FOR DIFFERENTIAL OPERATORS

Next we look how the condition (6.2) has to be modified in case of nonlinear operators in differential rings.

To this end let us consider the COLE-HOPF transformation in Example 5 of Section 6.2. Let $\mathbf{R}$ denote a differential ring with two derivations ∂_1, ∂_2 and let $\mathbf{R}^*$ denote the set of invertible elements of $\mathbf{R}$. $\mathbf{L}, \mathbf{T}$ and $\mathbf{L}^*$ are operators

$$\mathbf{L} : \mathbf{R} \to \mathbf{R}, \qquad \mathbf{T} : \mathbf{R}^* \to \mathbf{R}, \qquad \mathbf{L}^* : \mathbf{T}[\mathbf{R}^*] \to \mathbf{R}$$

defined by

$$\mathbf{L}u := \partial_1^2 u - \partial_2 u \tag{6.36}$$

$$\mathbf{T}u := u^{-1}\partial_1 u \tag{6.37}$$

$$\mathbf{L}^* v := \partial_1^2 v + 2v\partial_1 v - \partial_2 v \tag{6.38}$$

(cf. (6.18), (6.19), (6.20)). Using $\partial_j u^{-1} = -u^{-1}(\partial_j u)u^{-1}$ one can show that

$$(\mathbf{L}^* \circ \mathbf{T})u = u^{-1}\partial_1 \mathbf{L}u - u^{-1}(\mathbf{L}u)u^{-1}\partial_1 u \tag{6.39}$$

This equation gives rise to the following generalization of the right hand side product in (6.2). Instead of $\mathbf{T}^* : \mathbf{R}^* \to \mathbf{R}$ let

$$\mathbf{T}^* : \mathbf{R}^* \times \mathbf{R} \to \mathbf{R}$$

be defined by

$$\mathbf{T}^*(u, w) := u^{-1}\partial_1 w - u^{-1}wu^{-1}\partial_1 u \tag{6.40}$$

Further let us define a "new" operator product $\mathbf{T}^* \triangleleft \mathbf{L}$ by

$$(\mathbf{T}^* \triangleleft \mathbf{L})u := \mathbf{T}^*(u, \mathbf{L}u) \tag{6.41}$$

Thus (6.2) should be replaced by

$$\mathbf{L}^* \circ \mathbf{T} = \mathbf{T}^* \triangleleft \mathbf{L} \tag{6.42}$$

which is equivalent to (6.39). Note that $\mathbf{T}^*(u, o) = o$.

Of course (6.42) can be interpreted as a generalization of the very effective idea of transmutation of operators in the sense of R.W. CARROLL, R.P. GILBERT and H. BEGEHR (see [2]). Let us say that the pair of operators $(\mathbf{T}, \mathbf{T}^*)$ transmutes $\mathbf{L}$ into $\mathbf{L}^*$.

Operator relations (6.42) with sufficient $\mathbf{T}^*$ can be derived for all transformations mentioned in Section 6.2 (and much more) if we interpret those as matrix differential transformations.

Notes

1. AMS-No. 35A22, 35C05, 35L70, 15A54

Key words: nonlinear partial differential equations, differential rings, differential transformations, transmutations, systems of partial differential equations

References

[1] H. Begehr and R. P. Gilbert, Pseudohyperanalytic functions, Compl. Variabl. 9 (1988), 343-357

[2] H. Begehr and R. P. Gilbert, Transformations, transmutations and reproducing kernels, Vol.I and Vol.II, Pitman-Longman 1993 and 1994

[3] H. Florian and J. Püngel, Differentialtransformationen bei quasilinearen Gleichungen zweiter Ordnung. Contained in Partial differential equations with complex analysis (edited by H.Begehr & A.Jeffrey), Pitman research notes in mathematics series, vol. 262, 14-22, 1992

[4] H. Florian, J. Püngel and W. Tutschke, Solution of boundary value problems for non-linear partial differential equations using differential transformations, Anzeiger Österr. akad. Wiss., Abt.II, Math., Phys. u. Techn. Wiss., 131.Band, 1994,16p

[5] H. Florian, J. Püngel and W. Tutschke, Matrix differential transformations, Applicable analysis, in print

[6] R. P. Gilbert, Constructive methods for elliptic equations, Lecture Notes in Math. Vol. 365, Springer-Verlag Berlin- Heidelberg-New York, 1974

[7] J. Püngel and H. Florian, Differentialtransformationen von Lösungen partieller Differentialgleichungen, Compl. Variabl. 18 (1992), 125-134

[8] M. Kniepeiß and J. Püngel, Transformation of partial differential equations into ordinary differential equations by differential operators, in preparation

7 SOLUTION OF INITIAL VALUE PROBLEMS FOR PSEUDOHOLOMORPHIC FUNCTIONS IN CONICAL DOMAINS

A. O. Çelebi
and U. Yüksel

Dedicated to Professor Robert R. Gilbert on the occasion of his 65th birthday

The extension of the classical Cauchy-Kowalewski theorem for the initial value problem

$$\frac{\partial w}{\partial t} = C_0(t,z)\frac{\partial w}{\partial z} + A_0(t,z)w + D_0(t,z)$$

$$w(0,z) = w_0(z)$$

to the space of pseudoholomorphic function has been investigated by several authors. For example Tutschke and Withalm [2] considered the initial value problem

$$\frac{\partial w}{\partial t} = C_0(t,z)\frac{d_E w}{dz} + A_0(t,z)w + B_0(t,z)\overline{w}$$

$$w(0,z) = w_0(z)$$

where

$$\frac{d_E w}{dz} = \frac{\partial w}{\partial z} - A_E w + B_E \overline{w}$$

H. Florian et al. (eds.), Generalized Analytic Functions, 65–67.

$$lw = \frac{\partial w}{\partial \overline{z}} - a_E w - b_E \overline{w} = 0 \tag{7.1}$$

in which a_E, b_E, A_E, B_E are the characteristic coefficients belonging to the generating pair E in the sense of Bers [1]. Later Yüksel and Çelebi proved a Cauchy-Kowalewski Theorem for the initial value problem

$$\frac{\partial w}{\partial t} = Lw, \quad w(0,z) = w_0(z) \tag{7.2}$$

where

$$Lw = E_0(t,z)\frac{\partial}{\partial \overline{z}}\left(\frac{d_E w}{dz}\right) + F_0(t,z)\overline{\left(\frac{\partial}{\partial \overline{z}}\left(\frac{d_E w}{dz}\right)\right)}$$

$$+ C_0(t,z)\frac{d_E w}{dz} + G_0(t,z)\overline{\left(\frac{d_E w}{dz}\right)} + A_0(t,z)w + B_0(t,z)\overline{w} + D_0(t,z) \tag{7.3}$$

in the space $P_G(E)$ of pseudoholomorphic functions in the sense of Bers which are defined in the (bounded) domain G in the complex plane. Let us also note that for the particular choices of the coefficients E_0, F_0, G_0 and D_0 the problem (7.2) is reduced to one of the problems discussed in [2],[3] and [4].

This note is aimed at reporting the solution of the initial value problem (7.2) by the contraction mapping principle in conical domains. We assume that

(a) the characteristic coefficients a_E, b_E, A_E, B_E belong to $C^\lambda(\overline{G})$;

(b) for every t the operator l defined by (7.1) is associated to the operator L defined by (7.3);

(c) the coefficients $A_0, B_0, \cdots, G_0$ of L are defined and continuous in the conical domain

$$M = \{(t,z) : z \in G, 0 \leq t \leq \eta(s_0 - s(z))\}$$

and they belong to $C^\lambda(\overline{G}_s)$ provided $s < s_0 - \frac{t}{\eta}$ for fixed t;

(d) the initial function $w_0(z)$ belongs to $W_s := C^\lambda(\overline{G}_s) \cap P_{G_s(E)}, \quad 0 < s < s_0$ where the subdomains G_s exhaust the given domain G.

First we have shown that $W_*(M) := \{w : w \in W_s, ||w||_* < \infty\}$ equipped with *-norm is a Banach space, where $||w||_* = \sup_M ||w||_{s(z)} d^\sigma(t,z)$ with $d(t,z) = s_0 - s(z) - \frac{t}{\eta}$ and $s(z)$ is the (uniquely determined) index, $0 < s(z) < s_0$, such that z belongs to the boundary of $G_{s(z)}$.

Next we have obtained that the operator

$$Fw(t,z) := w_0(z) + \int_0^t Lw(t,z)dz$$

maps the space $W_*(M)$ into itself

Lastly we have proved that the initial value problem (7.2) is solvable in the conical domain M by the contraction mapping principle in case η satisfies an inequality of type

$$\eta < \frac{c_1 \sigma}{c_2(1+\sigma)e}$$

and the solution is uniquely determined in $W_*(M)$.

References

[1] L.Bers; Theory of Pseudonanalytic Functions, N.Y.Univ. 1953.

[2] W.Tutschke, C.Withalm; The Cauchy-Kowalewska Theorem for Pseudo-holomorphic Functions in the sense of L.Bers; Complex Var. Vol.1 (1983), 383-393.

[3] U. Yüksel; Solution of Cauchy-Kowalewski problems in a subspace of Pseudoholomorphic Functions in the sense of L.Bers, Complex Variables, Vol. 31 (1996), 1-8

[4] U.Yüksel; The Cauchy-Kowalewski Theorem for Pseudoholomorphic Functions in Conical Domains; Complex Variables, 1996, In Print.

[5] U.Yüksel, A.O. Çelebi; The Cauchy Kowalewski Theorem in the Space of Pseudoholomorphic functions; Complex Variables Vol. 29 (1996), 305-311.

8 THE CONTRACTION-MAPPING PRINCIPLE —SOME ASSOCIATED OPTIMIZATION PROBLEMS

W. Tutschke

Dedicated to Robert P. Gilbert on the occasion of his 65th birthday

Although the contraction-mapping principle is rather elementary and has been in use for a long time. its inherent strength still remains to be exploited. For instance, W. WALTER, in this journal, published an elementary proof of the CAUCHY-KOWALEVSKY Theorem based on the contraction-mapping principle a decade ago (see [4]). The present paper is aimed at showing that this principle is connected with some optimization problems as well.

Let $\mathcal{B}$ be a BANACH space with norm $\|\cdot\|$ and $\mathcal{A}$ a linear and bounded operator mapping $\mathcal{B}$ into itself. Consider the fixed-point problem

$$u = u_0 + \mathcal{A}\mathcal{F}u \tag{8.1}$$

where u_0 is fixedly chosen in $\mathcal{B}$ and $\mathcal{F}$ is a (non-linear) operator mapping $\mathcal{B}$ into itself. Suppose $\mathcal{F}$ is defined everywhere in $\mathcal{B}$ and LIPSCHITZ continuous with the LIPSCHITZ constant $L(d)$ in the ball $\|u - u_0\| \leq d$. Then in this ball the estimate $\|\mathcal{F}u\| \leq L(d) \cdot d + M$ is true implying that the operator defined by the right-hand side of (8.1) maps the ball into itself provided $\|\mathcal{F}u\| \, (L(d) \cdot d + M) \leq d$. Note that this condition implies contractivity, too (if $M = 0$, then the left-hand side has to be smaller than the right-hand side). Thus we have proved:

Lemma 1. *The smallest restriction of $\|\mathcal{A}\|$ is obtained if*

$$\frac{d}{dL(d) + M} \tag{8.2}$$

H. Florian et al. (eds.), Generalized Analytic Functions, 69–72.

is maximal.

In case $L(d)$ is differentiable and $L'(d) > 0$ everywhere, then the equation

$$d^2 = \frac{M}{L'(d)} \tag{8.3}$$

has to be satisfied necessarily. Supposing $L''(d) > 0$, additionally, this equation is uniquely solvable[1].

Of course, refining the above estimates under additional assumptions, still better estimates for the admissible operators can be expected[2]. Generally speaking, the estimates can be improved by replacing the ball with radius d by a polycylinder whose radii can be varied independently of each other. That way one can still better harmonize the set of definition of the operators with its properties. Many such investigations are carried out in S. GRAUBNER's dissertation [1]. As also done by S. GRAUBNER, this optimization procedure can be applied when solving boundary value problems for partial (complex) differential equations. In this case the polycylinders can be chosen in such a manner that the estimates for the functions themselves and for the derivatives are different from each other (see [2] and also S. GRAUBNER's paper [2] in this journal).

Another type of optimization problems connected with the contraction-mapping principle are optimizations of bounds for the derivatives with respect to parameters. Again, consider the operator equation (8.1). Suppose, however, that the operator $\mathcal{F}$ depends not only on the desired element u but also on a parameter m, i.e., $\mathcal{F}u$ is to be replaced by $\mathcal{F}(u, m)$. To be short, suppose that m is a real variable running in a closed interval $a \leq m \leq b$. Moreover, the following conditions are satisfied:

a) the underlying space $\mathcal{B}$ is a BANACH algebra,
b) the operator $\mathcal{A}$ and the differentiation with respect to the parameter m are interchangeable,
c) the derivative of the composition $\mathcal{F}(u(m), m)$ with respect to m is equal to $\partial_u \mathcal{F} \cdot \partial_m u + \partial_m \mathcal{F}$ (chain rule).

All of these assumptions are satisfied, for instance, in case the operator equation under consideration is the integral rewriting of an initial value problem for an ordinary differential equation whose right-hand side depends on a real parameter m and $\mathcal{B}$ is the BANACH algebra of continuous functions defined on a closed interval.

Now let $\mathcal{B}$ be a BANACH algebra whose elements are continuously differentiable with respect to a real parameter m, $a \leq m \leq b$. Introduce the norm

$$\|u\|_* = \max\left(\|u\|, \|\partial_m u\|\right).$$

Apply the contraction-mapping principle to to the above operator in the set

$$\left\{\|u - u_0\| \leq d, \quad \|\partial_m u\| \leq D\right\}.$$

The operator under consideration is contractive with respect to the $*$-norm and maps this set into itself in case the additional conditions

$$\begin{aligned} \|A\| &< \frac{1}{L_u(d+D)+M_u+L_m} \\ \|A\| &\leq \frac{D}{(L_ud+M_u)D+L_\mu d+M_m} \end{aligned}$$

are satisfied where L_u, M_u and L_m, M_m have the same meaning for $\partial_u f$ and $\partial_m f$ resp. as L, M for f have where L_u and L_m may depend on d. The right-hand sides of the last two inequalities coincide for

$$D = \frac{1}{2L_u}\left(-L_m + \sqrt{L_m^2 + 4(L_md + M_m)L_u}\right), \tag{8.4}$$

and the common value of these two bounds is

$$\frac{2}{L_m + \sqrt{L_m^2 + 4(L_md + M_m)L_u} + 2(L_ud + M_u)}. \tag{8.5}$$

Thus the following statement is true (cf. also [3]):

Lemma 2. *Choose any $d > 0$. Then largest possible bound for $\|A\|$ is given by (8.5).*

To sum up, the following statement has been proved:

Theorem *Suppose $L''(d) > 0$ and let d_* be the uniquely determined solution of the equation (8.3). Replace d by d_* in (8.2), (8.5) and (8.4). Then the largest possible bound for $\|A\|$ is the minimum of the quantities (8.2) and (8.5), and the desired fixed point satisfies the a-priori estimate*

$$\|\partial_m u\| \leq D.$$

Moreover, not only the successive approximations themselves but also their derivatives with respect to the parameter m converge in the norm of B.

Notes

1. The example $L'(d) = \dfrac{3}{1+d^4}$ shows that the condition $L'(d) > 0$ alone is not sufficient for unique solvability.

2. Consider, for instance, the initial value problem $y' = 1 + y^2$, $y(0) = 0$ in its integral rewriting. Then $\|A\|$ is the length ϱ of the existence interval, and the choice (8.3) leads to the maximal bound $\frac{1}{4}\sqrt{2}$ for ϱ. If $|y| \leq d$, then the right-hand side of the differential equation can be estimated by $K = 1 + d^2$ and thus the integral operator maps the interval under consideration into itself in case $K\varrho \leq d$, i.e., $\varrho \leq d/(1+d^2)$. Further, the operator is contractive if $\varrho < 1/2d$. Both restrictions of ϱ coincide for $d = 1$ and lead to the better bound $\frac{1}{2}$ for ϱ.

References

[1] S. GRAUBNER, Über die Optimierung von Polyzylindern bei Anwendung des Kontraktionsprinzips auf Differentialgleichungen. Thesis (Dr.rer.nat. dissertation) at FU Berlin, 1997.

[2] - , Über die Optimierung der Definitionsmenge von Operatoren bei der Anwendung des Kontraktionsprinzips auf Differentialgleichungen. This volume, the following Chapter 9.

[3] W. TUTSCHKE, Optimal bounds for the derivative with respect to a parameter and an application to partial complex differential equations. Contained in the Proceedings of the Peking Conference on "Integral equations and boundary value problems" (Beijing September 2-7, 1990) ed. by GUO-CHUN WEN and ZHEN ZHAO. World Scientific Publishing Co. 1991.

[4] W. WALTER, An elementary proof of the Cauchy-Kowalevsky theorem. Amer. Math. Monthly, vol. 92 (1985), 115-125.

9 ON THE OPTIMIZATION OF THE SET OF DEFINITION OF OPERATORS IN APPLICATIONS OF THE CONTRACTION-MAPPING PRINCIPLE TO DIFFERENTIAL EQUATIONS

S. Graubner

In case one reduces problems for differential equations (e.g., initial value problems for ordinary differential equations or boundary value problems for partial differential equations) to fixed-point problems, one has to solve related operator equations in polycylinders of suitably chosen function spaces.

In order to get necessary estimates of the operators, the right-hand sides of the given differential equations are supposed to be LIPSCHITZ continuous. Then the following problem arizes:

The larger the polycylinders, the larger the corresponding LIPSCHITZ constants, i.e., the estimates become worse. On the other hand, this can be avoided by choosing small polycylinders. This, however, leads to new restrictions of the length of the existence interval or one gets certain additional restrictions on the given initial data (such as initial values or boundary data).

In order to illustrate this problem, consider the operator defined by

$$U = u_0 + \mathcal{A}F(\cdot, u) \tag{9.1}$$

H. Florian et al. (eds.), Generalized Analytic Functions, 73–76.

where $\mathcal{A}$ is a linear and bounded operator with norm $\|\mathcal{A}\|$ and u belongs to the ball with radius d centred at u_0, i.e.,

$$\|u - u_0\| \leq d. \tag{9.2}$$

Note that a LIPSCHITZ condition of form

$$\|F(\cdot, u_2) - F(\cdot, u_1)\| \leq L\,\|u_2 - u_1\| \tag{9.3}$$

implies that the operator (9.1) is contractive provided

$$\|\mathcal{A}\|\,L < 1. \tag{9.4}$$

Denoting $\|F(\cdot, u_0)\|$ by M, one has

$$\begin{aligned} \|F(\cdot, u)\| &\leq \|F(\cdot, u) - F(\cdot, u_0)\| + \|F(\cdot, u_0)\| \\ &\leq Ld + M. \end{aligned} \tag{9.5}$$

Consequently, the operator (9.1) maps the polycylinder (9.2) into itself in case the condition

$$\|\mathcal{A}\|\,(Ld + M) \leq d \tag{9.6}$$

is satisfied. Notice that this condition (9.6) strengthens the condition (9.4) for contractions. Note that $d = (d_1, ..., d_n)$ if the Kdesired u is an n-tupel and if the set of definition of the operator (9.1) is the polycylinder

$$D = \left\{u = (u_1, , , .u_n) : \|u_j - u_{0j}\| \leq d_j\right\}.$$

In the paper [1] the optimal d is determined under different assumptions on the dependence of L on the radii of the polycylinder. The following simple Example 1 shows the behaviour of ordinary differential equations with respect to the optimization problem under consideration.

Example 1. Consider the initial value problem

$$y' = 1 + y^n \tag{9.7}$$

$$y(0) = 0 \tag{9.8}$$

where $n \geq 2$ is an integer. Consider the right-hand side of the differential equation in the interval

$$D = \{y : |y| \leq d\}.$$

Then $L = nd^{n-1}$ is a LIPSCHITZ constant, while $K = 1 + d^n$ is a bound for the absolute value of the right-hand side of the differential equation. Looking for the solution of (9.7), (9.8) in the interval $[0, \varrho]$, one gets the two conditions

$$\varrho\,(1 + d^n) \leq d \quad \text{and} \quad \varrho n d^{n-1} < 1,$$

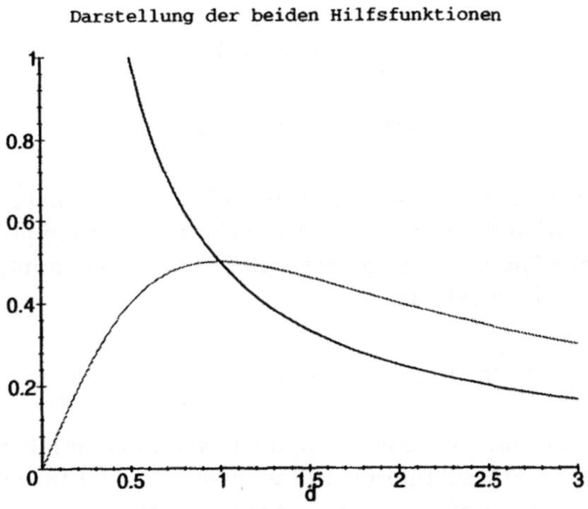

i.e., in dependence on the choice of d, the length ϱ of the interval of existence has to satisfy the inequalities

$$\varrho \leq \frac{d}{1+d^n} \quad \text{and} \quad \varrho < \frac{1}{nd^{n-1}}, \tag{9.9}$$

see the figure above. The point of intersection of the two curves leads to the optimal values

$$d_{opt} = \frac{1}{(n-1)^{\frac{1}{n}}} \quad , \qquad \varrho_{opt} = \frac{1}{n}(n-1)^{1-\frac{1}{n}}.$$

The second example shows that the method is applicable to partial differential equations, too:

Example 2. Consider an (elliptic) differential equation of type

$$\frac{\partial w}{\partial \bar{z}} = F\left(z, w, \frac{\partial w}{\partial z}\right) \tag{9.10}$$

in the unit disk G. We look for a solution $w = w(z)$ of the DIRICHLET boundary value problem, i.e., Re w and Im w are given on the boundary $|z| = 1$ and at one fixedly chosen point, resp. Note that every solution $w = w(z)$ of (9.10) satisfies an integral relation of form

$$w = \Phi + T_G F\left(\cdot, w, \frac{\partial w}{\partial z}\right)$$

where Φ is holomorphic and T_D is the well-known domain integral operator

$$(T_D h)(z) = -\frac{1}{\pi} \iint\limits_{|z|<1} \frac{h(\zeta)}{\zeta - z} d\xi d\eta, \quad \zeta = \xi + i\eta$$

(cf. [2]). Therefore, solutions of the differential equation (9.10) can be constructed as fixed point of the operator which can be defined by

$$W = \Phi + T_G F\left(\cdot, w, \frac{\partial w}{\partial z}\right). \tag{9.11}$$

Moreover, suitably choosing the boundary values of Φ, one can prescribe the boundary values of the images W and thus the boundary values of the fixed point, especially. In order to construct fixed points of the operator (9.11), one has to apply the contraction-mapping principle to the bicylinder

$$D = \left\{\left(w, \frac{\partial w}{\partial z}\right) : \|w - \Psi\|_\lambda \leq d_1, \quad \left\|\frac{\partial w}{\partial z} - \Psi'\right\|_\lambda \leq d_2\right\}$$

where Ψ is the holomorphic solution of the boundary value problem under consideration and $\|\cdot\|_\lambda$ means a HÖLDER-norm. In the dissertation [1] the radii d_1 and d_2 are varied and, using a change of coordinates, they are determined in such a way that one gets a least possible restriction of the permissible boundary values.

Provided the LIPSCHITZ constants depend linearly on the radii of the bicylinder, the optimal radii are determined by the position of two ellipses whose axes are parallel to the coordinate axes in the d_1, d_2-plane. This is also true for operator equations in general BANACH spaces as is also shown in [1].

References

[1] S. GRAUBNER, Über die Optimierung von Polyzylindern bei Anwendung des Kontraktionsprinzips auf Differentialgleichungen. Thesis (Dr.rer.nat. dissertation) at FU Berlin, 1997.

[2] I. N. VEKUA, Generalized Analytic Functions. Pergamon Press 1962.

10 GENERALIZED ANALYTIC FUNCTIONS IN MULTIDIMENSIONAL SPACES

E. Obolashvili

Dedicated to Professor Robert P. Gilbert on the occasion of his 65th birthday

Introduction

Several essential extensions of the generalized Cauchy–Riemann equations (GCRE), whose solutions are considered as in two as well as in multidimensional spaces e.g. [1], [2], [16], [4]... Methods of classical complex analysis are successfully used to investigate them. The very remarkable appreciation of complex analysis is given by H. Bergehr in [1]: "richness beauty and fascination of complex analysis originate from the coincidence of analytic, algebraic and geometrical methods... There is the branch which just recently became quite active: complex analysis in partial differential equations". Moreover complex analysis can be considered as the foundation of Clifford analysis,which represents one of the remarkable fields of modern mathematics. It has been found that almost all classical linear partial differential equations of mathematical physics are put in Clifford analysis and they can be obtained using only Dirac operator with various Clifford algebras without applications any physical laws. By this way GCRE in higher dimensional spaces can be constructed too [5], [12].

However in a three dimensional space the system

$$\begin{aligned} &\operatorname{div} U + (U \cdot A) = 0, \\ &\operatorname{grad} \varphi + \operatorname{rot} U + [U \times B] + C\varphi = 0, \end{aligned} \tag{10.1}$$

where U, A, B, C three-component vector-functions, $\varphi(x)$ scalar function, $x \in R^3$, was constructed in [12] as on the one hand three dimensional analogue of GCRE and

H. Florian et al. (eds.), Generalized Analytic Functions, 77–87.

on the other hand as generalization of Moisil-Theodorescu system. For the solutions (10.1) four-component -vector $\phi(U, \varphi)$, called generalized holomorphic vector, almost all properties of GAF of one complex variable were established too. In the case of $A = B = C \equiv 0$ (10.1) is Moisil-Theodorescu system written in the vectorial form. Before in the origin [10] this system was written in the matrix form. Besides it must be noted that as generalization of the potential system: $\operatorname{div} U = 0$, $\operatorname{rot} U = 0$ just mentioned system is considered in [9], but obviously the author did not notice that it is Moisil-Theodorescu system in vectorial form, there are cited neither [10] nor others.

I want to note with gratitude that the system (10.1) was mentioned in[7], [6] and there was noted that it has to be put in the context of algebras. Using Clifford algebra GCRE in a multidimensional space is given by B. Goldschmidt [5] who had only existence theorems but had not picked out such cases for which the Cauchy-kernel and representations can be written explicitly, besides he had neither hyperbolic nor parabolic cases. Natural multidimensional GCRE can be considered using Dirac operator and its modification in various Clifford algebras [12]. These will be elliptic, hyperbolic or parabolic systems according to what kind of multiplication rules in clifford analysis are used. In elliptic case the theory of regular functions as multidimensional holomorphic functions is given in [3], [7], [4], [6], ... As the aim of this paper is to consider GAF in a higher dimensional space we will consider only Clifford algebra $R_(n)$ $(n \geq 1)$ with the basis e_A, $A : (\alpha_1 \dots \alpha_k)$, $0 \leq \alpha_1 < \dots < \alpha_k \leq n$ and with the multiplication rules

$$e_0^2 = e_0, \quad e_j^2 = -e_0, \quad j = 1, \dots, n; \quad e_j e_k + e_k e_j = 0, \quad k \neq j = 1, 2, \dots, n, \tag{10.2}$$

where e_0 is identity element. Any element of this space can be represented as

$$u = \sum_A u_A e_A, \tag{10.3}$$

where u_A are real. u will be called vectorial if

$$u = \sum_{k=0}^{n} u_k e_k \tag{10.4}$$

In the case of $n \geq 2$ for u two conjugation are defined

$$\overline{u} = \sum_A u_A \overline{e}_A, \quad \widetilde{u} = \sum u_A \widetilde{e}_A, \tag{10.5}$$

$$\overline{e}_0 = e_0, \quad \overline{e}_j = \widetilde{e}_j = -e_j, \quad j = 1, \dots, n,$$
$$\overline{e}_A = \overline{e}_{\alpha_k} \dots \overline{e}_{\alpha_1}, \quad \widetilde{e}_A = \overline{e}_{\alpha_1} \dots \overline{e}_{\alpha_k}. \tag{10.6}$$

The product of any two elements is defined as

$$uv = \sum_{A,B} u_A v_B e_A e_B \tag{10.7}$$

For the vectorial element $u \neq 0$ the inverse element can be defined:

$$u^{-1} = \frac{\overline{u}}{|u|^2}, \quad u\overline{u} = \sum_0^n u_k^2 = |u|^2. \tag{10.8}$$

10.1 *H*-REGULAR FUNCTIONS

Let a domain $\Omega \in R^{n+1}$ $(n \geq 1)$ and a function $u(x) : \Omega \to R_{(n)}$, $x = \sum_0^n x_k e_k$, $u_A(x)$ be assumed of $C^1(\Omega)$ functions. Consider the modification of the Dirac operator [3]

$$\overline{\partial} = \sum_0^n \frac{\partial}{\partial x_k} e_k = \frac{\partial}{\partial x_0} e_0 + D, \quad \partial = \frac{\partial}{\partial x_0} e_0 - D, \tag{10.9}$$

where D is the Dirac operator. It is clear $\overline{\partial}$ considered in $R_{(1)}$ will be the classical Cauchy-Riemann operator $\frac{\partial}{\partial \overline{z}}$, $z = x_0 e_0 + x_1 e_1$. Many principal properties of holomorphic functions of one complex variable are established for the regular functions in $R_{(n)}$ [3], [7], [4], i.e. for the solutions of the equation

$$\overline{\partial} u = 0 \tag{10.10}$$

As $\partial\overline{\partial} = \Delta$ is Laplace operator in R^{n+1} each $u_A(x)$ will be harmonic function too.

The function $u(x)$ with values in $R_{(n)}$ is said to be h-regular in Ω if

$$Lu \equiv \overline{\partial} u + \overline{u} h = 0, \tag{10.11}$$

or

$$\overline{\partial} u + \overline{u} h = 0, \tag{10.12}$$

where $\overline{u}$ and $\widetilde{u}$ are defined by (10.5), $h(x)$ is given function with values in $R_{(n)}$. For $u \in R_{(1)}$, i.e. in two dimensional case these equations will be the generalized Cauchy-Riemann system

$$\frac{\partial u}{\partial \overline{z}} + B\overline{u} = 0 \tag{10.13}$$

Thus h-regular functions can be considered as GAF in higher dimensional space. Let consider $R_{(2)}$ and

$$u = u_0 e_o + u_1 e_1 + u_2 e_2 + u_{12} e_1 e_2, \quad h = h_0 e_0 + h_1 e_1 + h_2 e_2, \tag{10.14}$$

then (10.11) and (10.12) will be equivalent to the system

$$\begin{aligned} \operatorname{div} U + (H \cdot U) &= 0 \\ \operatorname{grad} \varphi + \operatorname{rot} U + [U \times H] \pm H\varphi &= 0, \end{aligned} \tag{10.15}$$

where the sign(+) one has for (10.11), and (−) for (10.12); $U = (u_0, -u_1, -u_2)$, $\varphi = -u_{12}$, $H = (h_0, -h_1, -h_2)$. In the case of constant vectorial h for the equation (10.11) 2^n-linearly independent fundamental solutions can be represented as [15]

$$u^A(r) = (\partial g)e_A - g(r)\widetilde{e}_A h, \quad A : (\alpha_1 \dots \alpha_k) \tag{10.16}$$

where g is fundamental solution of Helmholtz equation in R^{n+1}

$$\Delta g - |h|^2 g = 0, \tag{10.17}$$

defined as

$$g(r) = C_n \Big(\frac{1}{r}\frac{d}{dr}\Big)^{m-1} \psi(r), \quad r = |x|,$$

$\psi(r) = H_0^{(1)}(i|h|r)$ when $n = 2m-1$ and $\psi(r) = \frac{1}{r}e^{-|h|r|}$ when $n = 2m$, $H_0^{(1)}$ is Hanckel function zeroth order. For the equations (10.15) with (+) one has four linearly independent fundamental solutions, which are given in [12] too. But for the equation (10.12), i.e. also for (10.15) with the sign (−) all corresponding linearly independent fundamental solutions can not be defined explicitly. That is why the Cauchy-Kernel for (10.11) is represented explicitly while for (10.12) we can not. Thus in the sequel (10.11) equation will be considered.For it the adjoint equation

$$L^* v \equiv v\overline{\partial} - \overline{h}\widetilde{v} = 0$$

is defined. Its 2^n linearly independent fundamental solutions will be represented as

$$v^A = e_A(g\partial) + \overline{h}\widetilde{e}_A g.$$

It is easy to obtain for the operators Lu and L^*v

$$\mathrm{Re}[v(Lu) - (L^*v)u] = \mathrm{Re} \sum_{A,B} \sum_{j=0}^{n} \frac{\partial u_A v_B}{\partial x_j} e_A e_j e_B \tag{10.18}$$

Then by the classical well-known way one can obtain generalized Cauchy's integral theorem and formula, Borel-Pompeju representation, Liouvile's theorem, Cauchy type integral with explicit Cauchy kernel, Poincare-Bertrand formula and so on, [14].

Note that as partial case from (10.11) one can obtain generalized Riesz system[13]. Namely if $u(x)$, $h(x)$ are vectorial with values in $R_{(n)}$

$$u(x) = u_0 e_0 - \sum_1^n u_k e_k, \quad h(x) = h_0 e_0 - \sum_1^n h_k e_k,$$

then $\widetilde{u} = \overline{u}$ and (10.11) can be written in the form:

$$\sum_0^n \Big[\frac{\partial u_j}{\partial x_j} + h_j u_j\Big] = 0, \tag{10.19}$$

$$\frac{\partial u_j}{\partial x_k} - \frac{\partial u_k}{\partial x_j} - h_k u_j + h_j u_k = 0, \quad j, k = 0, 1, \dots n.$$

It can be considered also as multidimensional analogue of GCRE. For $n \geq 2$ it is overdetermined elliptic system. As the equations (10.11) has 2^n fundamental solutions (10.16) and among them only one is vectorial, one can conclude that (10.19) has only one fundamental solution which with $h =$ const has the form

$$u = \partial g - gh.$$

It is obvious Cauchy's integral formula for the solution of (10.19) will be just because if $u(u_o, u_1, \dots, u_n)$ is the solution of (10.19) then $v(u_o, u_1, \dots, u_n, 0, \dots, 0)$ will be the solution of the equation (10.11). That is why Liouville theorem will be true for the solutions of (10.19) too. Besides we had considered the system using notions of tensor analysis:

$$\sum_0^n \nabla^k u_k = \sum_0^n \nabla_k u^k = 0, \tag{10.20}$$
$$\nabla^j u^k - \nabla^k u^j = 0, \quad j, k = 0, 1, \dots, n,$$

where $u^j(x)$, $u_j(x)$ are contravariant and covariant components of a vector respectively in a curvilinear coordinate system $(x_0, x_1, \dots, x_n)$, ∇_j, ∇^j are covariant and contravariant derivatives respectively. As in the orthonormal coordinate system

$$\nabla_j = \nabla^j = \frac{\partial}{\partial x_j}, \tag{10.21}$$

(10.20) can be considered as Riesz system or multidimensional analogue of Cauchy-Riemann system in a Riemannean space. Let the metric quadratic form of the space be given as

$$ds^2 = g_{ik} dx_i dx_k, \quad g_{ik} = \delta_{ik} \exp[a_j x_j], \tag{10.22}$$

where δ_{ik} are Kroneker's symbols, $a_j (j = 0, 1, \dots, n)$ are real constants. Here Einstein's summation convention will be used omitting the summation sign agree upon adding over the repeated indices. Then (10.20) will have the form (10.19) with constant coefficients. Hence (10.19) can be considered as analogue of Cauchy-Riemann system in multidimensional Riemannean space with (10.22). But if $x_i (i = 0, 1, \dots, n)$ are considered as cartesian coordinates of the point it will be considered as multidimensional analogue of GCRE. The coordinate system in a multidimensional Riemannean space, for which metric tensor is defined by (10.22), may be regarded as analogue of the cartesian coordinates.

Note that (10.11) can be considered in the Clifford algebra with other multiplication rules in such a way that it will be hyperbolic system related to Maxwell equations of electromagnetic fields and Dirac equations of relativity quantum mechanics. Thus (10.11) represents in one Clifford algebra elliptic equations and in the other one hyperbolic equations. These questions will be regarded in the next paper.

10.2 BOUNDARY VALUE PROBLEMS

Georgian scientific mathematical school founded by N. Muskhelishvili and then considerable contribution done to its development by I. Vekua, has great traditions to solve boundary value problems for holomorphic and generalized holomorphic functions of one complex variable. Many of them are solved explicitly in quadratures.

In the classical complex analysis, as is well-known, there are two remarkable types of principal boundary value problems for analytic and GAF of one complex variable which have a lot of applications, namely 1.The boundary value problem of linear conjugation, i.e. Riemann bvp, 2.The Riemann-Hilbert boundary value problem.

Both these problems were posed for the first time by Riemann and now are investigated with sufficient completeness. Note that there exists wide classes of given functions and boundary curves for which the first problem for analytic functions is solved in quadratures, however second problem for them is solved explicitly in a much fewer cases. Unlike analytic functions for generalized analytic functions these problems are solved explicitly in very particular cases. The discussion and the historical survey can be found in [11], [17]. In higher dimensions for the equations (10.10), (10.11) the above indicated problems can be correspondly considered, but their solutions can be represented in quadratures only in more less cases. The main reason of this fact is that the product of two analytic functions is also analytic only in two dimensional case ,i.e. in $R_{(1)}$, but in $R_{(n)}(n > 1)$ it is not right, because for any functions $u, v \in R_{(n)}$

$$\overline{\partial}(uv) \neq (\overline{\partial}u)v + u(\overline{\partial}v). \tag{10.23}$$

If G is constant then it is easy to obtain:

$$\overline{\partial}(uG) = (\overline{\partial}u)G, \quad (Gu)\overline{\partial} = G(u\overline{\partial}). \tag{10.24}$$

That is why if u is left-regular:$\overline{\partial}u = 0$, then uG will be again left regular and if u is right regular: $u\overline{\partial} = 0$, then Gu will be right regular.

We are mainly interested in considering the correctly posed boundary value problems for GAF in higher dimensions whose solutions can be represented in quadratures.

Let $\Omega^+ \in R^{n+1}$ be bounded domain with the boundary S supposed to be piecewise Liapunov manifold. By Ω^- is denoted the complement domain of $\Omega^+ + S$ to the whole space R^{n+1}.

Define a piecewise left-regular $u(x)$ or right-regular $v(x)$ functions with values in $R_(n)$ vanishing at infinity with the conditions

$$u^+(x) = u^-(x)G + g(x), x \in S, \tag{10.25}$$

$$v^+(x) = Gv^-(x) + g(x), \tag{10.26}$$

where $G \neq 0$ is vectorial constant element of $R_{(n)}$, $g(x)$ is Holder continuous given function with values in $R_{(n)}$ $(n \geq 1)$. To solve this problem one can use exactly the same way as for piecewise holomorphic functions of one complex variable [11].

Defining canonical function

$$\chi(x) = \begin{cases} G, & x \in \Omega^+, \\ e_0, & x \in \Omega^-, \end{cases}$$

i.e. $\chi^+(x) = \chi^-(x)G$ for (10.25) and $\chi^+(x) = G\chi^-(x)$ for (10.26), and as G^{-1} can be defined by (10.8), $u(x)G^{-1}$,i.e. $u(x)[\chi(x)]^{-1}$ will be the solution of (10.25) and $[\chi(x)]^{-1}v(x)$ will be the solution of (10.26), that is why the corresponding solution will be represented explicitly by Cauchy type integral.

Let D^+ be a half-space $x_n > 0$ and D_- be $x_n < 0$, S be the plane $x_n = 0$. Then for the equation (10.25) Riemann-Hilbert problem will be formulated as: To define in D^+ a left-regular function $u(x)$ vanishing at infinity by the conditions

$$\mathrm{Re}[u(x)Ge_A] = f_A \quad \text{for} \quad x_n = 0, \quad n \geq 1, \tag{10.27}$$

where A takes 2^{n-1} different values from $(\alpha_1 \ldots \alpha_k)$, $0 \leq \alpha_1, \ldots, \alpha_n \leq n$, f_A given scalar functions, i.e. we have 2^{n-1} real conditions on the boundary, $G \neq 0$ is again vectorial constant element of $R_{(n)}$. These boundary conditions will be correspondly changed to solve explicitly the problem for right regular functions. As for the holomorphic functions of one complex variable, the solution of (10.27) can be easily reduced to the solution of Riemann bvp (10.25).

In order to obtain explicit solution of Riemann and Riemann-Hilbert boundary value problems for h-regular functions,their formulation must be a little modified. The point is that if u is h-regular and G is constant element in $R_{(n)}$ then $u(x)G$ is not h-regular.

Let $u(x)$ with values in $R_{(n)}$ be the solution of the equations

$$\begin{aligned} \overline{\partial} u + \widetilde{u}h = 0, \quad & x \in \Omega^+, \\ \overline{\partial} u + \widetilde{u}h^1 = 0, \quad & x \in \Omega^- \end{aligned} \tag{10.28}$$

where $h^1 = \overline{G}hG^{-1}$; G and h are supposed to be vectorial in $R_{(n)}$, then h^1 will be vectorial too and $|h^1| = |h|$.

Define $u(x)$ vanishing at infinity with the boundary condition

$$u^+(x) = u^-(x)G + g(x), \quad x \in S, \quad G \neq 0, \tag{10.29}$$

$g(x)$ is given Holder continuous function with values in $R_{(n)}$.

If u is right h-regular function the problem which can be solved explicitly will be posed as

To find the solution vanishing at infinity of equations

$$\begin{aligned} u\overline{\partial} + h\widetilde{u} = 0, \quad & x \in \Omega^+ \\ u\overline{\partial} + h^1\widetilde{u} = 0, \quad & x \in \Omega^- \end{aligned} \tag{10.30}$$

where $h^1 = G^{-1}hG$; h, G are vectorial constants, with the condition

$$u^+ = Gu^- + g. \tag{10.31}$$

Defining correspondly new h-regular function $v(x)$, these both cases will be easily reduced to the problem with the boundary condition

$$v^{+} - v^{-} = g,$$

and hence, $u(x)$ will be uniquely represented by Cauchy-type integral.

Let D^{+} be a half-space $x_n > 0$, D^{-} be $x_n < o$, Riemann-Hilbert problem: define h-regular function $u(x)$ in D^{+}, vanishing at infinity and satisfying 2^{n-1} real conditions

$$\mathrm{Re}[u(x)Ge_A] = f_A \quad \text{for} \quad x_n = 0 \tag{10.32}$$

where A takes 2^{n-1} different values from $(\alpha_1\alpha_2\ldots\alpha_k)$, $0 \le \alpha_1 < \alpha_2 < \cdots < \alpha_k \le n$, f_A given scalar functions.

Let $\Omega^{+} \in D^{+}$ with the boundary S, Ω^{-} be the complement domain of $\Omega^{+} + S$ to the half-space.

The compound boundary value problem: Let $u(x)$ be the solution of (10.28) vanishing at infinity. Define $u(x)$ by the conditions:

$$u^{+}(x) = u^{-}(x)G + g(x), \quad x \in S$$
$$\mathrm{Re}[u(x)e_A] = f_A(x_0, \ldots, x_{n-1}), \quad x_n = 0,$$

where $G \neq 0$, g, 2^{n-1} functions f_A, A are defined as in (10.32).

Both these problems can be also solved by classical way [15]

Now note that in the case $n \geq 2$, for some equations though with constant coefficients Dirichle problem is correctly posed in one domain of the space and in the other domain Cauchy's problem is correctly posed. In $R_{(n)}$ i.e. for elliptic equations such situations can not have place, that is why they can be called properly elliptic equations. But in $R_{(n,n-1)}$ $(n \geq 2)$ for which one has

$$e_j^2 = -e_0, \quad j = 1, \ldots, n-1, \quad e_n^2 = e_0,$$

if we consider for instance hyperbolic-regular equations

$$\bar{\partial} u = 0, \tag{10.33}$$

we can define the corresponding domains with above indicated properties. Let $x_0, \ldots,$ x_{n-1} be space variables, $x_n \equiv t$ time variable. They say that the manifold $t = 0$, with respect to the equation (10.33) has the space form and the manifold $x_k = 0$ with fixed k from $(0, 1, \ldots, n-1)$, has the time form. J.Hadamar had noted in [8] that Cauchy's problem on the time manifold for the wave equation is not correctly posed. We will see that it can be modified in such a way that one get correctly posed problems. As $u(x,t)$ at the same time is the solution of the wave equation

$$\Delta u = \frac{\partial^2 u}{\partial t^2} \tag{10.34}$$

where Δ is Laplace operator with respect to the variables $x_0, \ldots, x_{n-1}$, it is sufficient to consider the problems for (10.34). First note that from the characteristic equation of (10.34) follows

$$\xi_{n-1}^2 = \tau^2 - \sum_0^{n-2} \xi_k^2,$$

thus this equation with respect to ξ_{n-1} will have two real different solutions if

$$\lambda^2 \equiv \tau^2 - \sum_0^{n-2} \xi_k^2 > 0 \tag{10.35}$$

and will have two conjugate complex solutions if

$$\mu^2 \equiv \sum_0^{n-2} \xi_k^2 - \tau^2 > 0. \tag{10.36}$$

Consider (10.34) for $x_{n-1} \geq 0$, $x(x_0, \dots, x_{n-2}) \in R^{n-1}$, and as this equation does not change if t will be changed by $-t$, we will suppose $t \in R$.

Problem. Find the bounded solution of (10.34) in this domain with the conditions like the Cauchy data if we have (10.35):

$$u(x, 0, t) = \varphi(x, t), \quad \frac{\partial u}{\partial x_{n-1}} = \psi(x, t), \text{ for } x_{n-1} = 0 \tag{10.37}$$

or only one condition like the Dirichlet datum if we have (10.36);

$$u(x, 0, t) = \varphi(x, t), \tag{10.38}$$

where the functions φ, ψ must satisfy the conditions given below.

Solution. By Fourier integral transformation FIT of (10.34) with respect to the variables $x_0, \dots, x_{n-2}, t$, one can get:

$$\frac{d^2 \widehat{u}}{dx_{n-1}^2} - (|\xi|^2 - \tau^2)\widehat{u} = 0, \quad \xi(\xi_0, \dots, \xi_{n-2}).$$

This equation will have two linearly independent bounded solutions if (10.35) condition is supposed and then

$$\widehat{u} = c_1 \cos \lambda x_{n-1} + c_2 \sin \lambda x_{n-1}, \quad x_{n-1} > 0,$$

but if (10.36) condition is supposed then it will have one linearly independent solution vanishing at infinity

$$\widehat{u} = c e^{-\mu x_{n-1}}, \quad x_{n-1} > 0, \quad \mu > 0,$$

These representations show how many conditions for $x_{n-1} = 0$ must be given. Hence, by force of (10.37) one can get with the condition (10.35):

$$\widehat{u}(\xi, x_{n-1}, \tau) = \begin{cases} \widehat{\varphi}(\xi, \tau) \cos \lambda x_{n-1} + \frac{\widehat{\psi}(\xi, \tau)}{\lambda} \sin \lambda x_{n-1}, & \text{for } \tau^2 - |\xi|^2 > 0, \\ 0, \text{ for } \tau^2 - |\xi|^2 \leq 0. \end{cases} \tag{10.39}$$

But in the case (10.36) the solution of (10.38) will be

$$\widehat{u}(\xi, x_{n-1}, \tau) = \begin{cases} 0, & \tau^2 - |\xi|^2 \geq 0, \\ \widehat{\varphi}(\xi, \tau) e^{-\mu x_{n-1}}, & \text{for } \tau^2 - |\xi|^2 > 0. \end{cases} \tag{10.40}$$

(10.39) means that FIT of the given functions φ and ψ in (10.37) must be zero outside of the domain $\tau^2 > |\xi|^2$ and (10.40) means that FIT of $\varphi(x,t)$ in (10.38) must be zero outside of $\tau^2 < |\xi|^2$. Thus the domain for the Cauchy's data must be correspondly defined. Then by force of inverse theorem for FIT with the help (10.39) and (10.40) the solutions of the problem with the conditions (10.37) or (10.38) will be represented correspondly. Note that in place of (10.38) one can consider Neumann condition.

For the equations (10.33) (10.28) analogue problems can be considered too. In the case of (10.35) on the boundary must be given 2^n conditions and in the case of (10.36) 2^{n-1} conditions must be given.

I should like to express my appreciation to Technical University in Graz, especially to professor Wolfgang Tutschke for the remarkable conference, for the invitation, for the financial support and for the hospitality offered me during my stay in Graz-the wonderful place of the fairyland Austria.

References

[1] Begehr H., Complex analytic methods for PDE. World scientific, Singapore, London, 1994.

[2] Bojarski B., Old and new on Beltrami equations. Functional analytic methods in complex analysis and applications to PDE. World scientific. Trieste 1990.

[3] Brackx F.,Delanghe R., Sommen F., Clifford Analysis, Pitman, London, 1982.

[4] Gilbert R., Buchanan J.,First order systems a function theoretic approach. Academic Press, New York. 1983.

[5] Goldschmidt B., Properties of generalized analytic vectors in R^n. Math. Nachr. 103, 1981, 245-254.

[6] Gürlebeck K., Sprössig W., Quaternionic analysis and elliptic boundary value problems. Akademie-Verlag, Berlin, 1989.

[7] Habetha K. Function theory in algebras. Complex Analysis, Methods, trends and applications. Akademie Verlag, Berlin, 1985, 225-237.

[8] Hadamard J., Le problem de Cauchy. Paris. 1932.

[9] Krein S.G. Linear Equations in Banach space, "Nauka", Moskow, 1971. (russian).

[10] Moisil G., Teodorescu N., Functions Holomorphes dans l'espace. Mathematica, 5, 1931, 142-150.

[11] Muskhelishvili N., Singular integral Equations, Noordhoff Groningen 1953.

[12] Obolashvili E., Some *PDE* in Clifford Analysis. Banach Center Pub. 37(1996) 251-257.

[13] Obolashvili E., Space generalized holomorphic vectors. Diff. urav. XI, 1, 1975, 108-115 (russian).

[14] Obolashvili E., Partial differential equations in Clifford Analysis. Longman (to appear).

[15] Obolashvili E.,Boundary and Initial Value Problems in Clifford Analysis. Proceedings of the symposium in Seiffen, edited by W. Sprössig and K. Gürlebeck. 1996. 145-152.

[16] Tutschke W., Solutions of Initial value problems in classes of generalized analytic functions. Leipzig, Teubner-Text zur Math. 1989.

[17] Vekua I., Generalized Analytic functions. Pergamon Press. Oxford 1962.

11 OVERDETERMINED SYSTEMS OF SECOND ORDER ELLIPTIC EQUATIONS IN SEVERAL COMPLEX VARIABLES

H. Begehr and A. Dzhuraev

Dedicated to Robert P. Gilbert on the occasion of his 65th birthday

In complex analysis there exist two basic elliptic second order differential operators having quite different properties. One is the well-known Laplace operator $\partial^2/\partial z\partial\overline{z}$ which is strongly elliptic. The second one is the Bitsadze operator $\partial^2/\partial\overline{z}\partial\overline{z}$, see [7]. It is just elliptic but not strongly elliptic. A substantial difference is that while the Dirichlet problem for the Laplace equation $\partial^2 u/\partial z\partial\overline{z} = 0$ is uniquely solvable it is not for the Bitsadze equation $\partial^2 u/\partial\overline{z}^2 = 0$. For example

$$u_k(z) = (1-|z|^2)\, z^k\,, \quad k \in \mathbb{N}_0\,,$$

are infinitely many linearly independent solutions to the Bizadse equation vanishing on the boundary $|z| = 1$ of the unit disc $\mathbb{D} := \{|z| < 1\}$ in the complex plane $\mathbb{C}$.

The same situation occurs in several complex variables where instead of just one equation overdetermined systems appear. The system of n^2 equations

$$u_{\overline{z_k} z_\ell} = 0\,, \quad 1 \le k,\ell \le n\,, \tag{11.1}$$

the solution of which are called pluriharmonic is strongly elliptic. The system of $n(n+1)/2$ equations

$$u_{\overline{z_k}\,\overline{z_\ell}} = 0\,, \quad 1 \le k \le \ell \le n\,, \tag{11.2}$$

H. Florian et al. (eds.), Generalized Analytic Functions, 89–109.

is not strongly elliptic. Its solutions are the pluriholomorphic functions. Because any pluriharmonic function is harmonic i.e. satisfies

$$u_{\overline{z}_k z_k} = 0\,, \quad 1 \leq k \leq n\,, \tag{11.3}$$

it is evident that the Dirichlet problem will be uniquely solvable if it is solvable. To show that this does not hold for pluriholomorphic functions consider the unit ball $B := \{|z| < 1\}$ in $\mathbb{C}^n$. All the functions

$$u_k(z) = (1 - |z|^2)\, z^k := (1 - \sum_{k=1}^{n} |z_k|^2) \prod_{\nu=1}^{n} z_\nu^{k_\nu}\,, \quad k = (k_1, \dots, k_n) \in \mathbb{N}_0^n\,,$$

are pluriholomorphic and vanish on $\partial B := \{|z| = 1\}$. Hence, there are in general infinitely many linearly independent pluriholomorphic functions satisfying homogeneous Dirichlet boundary conditions. Contrary to the one–dimensional case for several complex variables boundary value problems are not unconditionally solvable, see [1,2,3,4,8,10,11]. In order to solve some boundary value problem therefore at first solvability conditions have to be found out. In case they are satisfied then one can look for solutions.

Here some boundary value problems for inhomogeneous systems will be studied which are related to the above mentioned homogeneous systems. They are considered for the unit ball and partially also in the unit polydisc $\mathbb{D} = \times_{k=1}^{n} \{|z_k| < 1\}$.

11.1 DIRICHLET PROBLEM FOR SIMPLE HOMOGENEOUS EQUATIONS

Analytic functions φ in the unit ball B which are continuous in the closure $\overline{B} = B \cup \partial B = \{|z| \leq 1\}$ can be represented by the Cauchy formula

$$\varphi(z) = \int_{\partial B} \varphi(\zeta) \frac{d\sigma(\zeta)}{(1 - \langle z, \zeta \rangle)^n}\,, \quad \langle z, \zeta \rangle := \sum_{k=1}^{n} z_k \overline{\zeta_k}\,, \quad z \in B\,. \tag{11.4}$$

Here σ is the rotation–invariant positive Borel measure on ∂B with respect to the orthogonal group $\mathcal{O}(2n)$ of all isometries of $\mathbb{R}^{2n}$ fixing the origin normalized by $\sigma(\partial B) = 1$. For the boundary values of this Cauchy integral Plemelj–kind formulas hold, see [12]. Thus the following is valid.

Lemma A *The function $f \in C(\partial B) \cap L_2(\partial B)$ is the boundary value on ∂B of a holomorphic function in B if and only if*

$$\int_{\partial B} f(\eta) \frac{d\sigma(\eta)}{(1 - \langle \zeta, \eta \rangle)^n} = \frac{1}{2} f(\zeta)\,, \quad \zeta \in \partial B\,. \tag{11.5}$$

Here the integral is understood as a Cauchy principle value integral.

Definition *Let P_h denote the set of functions on ∂B which are boundary values* *of pluriharmonic functions in B.*

In [10] the following is shown.

Lemma B *The function $f \in C(\partial B) \cap L_2(\partial B)$ belongs to P_h if and only if*

$$\int_{\partial B} f(\eta) \operatorname{Re} \frac{1}{(1-\langle \zeta, \eta \rangle)^n} \, d\sigma(\eta) = \frac{1}{2} \int_{\partial B} f(\eta) \, d\sigma(\eta) \,, \quad \zeta \in \partial B \,. \tag{11.6}$$

With these results the Dirichlet problem can be solved for pluriharmonic functions.

Theorem 1 *For given $\gamma \in C(\partial B) \cap L_2(\partial B)$ there exists a uniquely defined pluriharmonic function in B satisfying the Dirichlet condition $u = \gamma$ on ∂B if and only if γ satisfies condition (11.6). The solution is*

$$u(z) = \int_{\partial B} \gamma(\zeta) \left(\frac{1}{(1-\langle z, \zeta \rangle)^n} + \frac{1}{(1-\langle \zeta, z \rangle)^n} - 1 \right) d\sigma(\zeta) \,, \quad z \in B \,. \tag{11.7}$$

Proof As can be seen from (11.1) any pluriharmonic function u can be represented by two analytic functions φ and ψ as $u = \varphi + \overline{\psi}$. Hence, the Dirichlet condition on ∂B can be written as

$$\varphi + \overline{\psi} = \gamma \quad \text{on} \quad \partial B \,.$$

Applying the Cauchy formula

$$\varphi(z) = \int_{\partial B} \gamma(\zeta) \frac{d\sigma(\zeta)}{(1-\langle z, \zeta \rangle)^n} - C_1 \,, \quad \psi(z) = \int_{\partial B} \overline{\varphi(\zeta)} \frac{d\sigma(\zeta)}{(1-\langle z, \zeta \rangle)^n} - C_2 \,,$$

with

$$\begin{aligned} C_1 &= \int_{\partial B} \overline{\psi(\zeta)} \frac{d\sigma(\zeta)}{(1-\langle z, \zeta \rangle)^n} = \overline{\int_{\partial B} \frac{\psi(\zeta)\, d\sigma(\zeta)}{(1-\langle \zeta, z \rangle)^n (1-\langle w, \zeta \rangle)^n}\bigg|_{w=0}} = \overline{\psi(0)} \,, \\ C_2 &= \int_{\partial B} \overline{\varphi(\zeta)} \frac{d\sigma(\zeta)}{(1-\langle z, \zeta \rangle)^n} = \overline{\varphi(0)} = \int_{\partial B} (\overline{\gamma(\zeta)} - \psi(\zeta)) \, d\sigma(\zeta) \\ &= \int_{\partial B} \overline{\gamma(\zeta)} \, d\sigma(\zeta) - \psi(0) \,. \end{aligned}$$

Thus

$$u(z) = \int_{\partial B} \gamma(\zeta) \left(\frac{1}{(1-\langle z, \zeta \rangle)^n} + \frac{1}{(1-\langle \zeta, z \rangle)^n} - 1 \right) d\sigma(\zeta) \,, \quad z \in B \,.$$

Passing with z to the boundary point η from the Plemelj formula

$$u^+(\eta) = \gamma(\zeta) + \int_{\partial B} \gamma(\zeta) (2 \operatorname{Re} \frac{1}{(1-\langle \eta, \zeta \rangle)^n} - 1) d\sigma(\zeta) \,, \quad \eta = \partial B \,,$$

follows. This coincides with γ if and only if γ satisfies (11.6).

Remark For $n = 1$ (11.6) is an identity rather than a condition because then

$$2\,\mathrm{Re}\,\frac{1}{1-\langle\zeta,\eta\rangle} = \frac{1}{1-\zeta\overline{\eta}} + \frac{1}{1-\overline{\zeta}\eta} = \frac{2-2\,\mathrm{Re}\,\zeta\overline{\eta}}{|1-\zeta\overline{\eta}|^2} = 1\,.$$

Theorem 2 *For given $\gamma \in C(\partial B) \cap L_2(\partial B)$ the Dirichlet problem $u = \gamma$ on ∂B is solvable for pluriholomorphic functions if γ satisfies the conditions*

$$\int_{\partial B} \eta_k \gamma(\eta)\,\frac{d\sigma(\eta)}{(1-\langle\zeta,\eta\rangle)^n} = \frac{\zeta_k}{2}\,\gamma(\zeta)\,, \quad 1 \le k \le n\,,\ \zeta \in \partial B\,. \tag{11.8}$$

The general solution then is

$$u(z) = \int_{\partial B} \langle\eta,z\rangle\gamma(\eta)\,\frac{d\sigma(\eta)}{(1-\langle z,\eta\rangle)^n} + \sum_{0\le|k|} a_k z^k (1-|z|^2) \tag{11.9}$$

with arbitrary complex constants a_k, $k = (k_1,\dots,k_n) \in \mathbb{N}_0^n$, $|k| = \sum_{\nu=1}^n k_\nu$, $z^k = \prod_{\nu=1}^n z_\nu^{k_\nu}$.

Proof Any pluriholomorphic function u is representable as

$$u(z) = \varphi_0(z) + \langle\varphi(z),z\rangle = \varphi_0(z) + \sum_{k=1}^n \varphi_k(z)\,\overline{z_k}$$

with arbitrary analytic functions φ_k, $0 \le k \le n$, in B, see [2], 4.3. Hence, the Dirichlet problem can be written as

$$\sum_{k=1}^n \overline{\zeta_k}(\zeta_k\varphi_0(\zeta) + \varphi_k(\zeta)) = \sum_{k=1}^n \overline{\zeta_k}\zeta_k\gamma(\zeta)\,, \quad \zeta \in \partial B\,.$$

This holds, obviously, if

$$\zeta_k\gamma(\zeta) = \zeta_k\varphi_0(\zeta) + \varphi_k(\zeta)\,, \quad 1 \le k \le n\,,\ \zeta \in \partial B\,.$$

Hence, $\zeta_k\gamma(\zeta)$ is the boundary value on ∂B of the holomorphic function $z_k\varphi_0(z) + \varphi_k(z)$ in B. By Lemma A this is the case if and only if (11.8) holds.

These equations for $1 \le k \le n$ imply

$$\int_{\partial B} \langle\eta,\zeta\rangle\,\gamma(\eta)\,\frac{d\sigma(\eta)}{(1-\langle\zeta,\eta\rangle)^n} = \frac{1}{2}\,\gamma(\zeta)\,, \quad \zeta \in \partial B\,. \tag{11.8'}$$

That

$$u_0(z) := \int_{\partial B} \langle\eta,z\rangle\,\gamma(\eta)\,\frac{d\sigma(\eta)}{(1-\langle z,\eta\rangle)^n}$$

is a particular solution to the Dirichlet problem follows from the Cauchy formula applied to $z_k\varphi_0(z) + \varphi_k(z)$, $1 \le k \le n$, and the Plemelj formula together with (11.8).

11.2 NEUMANN PROBLEM FOR SIMPLE HOMOGENEOUS EQUATIONS

Here $\partial/\partial\nu(\zeta)$ denotes the outward normal derivative on ∂B in the point ζ. Because the Neumann condition $\partial u/\partial\nu = \gamma$ on ∂B does not determine u uniquely in general this condition is accompanied by the normalization condition

$$\int_{\partial B} u(\zeta)\, d\sigma(\zeta) = 0\,. \tag{11.10}$$

It can easily be seen that the homogeneous Neumann problem for pluriharmonic functions is only trivially solvable. If a pluriharmonic function $u = \varphi + \overline{\psi}$ would satisfy on ∂B

$$\frac{\partial u(\zeta)}{\partial\nu(\zeta)} = 0\,, \quad \text{i.e.} \sum_{k=1}^{n}\left\{\zeta_k\frac{\partial u}{\partial\zeta_k} + \overline{\zeta_k}\frac{\partial u}{\partial\overline{\zeta_k}}\right\} = 0$$

then

$$\sum_{k=1}^{n}\left\{\zeta_k\varphi_{\zeta_k}(\zeta) + \overline{\zeta_k}\,\overline{\psi_{\zeta_k}(\zeta)}\right\} = 0 \quad \text{on } \partial B\,.$$

This means for the analytic functions

$$\mathrm{Re}\sum_{k=1}^{n}\zeta_k(\varphi_{\zeta_k}(\zeta) + \psi_{\zeta_k}(\zeta)) = 0\,, \quad \mathrm{Im}\sum_{k=1}^{n}\zeta_k(\varphi_{\zeta_k}(\zeta) - \psi_{\zeta_k}(\zeta)) = 0 \text{ on } \partial B$$

so that

$$\sum_{k=1}^{n} z_k\varphi_{z_k}(z) = 0\,, \quad \sum_{k=1}^{n} z_k\psi_{z_k}(z) = 0 \quad \text{in } B\,.$$

Hence, $\varphi_{z_k}(z) = 0$, $\psi_{z_k}(z) = 0$, $1 \le k \le n$, i.e. φ and ψ and hence u are constant. Because of the normalization the constant u vanishes.

As the Dirichlet problem also the homogeneous Neumann problem for pluriholomorphic functions has infinitely many linearly independent solutions. They are

$$u_k(z) = (1 - \frac{|k|}{|k|+2}|z|^2)\, z^k\,, \quad k \in \mathbb{N}_0^n\,.$$

Obviously, they are pluriholomorphic. Moreover, on ∂B

$$\begin{aligned}
\partial u_k/\partial\nu &= \sum_{\nu=1}^{n}\{\zeta_\nu u_{k\zeta_\nu} + \overline{\zeta_\nu} u_{k\overline{\zeta_\nu}}\} \\
&= \sum_{\nu=1}^{n}\{k_\nu(1 - \frac{|k|}{|k|+2}|\zeta|^2)\,\zeta^k - \frac{|k|}{|k|+2}\zeta_\nu\overline{\zeta_\nu}\zeta^k - \frac{|k|}{|k|+2}\overline{\zeta_\nu}\zeta_\nu\zeta^k\} \\
&= 0\,.
\end{aligned}$$

Theorem 3 *The normalized Neumann problem for pluriharmonic functions in the unit ball is solvable if and only if the right–hand side γ satisfies (11.6) and*

$$\int_{\partial B} \gamma(\eta)\, d\sigma(\eta) = 0\,. \tag{11.11}$$

The solution then is uniquely given as

$$\begin{aligned} u(z) &= \int_{\partial B} \gamma(\zeta)\Big[\sum_{k=1}^{n-1} \mathrm{Re}\, \frac{2}{k(1-\langle z,\zeta\rangle)^k} + \log \frac{1}{|1-\langle z,\zeta\rangle|^2}\Big] d\sigma(\zeta) \\ &\quad - \int_{\partial B}\int_{\partial B} \gamma(\zeta)\Big[\sum_{k=1}^{n-1} \mathrm{Re}\, \frac{2}{k(1-\langle \eta,\zeta\rangle)^k} + \log \frac{1}{|1-\langle \eta,\zeta\rangle|^2}\Big] d\sigma(\zeta)\, d\sigma(\eta)\,. \end{aligned}$$

Proof The boundary condition is for $u = \varphi + \overline{\psi}$

$$\sum_{k=1}^{n} \{\zeta_k \varphi_{\zeta_k}(\zeta) + \overline{\zeta_k}\, \overline{\psi_{\zeta_k}(\zeta)}\} = \gamma(\zeta) \quad \text{on } \partial B\,.$$

From the Cauchy formula

$$\sum_{k=1}^{n} z_k \varphi_{z_k}(z) = \int_{\partial B} \gamma(\zeta)\, \frac{d\sigma(\zeta)}{(1-\langle z,\zeta\rangle)^n}\,, \quad z \in B\,,$$

follows because

$$\int_{\partial B} \zeta_k \psi_{\zeta_k}(\zeta)\, \frac{d\sigma(\zeta)}{(1-\langle \zeta,z\rangle)^n} = \int_{\partial B} \frac{\zeta_k \psi_{\zeta_k}(\zeta)}{(1-\langle \zeta,z\rangle)^n}\, \frac{d\sigma(\zeta)}{(1-\langle w,\zeta\rangle)^n}\Big|_{w=0} = 0\,.$$

In the same way

$$\sum_{k=1}^{n} z_k \psi_{z_k}(z) = \int_{\partial B} \overline{\gamma(\zeta)}\, \frac{d\sigma(\zeta)}{(1-\langle z,\zeta\rangle)^n}\,, \quad z \in B\,,$$

follows. Obviously, by choosing $z = 0$, we see (11.11) holds. Hence, rewriting

$$\begin{aligned} \sum_{k=1}^{n} z_k \varphi_{z_k}(z) &= \int_{\partial B} \gamma(\zeta)\Big(\frac{1}{(1-\langle z,\zeta\rangle)^n} - 1\Big) d\sigma(\zeta) \\ &= \int_{\partial B} \gamma(\zeta)\, \frac{\frac{1}{1-\langle z,\zeta\rangle^n} - 1}{\frac{1}{1-\langle z,\zeta\rangle} - 1}\, \frac{\langle z,\zeta\rangle}{1-\langle z,\zeta\rangle}\, d\sigma(\zeta) \\ &= \int_{\partial B} \gamma(\zeta) \sum_{k=1}^{n} \frac{\langle z,\zeta\rangle}{(1-\langle z,\zeta\rangle)^k}\, d\sigma(\zeta) \end{aligned}$$

we see that

$$\phi(z) := \varphi(z) - \int_{\partial B} \gamma(\zeta)\Big[\sum_{k=1}^{n-1} \frac{1}{k(1-\langle z,\zeta\rangle)^k} + \log\frac{1}{1-\langle z,\zeta\rangle}\Big] d\sigma(\zeta)$$

satisfies

$$\sum_{\ell=1}^{n} z_\ell \phi_{z_\ell}(z) = 0\,.$$

Therefore ϕ is constant i.e. $\phi(z) = \varphi(0)$ because of (11.11). In the same way

$$\psi(z) = \psi(0) + \int_{\partial B} \overline{\gamma(\zeta)}\Big[\sum_{k=1}^{n-1} \frac{1}{k(1-\langle z,\zeta\rangle)^k} + \log\frac{1}{1-\langle z,\zeta\rangle}\Big] d\sigma(\zeta)$$

can be deduced. Thus u has the desired form where the constant $\varphi(0) + \overline{\psi(0)}$ follows from the normalization (11.10).

Theorem 4 *If γ satisfies (11.8) and*

$$\sum_{k=1}^{n-1} \int_{\partial B}\int_{\partial B} \langle \eta,\zeta\rangle \gamma(\eta) \frac{d\sigma(\eta)\, d\sigma(\zeta)}{(1-\langle \zeta,\eta\rangle)^{n-k}} = 0 \tag{11.12}$$

the Neumann problem is solvable for pluriharmonic functions. The general solution then is

$$u(z) = \frac{1}{n-1}\sum_{k=1}^{n-1} \int_{\partial B} \langle \zeta, z\rangle \gamma(\zeta) \frac{d\sigma(\zeta)}{(1-\langle z,\zeta\rangle)^k} + \sum_{0<|k|} a_k\Big(1 - \frac{|k|}{|k|+2}|z|^2\Big) z^k$$

with arbitrary complex constants a_k, $k \in \mathbb{N}_0^n$.

Proof The boundary condition is for $u = \varphi_0 + \sum_{k=1}^{n} \varphi_k \overline{z_k}$

$$\sum_{k=1}^{n} \{\zeta_k \Big[\varphi_{0\zeta_k} + \sum_{\ell=1}^{n} \varphi_{\ell \zeta_k} \overline{\zeta_\ell}\Big] + \overline{\zeta_k}\varphi_k\} = \gamma \quad \text{on } \partial B$$

or

$$\sum_{k=1}^{n} \overline{\zeta_k}\Big\{\zeta_k \sum_{\ell=1}^{n} \zeta_\ell \varphi_{0\zeta_\ell} + \sum_{\ell=1}^{n} \zeta_\ell \varphi_{k\zeta_\ell} + \varphi_k\Big\} = \gamma(\zeta) = \sum_{k=1}^{n} \overline{\zeta_k}\zeta_k \gamma \text{ on } \partial B\,.$$

This equality holds if

$$\zeta_k \sum_{\ell=1}^{n} \zeta_\ell \varphi_{0\zeta_\ell} + \sum_{\ell=1}^{n} \zeta_\ell \varphi_{k\zeta_\ell} + \varphi_k = \zeta_k \gamma \quad \text{on } \partial B\,,\ 1 \le k \le n\,.$$

Hence, $\zeta_k\gamma$ is the boundary value of an analytic function in B. By Lemma A this is the case if and only if

$$\int_{\partial B} \eta_k \gamma(\eta) \frac{d\sigma(\eta)}{(1-\langle \zeta,\eta\rangle)^n} = \frac{1}{2}\zeta_k\gamma(\zeta)\,, \quad \zeta \in \partial B\,.$$

From the Cauchy formula

$$z_k \sum_{\ell=1}^{n} z_\ell \varphi_{0z_\ell}(z) + \sum_{\ell=1}^{n} z_\ell \varphi_{kz_\ell}(z) + \varphi_k(z) = \int_{\partial B} \zeta_k \gamma(\zeta) \frac{d\sigma(\zeta)}{(1-\langle z,\zeta\rangle)^n}, \; z \in B, \; 1 \le k \le n,$$

follows. This is a system of differential equations for the φ_k, $1 \le k \le n$. Rewriting it as

$$\varphi_k(z) + \sum_{\ell=1}^{n} z_\ell \varphi_{kz_\ell}(z) = \int_{\partial B} \zeta_k \gamma_0(\zeta) \frac{d\sigma(\zeta)}{(1-\langle z,\zeta\rangle)^n}, \; z \in B, \; 1 \le k \le n,$$

with

$$\gamma_0(\zeta) := \gamma(\zeta) - \sum_{\ell=1}^{n} \zeta_\ell \varphi_{0\zeta_\ell}(\zeta), \quad \zeta \in \partial B,$$

where the analyticity of φ_0 is observed,

$$\varphi_k(z) = \frac{1}{n-1} \int_{\partial B} \frac{\zeta_k \gamma_0(\zeta)}{\langle z,\zeta\rangle} \Big[\frac{1}{(1-\langle z,\zeta\rangle)^{n-1}} - 1\Big] d\sigma(\zeta), \quad z \in B, \; 1 \le k \le n,$$

is seen. In order to find a particular solution to the inhomogeneous problem φ_0 is chosen as the constant 0.

Thus

$$\begin{aligned} u_0(z) &= \sum_{k=1}^{n} \overline{z_k} \varphi_k(z) = \frac{1}{n-1} \int_{\partial B} \frac{\langle \zeta, z\rangle}{\langle z,\zeta\rangle} \gamma(\zeta) \Big[\frac{1}{(1-\langle z,\zeta\rangle)^{n-1}} - 1\Big] d\sigma(\zeta) \\ &= \frac{1}{n-1} \sum_{k=1}^{n-1} \int_{\partial B} \langle \zeta, z\rangle \gamma(\zeta) \frac{1}{(1-\langle z,\zeta\rangle)^k} d\sigma(\zeta). \end{aligned} \tag{11.13}$$

Obviously, it is polyholomorphic. Moreover,

$$\begin{aligned} \sum_{k=1}^{n} \{z_k u_{0z_k} + \overline{z_k} u_{0\overline{z_k}}\} &= \frac{1}{n-1} \int_{\partial B} \frac{\langle \zeta, z\rangle}{\langle z,\zeta\rangle} \gamma(\zeta) \\ &\times \Big[1 - \frac{1}{(1-\langle z,\zeta\rangle)^{n-1}} + \frac{(n-1)\langle z,\zeta\rangle}{(1-\langle z,\zeta\rangle)^n} + \frac{1}{(1-\langle z,\zeta\rangle)^n} - 1\Big] d\sigma(\zeta) \\ &= \int_{\partial B} \langle \zeta, z\rangle \gamma(\zeta) \frac{d\sigma(\zeta)}{(1-\langle z,\zeta\rangle)^n} \end{aligned}$$

which for $z \in \partial B$ because of (11.8) and ∂B coincides with $\gamma(z)$.

The general solution then is

$$u(z) = u_0(z) + \sum_{0<|k|} a_k \Big(1 - \frac{|k|}{|k|+2} |z|^2\Big) z^k.$$

That the normalization

$$\int_{\partial B} u(z)\, d\sigma(z) = 0$$

holds follows from (11.12) and

$$\int_{\partial B} \zeta^k d\sigma(\zeta) = 0\,, \quad \text{for } 0 < |k|\,.$$

11.3 SOME WELL–POSED PROBLEMS FOR SIMPLE HOMOGENEOUS EQUATIONS

The Dirichlet and the Neumann problem were seen to be only well–posed for pluriharmonic but not for pluriholomorphic functions. Here some other well–posed boundary value problems will be introduced, see [11]. One example is the following boundary condition

$$\operatorname{Re} u(\zeta) = \gamma(\zeta)\,, \quad \operatorname{Re}\langle \operatorname{grad}_{\overline{\zeta}} u(\zeta), \zeta\rangle = \delta(\zeta) \text{ on } \partial B \tag{11.14}$$

together with the normalization

$$\operatorname{Im} \int_{\partial B} u(\zeta)\, d\sigma(\zeta) = 0\,. \tag{11.15}$$

Here

$$\operatorname{grad}_{\overline{z}} u = (u_{\overline{z_1}}, u_{\overline{z_2}}, \ldots, u_{\overline{z_n}})\,.$$

Theorem 5 *Problem (11.14), (11.15) is solvable for pluriharmonic functions if and only if γ and δ satisfy (11.6) and, moreover,*

$$\int_{\partial B} \delta(\zeta)\, d\sigma(\zeta) = 0\,. \tag{11.16}$$

Then the solution is uniquely given as

$$\begin{aligned} u(z) &= \int_{\partial B} \gamma(\zeta)\Big[\frac{2}{(1-\langle z,\zeta\rangle)^n} - 1\Big] d\sigma(\zeta) \\ &= -4i \operatorname{Im} \int_{\partial B} \delta(\zeta)\Big[\sum_{k=1}^{n-1} \frac{1}{k(1-\langle z,\zeta\rangle)^k} + \log\frac{1}{1-\langle z,\zeta\rangle}\Big] d\sigma(\zeta) \\ &\quad -2i \operatorname{Im} \int_{\partial B}\int_{\partial B} \Big[\frac{\gamma(\zeta)}{(1-\langle \eta,\zeta\rangle)^n} \\ &\quad -2\delta(\zeta)\Big[\sum_{k=1}^{n-1} \frac{1}{k(1-\langle \eta,\zeta\rangle)^k} + \log\frac{1}{1-\langle \eta,\zeta\rangle}\Big] d\sigma(\zeta)\, d\sigma(\eta)\,. \end{aligned} \tag{11.17}$$

Proof Inserting the representation $u = \varphi + \overline{\psi}$ for pluriharmonic functions in the boundary conditions gives the Schwarz conditions for analytic functions

$$\operatorname{Re}\{\varphi(\zeta) + \psi(\zeta)\} = \gamma(\zeta)\,, \quad \operatorname{Re}\sum_{k=1}^{n} \zeta_k \psi_{z_k}(\zeta) = \delta(\zeta) \text{ on } \partial B\,.$$

Because the real part of an analytic function is pluriharmonic, Theorem 1 may be applied twice giving

$$\mathrm{Re}\,\{\varphi(z)+\psi(z)\} = \mathrm{Re}\int_{\partial B}\gamma(\zeta)\Big[\frac{2}{(1-\langle z,\zeta\rangle)^n}-1\Big]d\sigma(\zeta)\,,\quad z\in B\,,$$

$$\mathrm{Re}\,\{\sum_{k=1}^{n} z_k\psi_{z_k}(z)\} = \mathrm{Re}\int_{\partial B}\delta(\zeta)\Big[\frac{2}{(1-\langle z,\zeta\rangle)^n}-1\Big]d\sigma(\zeta)\,,\quad z\in B\,.$$

Hence,

$$\begin{aligned}\varphi(z)+\psi(z) &= \int_{\partial B}\gamma(\zeta)\Big[\frac{2}{(1-\langle z,\zeta\rangle)^n}-1\Big]d\sigma(\zeta)+ic_1\,,\\ \sum_{k=1}^{n} z_k\psi_{z_k}(z) &= \int_{\partial B}\delta(\zeta)\Big[\frac{2}{(1-\langle z,\zeta\rangle)^n}-1\Big]d\sigma(\zeta)+ic_2\,,\end{aligned}$$

where $c_2 = 0$ and (11.16) has to be satisfied. Rewriting the last relation by the identity

$$\begin{aligned}&\frac{1}{(1-\langle z,\zeta\rangle)^n}-1\\ &\quad=\langle z,\zeta\rangle\sum_{k=1}^{n}\frac{1}{(1-\langle z,\zeta\rangle)^k}=\sum_{k=1}^{n} z_k\frac{\partial}{\partial z_k}\Big[\sum_{\ell=1}^{n-1}\frac{1}{(\ell(1-\langle z,\zeta\rangle)^\ell}+\log\frac{1}{1-\langle z,\zeta\rangle}\Big]\end{aligned}$$

as

$$\begin{aligned}0 &= \sum_{k=1}^{n} z_k\psi_{z_k}(z)-2\int_{\partial B}\delta(\zeta)\Big[\frac{1}{(1-\langle z,\zeta\rangle)^n}-1\Big]d\sigma(\zeta)\\ &= \sum_{k=1}^{n} z_k\frac{\partial}{\partial z_k}\Big[\psi(z)-2\int_{\partial B}\delta(\zeta)\Big(\sum_{\ell=1}^{n-1}\frac{1}{\ell(1-\langle z,\zeta\rangle)^\ell}+\log\frac{1}{1-\langle z,\zeta\rangle}\Big)d\sigma(\zeta)\Big]\end{aligned}$$

one sees that the analytic function in square brackets is homogeneous of degree zero, which means constant in B,

$$\psi(z) = 2\int_{\partial B}\delta(\zeta)\Big[\sum_{k=1}^{n-1}\frac{1}{k(1-\langle z,\zeta\rangle)^k}+\log\frac{1}{1-\langle z,\zeta\rangle}\Big]d\sigma(\zeta)+c\,.$$

Thus

$$\begin{aligned} u(z) &= \varphi(z) + \overline{\psi(z)} = \varphi(z) + \psi(z) - 2i \,\mathrm{Im}\, \psi(z) \\ &= \int_{\partial B} \gamma(\zeta) \Big[\frac{2}{(1 - \langle z, \zeta \rangle)^n} - 1 \Big] d\sigma(\zeta) \\ &\quad -4i \,\mathrm{Im} \int_{\partial B} \delta(\zeta) \Big[\sum_{k=1}^{n-1} \frac{1}{k(1 - \langle z, \zeta \rangle)^k} + \log \frac{1}{1 - \langle z, \zeta \rangle} \Big] d\sigma(\zeta) \\ &\quad + i(c_1 - 2 \,\mathrm{Im}\, c) . \end{aligned}$$

The imaginary constant is determined through (11.15) leading to (11.17).

Problem (11.14) cannot be well–posed for pluriholomorphic functions because (11.14) just impose two real conditions while in general pluriholomorphic functions are determined by $n+1$ holomorphic functions

$$u = \varphi_0 + \sum_{k=1}^{n} \overline{z_k} \varphi_k = \varphi_0 + \langle \varphi, z \rangle .$$

Thus (11.14) is underdetermined. But for the subclass S_p of pluriholomorphic functions being representable by two analytic functions φ and ψ in the form

$$u = \varphi + |z|^2 \psi \tag{11.18}$$

(11.14) is well–posed.

Theorem 6 *Necessary and sufficient for problem (11.14), (11.15) for pluriholomor phic functions from S_p in the unit ball B to be solvable are that γ and δ satisfy (11.6). Under the additional normalization condition*

$$\mathrm{Im} \int_{\partial B} \langle \mathrm{grad}_{\overline{\zeta}} u(\zeta), \zeta \rangle \, d\sigma(\zeta) = 0 \tag{11.19}$$

the uniquely given solution is

$$\begin{aligned} u(z) &= \int_{\partial B} [\gamma(\zeta) - (1 - |z|^2)\, \delta(\zeta)] \Big[\frac{2}{(1 - \langle z, \zeta \rangle)^n} - 1 \Big] d\sigma(\zeta) \\ &\quad -2i \,\mathrm{Im} \int_{\partial B} \int_{\partial B} [\gamma(\zeta) - (1 - |z|^2)\, \delta(\zeta)] \frac{d\sigma(\zeta)\, d\sigma(\eta)}{(1 - \langle \eta, \zeta \rangle)^n} , \quad z \in B . \end{aligned} \tag{11.20}$$

Proof Inserting (11.18) in (11.14) gives

$$\mathrm{Re}\, \{\varphi + \psi\} = \gamma , \quad \mathrm{Re}\, \psi = \delta \quad \text{on } \partial B .$$

Again as in the preceding proof

$$(\varphi+\psi)(z) = \int_{\partial B}\gamma(\zeta)\Big[\frac{2}{(1-\langle z,\zeta\rangle)^n}-1\Big]d\sigma(\zeta)+ic_1\,,$$

$$\psi(z) = \int_{\partial B}\gamma(\zeta)\Big[\frac{2}{(1-\langle z,\zeta\rangle)^n}-1\Big]d\sigma(\zeta)+ic_2\,.$$

From (11.19)

$$0=\mathrm{Im}\int_{\partial B}\psi(\zeta)\,d\sigma(\zeta)=2\,\mathrm{Im}\int_{\partial B}\int_{\partial B}\delta(\zeta)\frac{d\sigma(\zeta)\,d\sigma(\eta)}{1-\langle\eta,\zeta\rangle)^n}+c_2\,,$$

which determines c_2. Now

$$u=\varphi+\psi+(|z|^2-1)\psi$$

so that

$$u(z)=\int_{\partial B}[\gamma(\zeta)+(|z|^2-1)\,\delta(\zeta)]\Big[\frac{2}{(1-\langle z,\zeta\rangle)^n}-1\Big]d\sigma(\zeta)+i(c_1+c_2(|z|^2-1))\,.$$

From (11.15)

$$2\,\mathrm{Im}\int_{\partial B}\int_{\partial B}\gamma(\zeta)\frac{d\sigma(\zeta)}{(1-\langle\eta,\zeta\rangle)^n}+c_1=0$$

follows, and hence (11.20).

Next a well-posed problem for the entire class of pluriholomorphic functions is considered.

Theorem 7 *There exists a uniquely defined pluriholomorphic function u in B satisfying*

$$\mathrm{Re}\,u=\gamma_0\,,\quad \mathrm{Re}\,u_{\overline{\zeta_k}}=\gamma_k\,,\quad 1\le k\le n,\ \text{ on } \partial B \tag{11.21}$$

and the normalization conditions

$$\mathrm{Im}\int_{\partial B}u(\zeta)\,d\sigma(\zeta)=0\,,\quad \mathrm{Im}\int_{\partial B}u_{\overline{\zeta_k}}(\zeta)\,d\sigma(\zeta)=0\,,\ 1\le k\le n\,, \tag{11.22}$$

if and only if γ_k, $1\le k\le n+1$, satisfy condition (11.6) where with $\gamma:=(\gamma_1,\dots,\gamma_n)$

$$\gamma_{n+1}(\zeta) := \gamma_0(\zeta)-\mathrm{Re}\langle\gamma(\zeta),\zeta\rangle-2\int_{\partial B}\mathrm{Im}\,\{\gamma(\eta),\zeta\rangle\mathrm{Im}\frac{1}{(1-\langle\eta,\zeta\rangle)^n}\,d\sigma(\eta)$$
$$+2\int_{\partial B}\int_{\partial B}\mathrm{Im}\,\langle\gamma(\eta),\zeta\rangle\,\mathrm{Im}\frac{1}{(1-\langle\eta,\xi\rangle)^n}\,d\sigma(\eta)\,d\sigma(\xi)\,.$$

The solution then is given as

$$u(z) = \int_{\partial B} [\gamma_{n+1}(\zeta) + \langle \gamma(\zeta), z\rangle]\Big[\frac{2}{(1-\langle z,\zeta\rangle)^n} - 1\Big] d\sigma(\zeta) \\ -2i\Big(c_{k+1} + \sum_{k=1}^{n} c_k \overline{z_k}\Big) \tag{11.23}$$

with

$$c_k := \operatorname{Im} \int_{\partial B}\int_{\partial B} \gamma_k(\zeta)\frac{d\sigma(\zeta)\,d\sigma(\eta)}{(1-\langle \eta,\zeta\rangle)^n}, \quad 1 \le k \le n+1 .$$

Proof Inserting $u = \varphi_0 + \sum_{k=1}^{n} \overline{z_k}\varphi_k$ into the second term of (11.21) gives

$$\operatorname{Re}\varphi_k = \gamma_k , \quad 1 \le k \le n, \text{ on } \partial B .$$

Together with (11.22) this gives as in the preceding proofs

$$\varphi_k(z) = \int_{\partial B} \gamma_k(\zeta)\Big[\frac{2}{(1-\langle z,\zeta\rangle)^n} - 1\Big] d\sigma(\zeta) - 2i \operatorname{Im} \int_{\partial B}\int_{\partial B} \gamma_k(\eta)\frac{d\sigma(\zeta)\,d\sigma(\eta)}{(1-\langle \eta,\zeta\rangle)^n} .$$

From the first condition in (11.21)

$$\operatorname{Re}\varphi_0(\zeta) = \gamma_0(\zeta) - \operatorname{Re}\sum_{k=1}^{n} \overline{\zeta_k}\varphi_k(\zeta) = \gamma_0(\zeta) - \sum_{k=1}^{n}\Big\{\operatorname{Re}\zeta_k\gamma_k(\zeta) + \operatorname{Im}\zeta_k \operatorname{Im}\varphi_k(\zeta)\Big\} = \gamma_{n+1}$$

follows. Thus, again

$$\varphi_0(\zeta) = \int_{\partial B} \gamma_{n+1}(\zeta)\Big[\frac{2}{(1-\langle z,\zeta\rangle)^n} - 1\Big] d\sigma(\zeta) - 2i \operatorname{Im} \int_{\partial B}\int_{\partial B} \gamma_{n+1}(\zeta)\frac{d\sigma(\zeta)\,d\sigma(\eta)}{(1-\langle \eta,\zeta\rangle)^n}$$

so that (11.22) follows.

11.4 INHOMOGENEOUS PLURIHOLOMORPHIC SYSTEM IN THE UNIT BALL

Inhomogeneous systems of the form

$$u_{\overline{z_k} z_\ell} = f_{k\ell} \quad \text{and} \quad u_{\overline{z_k}\,\overline{z_\ell}} = f_{k\ell} , \quad 1 \le k \le n,$$

can be treated as in the case of one complex variable see e.g. [9], [5], by using a Pompeiu operator [2]. Because this system consists of n equation for just one unknown some compatibility conditions have to be satisfied in order that these overdetermined systems are solvable. Here the second system is treated.

Lemma 1 *Let the $f_{k\ell} \in C(\overline{B}) \cap C^1(B)$ satisfy the compatibility conditions*

$$f_{k\ell} = f_{\ell k} , \quad f_{k\ell \overline{z_j}} = f_{j\ell \overline{z_k}} , \quad 1 \le j,k,\ell \le n . \tag{11.24}$$

Then

$$u_0(z) := -TF(z) + \sum_{k=1}^{n} \overline{z_k} Tf^{(k)}(z) \tag{11.25}$$

is a particular solution to the inhomogenous system

$$w_{\overline{z_k}\,\overline{z_\ell}} = f_{k\ell} \quad \text{in } B$$

where

$$f^{(k)} := (f_{1k}, f_{2k}, \ldots, f_{nk}), \quad F := (F_1, \ldots, F_n),\ F_k := \sum_{\ell=1}^{n} \overline{z_\ell} f_{k\ell},\ 1 \le k \le n,$$

$$\begin{aligned} TF(z) &:= \int_{\partial B} TF(\zeta) \frac{1 - \langle \zeta, z\rangle}{|\zeta - z|^{2n}} d\sigma(\zeta) - \frac{1}{n} \int_B \frac{\langle F(\zeta), \zeta - z\rangle}{|\zeta - z|^{2n}} d\nu(\zeta), \quad z \in B, \\ TF(\zeta) &:= \frac{1}{n} \int_B \frac{\langle F(\eta), \zeta - \eta\rangle d\nu(\eta)}{(1 - \langle \zeta, \eta\rangle)^n (1 - \langle \eta, \zeta\rangle)}, \quad \zeta \in \partial B. \end{aligned} \tag{11.26}$$

Remark ν is the Lebesgue measure in $\mathbb{C}^n \cong \mathbb{R}^{2n}$ normalized by $\nu[B] = 1$. As σ also ν is rotation invariant and both measures, see [13], are related through

$$\int_{\mathbb{C}^n} f d\nu = 2n \int_0^\infty r^{2n-1} \int_{\partial B} f(r, \zeta)\, d\sigma(\zeta)\, dr.$$

Proof Because $F_{k\overline{z_\ell}} = F_{\ell\overline{z_k}}$, and $f_{k\ell\overline{z_j}} = f_{j\ell\overline{z_k}}$, $1 \le j, k, \ell \le n$, from [13] we know

$$\frac{\partial TF}{\partial \overline{z_k}} = F_k, \ 1 \le k \le n; \quad \frac{\partial Tf^{(\ell)}}{\partial \overline{z_k}} = f_{k\ell}, \ 1 \le k, \ell \le n,$$

so that

$$\begin{aligned} u_{0\overline{z_k}} &= -F_k + \sum_{\ell=1}^{n} \overline{z_\ell} f_{k\ell} + Tf^{(k)} = Tf^{(k)}, \quad 1 \le k \le n, \\ u_{0\overline{z_k}\,\overline{z_\ell}} &= f_{k\ell} = f_{\ell k}. \end{aligned}$$

Any boundary value problem for solutions to the inhomogeneous pluriholomorphic system can be reduced to the same kind of problem for pluriholomorphic functions in the following way. Let e.g. (11.21), (11.22) be prescribed. Then the function $u - u_0$ is pluriholomorphic and satisfies

$$\text{Re}\,(u - u_0) = \gamma - \text{Re}\, u_0,$$

$$\text{Re}\,(u - u_0)_{\overline{\zeta_k}} = \gamma_k - \text{Re}\, u_{0\overline{\zeta_k}}, \ 1 \le k \le n,$$

on ∂B and

$$\operatorname{Im}\int_{\partial B}(u-u_0)(\zeta)\,d\sigma(\zeta)) = -\operatorname{Im}\int_{\partial B} u_0(\zeta)\,d\sigma(\zeta)\,,$$

$$\operatorname{Im}\int_{\partial B}(u-u_0)_{\overline{\zeta_k}}(\zeta)\,d\sigma(\zeta) = -\operatorname{Im}\int_{\partial B} u_{0\overline{\zeta_k}}(\zeta)\,d\sigma(\zeta)\,,\ 1\le k\le n\,,$$

where the right–hand sides are known. In principle the solution $u-u_0$ is given via (11.22) where the new right–hand sides have to be inserted and the constants have to be modified according to the new side conditions. Expressing u_0 then through $f_{k\ell}$ in the above form leads to u. For details see [2], 4.4.

11.5 SIMPLE BOUNDARY VALUE PROBLEMS IN THE UNIT POLYDISC

The Cauchy formula for analytic functions φ in the polydisc $\mathbb{D}$ of $\mathbb{C}^n$ being continuous on the distinguished boundary $\partial_0\mathbb{D} := \{z : |z_k| = 1,\ 1\le k\le n\}$ is

$$\varphi(z) = \frac{1}{(2\pi i)^n}\int_{\partial_0\mathbb{D}}\varphi(\zeta)\,\frac{d\zeta}{\zeta - z}\,,\qquad \frac{d\zeta}{\zeta - z} := \prod_{k=1}^{n}\frac{d\zeta_k}{\zeta_k - z_k}\,.$$

It can be reformulated as

$$\varphi(z) = \frac{1}{(2\pi i)^n}\int_{\partial_0\mathbb{D}}\operatorname{Re}\varphi(\zeta)\Big[\frac{2\zeta}{\zeta - z} - 1\Big]\frac{d\zeta}{\zeta} + i\operatorname{Im}\varphi(0)\,,\ \ z\in\mathbb{D}\,,$$

where

$$\frac{\zeta}{\zeta - z} := \prod_{k=1}^{n}\frac{\zeta_k}{\zeta_k - z_k}\,,\qquad \frac{d\zeta}{\zeta} := \prod_{k=1}^{n}\frac{d\zeta_k}{\zeta_k}\,,$$

see [1,2]. This Schwarz formula for $n=1$ is related to the Poisson formula as the real part of the Schwarz kernel coincides with the Poisson kernel,

$$\operatorname{Re}\frac{2\zeta}{\zeta - z} - 1 = \frac{\zeta}{\zeta - z} + \frac{\overline{\zeta}}{\overline{\zeta - z}} - 1 = \operatorname{Re}\frac{\zeta + z}{\zeta - z} = \frac{|\zeta|^2 - |z|^2}{|\zeta - z|^2}\,.$$

For $n > 1$ we have

$$\begin{aligned}\operatorname{Re}\frac{2\zeta}{\zeta - z} - 1 &= \frac{\zeta}{\zeta - z} + \frac{\overline{\zeta}}{\overline{\zeta - z}} - 1 = \prod_{k=1}^{n}\frac{\zeta_k}{\zeta_k - z_k} + \prod_{k=1}^{n}\frac{\overline{\zeta_k}}{\overline{\zeta_k - z_k}} - 1\\ &= \prod_{k=1}^{n}\Big[\frac{\zeta_k}{\zeta_k - z_k} + \frac{\overline{\zeta_k}}{\overline{\zeta_k - z_k}} - 1\Big] - K(z,\zeta)\end{aligned}$$

with

$$K(z,\zeta) = \sum_{k=1}^{n-1}\Big[\Big(\prod_{\nu=1}^{k}\frac{\zeta_\nu}{\zeta_\nu - z_\nu} - 1\Big)\frac{\overline{\zeta_{k+1}}}{\overline{\zeta_{k+1} - z_{k+1}}}$$

$$+\Big(\prod_{\nu=1}^{k}\frac{\overline{\zeta_\nu}}{\overline{\zeta_\nu - z_\nu}} - 1\Big)\frac{\zeta_{k+1}}{\zeta_{k+1} - z_{k+1}}\Big] \prod_{\nu=k+2}^{n}\Big[\frac{\zeta_\nu}{\zeta_\nu - z_\nu} + \frac{\overline{\zeta_\nu}}{\overline{\zeta_\nu - z_\nu}} - 1\Big].$$

Lemma C *Necessary and sufficient for the Schwarz problem*

$$\operatorname{Re}\varphi = \gamma \quad \text{on } \partial_0 \mathbb{D}, \quad \operatorname{Im}\varphi(0) = c,$$

for analytic functions in $\mathbb{D}$ *is*

$$\frac{1}{(2\pi i)^n}\int_{\partial_0 \mathbb{D}} \gamma(\zeta)\, K(z,\zeta)\, \frac{d\zeta}{\zeta} = 0, \quad z \in \mathbb{D}. \tag{11.27}$$

The unique solution then is given by the Schwarz integral

$$\varphi(z) = \frac{1}{(2\pi i)^n}\int_{\partial_0 \mathbb{D}} \gamma(\zeta)\Big[\frac{2\zeta}{\zeta - z} - 1\Big]\frac{d\zeta}{\zeta} + ic. \tag{11.28}$$

For details, see [1,2]. Obviously, in case $n = 1$ there is no solvability condition for this Schwarz problem.

This result is the basis of the theory of Riemann–Hilbert boundary value problems for analytic functions in polydiscs. This problem is particularly interesting in polydomains because here the distinguished boundary is not a simply connected set as e.g. ∂B is. Therefore here the phenomenon of index arises making the Riemann–Hilbert problem more interesting. Again we refer to [2], 5.

In order to treat the inhomogeneous Cauchy–Riemann system

$$w_{\overline{z_k}} = f_k, \quad f_{k\overline{z_\ell}} = f_{\ell\overline{z_k}}, \quad 1 \le k, \ell \le n, \text{ in } \mathbb{D}$$

rather than analytic functions the Pompeiu operators

$$T_k f(z) := -\frac{1}{\pi}\int_{|\zeta_k|<1} f(z_1, \dots, \zeta_k, \dots, z_n)\, \frac{d\zeta_k}{\zeta_k - z_k}, \quad 1 \le k \le n, \tag{11.29}$$

are used. A particular solution of the inhomogeneous Cauchy–Riemann system is inductively seen to be given as

$$w_0 = \sum_{\nu=1}^{n}(-1)^{\nu+1} \sum_{1\le k_1<\cdots<k_\nu\le n} T_{k_\nu}\cdots T_{k_1} f_{k_1\overline{\zeta_{k_2}}\cdots\overline{\zeta_{k_\nu}}} \tag{11.30}$$

where the second sum is taken over all ordered ν–tuples $\{k_1, \dots, k_\nu\} \subset \{1, \dots, n\}$, $k_1 < k_2 < \cdots < k_\nu$.

The Schwarz problem for the inhomogeneous Cauchy–Riemann system can be reduced to the one for analytic functions by considering $w - w_0$ satisfying

$$\mathrm{Re}\,(w - w_0) = \gamma - \mathrm{Re}\, w_0 \ \text{ on } \partial_0 \mathbb{D},\ \mathrm{Im}\,(w - w_0)(0) = c - \mathrm{Im}\, w_0(0)\,.$$

Inserting this right–hand side into (11.27) and (11.28) leads to the solvability condition and the solution. Here the main problem is to reformulate the iterated integrals involving he $f_{k\ell}$'s.

Theorem 8 *Necessary and sufficient for the Schwarz problem in $\mathbb{D}$ for the inhomogeneous Cauchy–Riemann system satisfying the compatibility conditions is*

$$\begin{aligned}\frac{1}{(2\pi i)^n}&\int_{\partial_0 \mathbb{D}} \gamma(\zeta)\, K(z,\zeta)\, \frac{d\zeta}{\zeta} \\ &= \mathrm{Re}\Big\{\sum_{\nu=2}^{n}\sum_{\lambda=1}^{\nu-1} \sum_{\substack{1\le k_1<\cdots<k_\lambda\le n\\ 1\le k_{\lambda+1}<\cdots<k_\nu\le n}} \frac{(-1)^\nu}{\pi^\nu}\int_{\mathbb{D}_{k_1}}\cdots\int_{\mathbb{D}_{k_\nu}} f_{k_1\overline{\zeta_{k_2}}\cdots\overline{\zeta_{k_\lambda}}\zeta_{k_{\lambda+1}}\cdots\zeta_{k_\nu}}(\zeta) \\ &\quad\times \prod_{\tau=1}^{\lambda}\frac{\overline{z_{k_\tau}}}{1-\overline{z_{k_\tau}}\zeta_{k_\tau}} \prod_{\tau=\lambda+1}^{\nu}\frac{z_{k_\tau}}{1-z_{k_\tau}\overline{\zeta_{k_\tau}}} \prod_{\tau=1}^{\nu} d\xi_{k_\tau}d\eta_{k_\tau}\Big\}.\end{aligned} \tag{11.31}$$

The solution then is

$$\begin{aligned} w(z) &= \frac{1}{(2\pi i)^n}\int_{\partial_0 \mathbb{D}} \gamma(\zeta)\Big[\frac{2\zeta}{\zeta - z} - 1\Big]\frac{d\zeta}{\zeta} \\ &\quad+ \sum_{\nu=1}^{n}\sum_{1\le k_1<\cdots<k_\nu\le n}\frac{(-1)^\nu}{\pi^\nu}\int_{\mathbb{D}_{k_1}}\cdots\int_{\mathbb{D}_{k_\nu}}\Big\{\overline{f_{k_1\overline{\zeta_{k_2}}\cdots\overline{\zeta_{k_\nu}}}(\zeta)}\prod_{\tau=1}^{\nu}\frac{z_{k_\tau}}{1-z_{k_\tau}\overline{\zeta_{k_\tau}}} \\ &\qquad\qquad -.(-1)^\nu f_{k_1\overline{\zeta_{k_2}}\cdots\overline{\zeta_{k_\nu}}}(\zeta)\prod_{\tau=1}^{\nu}\frac{1}{\zeta_{k_\tau}-z_{k_\tau}}\Big\}\prod_{\tau=1}^{\nu}d\xi_{k_\tau}d\eta_{k_\tau} \\ &\quad+ \sum_{\nu=2}^{n}\sum_{\lambda=1}^{\nu-1}\sum_{\substack{1\le k_1<\cdots<k_\lambda\le n\\ 1\le k_{\lambda+1}<\cdots<k_\nu\le n}}\frac{(-1)^\lambda}{\pi^\nu}\int_{\mathbb{D}_{k_1}}\cdots\int_{\mathbb{D}_{k_\nu}}\overline{f_{k_1\overline{\zeta_{k_2}}\cdots\overline{\zeta_{k_\lambda}}\zeta_{k_{\lambda+1}}\cdots\zeta_{k_\nu}}(\zeta)} \\ &\quad\times\prod_{\tau=1}^{\lambda}\frac{z_{k_\tau}}{1-z_{k_\tau}\overline{\zeta_{k_\tau}}}\prod_{\tau=\lambda+1}^{\nu}\frac{1}{\zeta_{k_\tau}-z_{k_\tau}}\prod_{\tau=1}^{\nu}d\xi_{k_\tau}d\eta_{k_\tau} \\ &\quad+ ic\,,\end{aligned} \tag{11.32}$$

see [1,2].

This result can be used to solve the Dirichlet problem for the inhomogeneous pluriharmonic system in $\mathbb{D}$,

$$u_{\overline{z_k} z_\ell} = f_{k\ell}\,, \quad f_{k\ell \overline{z_i}} = f_{i\ell \overline{z_k}}\,, \ f_{k\ell z_j} = f_{kj z_\ell}, \ 1 \leq i,j,k,\ell \leq n\,, \text{ in } \mathbb{D}\,,$$

$$u = \gamma \quad \text{on } \partial_0 \mathbb{D}\,. \tag{11.33}$$

At first a particular solution to the inhomogeneous system is sought. Denoting $u_{\overline{z_k}} =: w_k$, $1 \leq k \leq n$, we get the n systems

$$w_{k z_\ell} = f_{k\ell}\,, \quad 1 \leq \ell \leq n, \text{ in } \mathbb{D}\,,$$

each of which can be treated as an inhomogeneous Cauchy–Riemann system in $\overline{z}$ rather than in z. The general solution is of the form

$$w_k = \sum_{\mu=1}^{n} (-1)^{\mu+1} \sum_{1 \leq \ell_1 \cdots < \ell_\mu \leq n} \overline{T_{\ell_\mu}} \cdots \overline{T_{\ell_1}} f_{k\ell_1 \zeta_{\ell_2} \cdots \zeta_{\ell_\mu}} + \overline{\psi_k} =: F_k \tag{11.34}$$

with arbitrary analytic function ψ_k. Choosing ψ_k in such a way that the inhomogeneous Cauchy–Riemann system

$$u_{\overline{z_k}} = F_k\,, \quad 1 \leq k \leq n\,, \text{ in } \mathbb{D}\,,$$

satisfies the compatibility conditions

$$F_{k \overline{z_\ell}} = F_{\ell \overline{z_k}}\,, \quad 1 \leq k,\ell \leq n\,, \text{ in } \mathbb{D}$$

a particular solution u_0 to the inhomogeneous pluriholomorphic system is found as

$$\begin{aligned} u_0 &= \sum_{\nu=1}^{n} (-1)^{\nu+1} \sum_{1 \leq k_1 < \cdots < k_\nu \leq n} T_{k_\nu} \cdots T_{k_1} F_{k_1 \overline{\zeta_{k_2}} \cdots \overline{\zeta_{k_\nu}}} \qquad (11.35) \\ &= \sum_{\nu,\mu=1}^{n} (-1)^{\mu+\nu} \sum_{\substack{1 \leq k_1 < \cdots < k_\nu \leq n \\ 1 \leq \ell_1 < \cdots < \ell_\mu \leq n}} T_{k_\nu} \cdots T_{k_1} \overline{T_{\ell_\mu}} \cdots \overline{T_{\ell_1}} f_{k_1 \ell_1 \overline{\zeta_{k_2}} \cdots \overline{\zeta_{k_\nu}} \zeta_{\ell_2} \cdots \zeta_{\ell_\mu}} \\ &\quad + \sum_{\nu=1}^{n} (-1)^{\nu+1} \sum_{1 \leq k_1 < \cdots < k_\nu \leq n} T_{k_\nu} \cdots T_{k_1} \overline{\psi_{k_1 \zeta_{k_2} \cdots \zeta_{k_\nu}}}\,. \end{aligned}$$

Remark

$$u_{0 \overline{z_{k_1}} z_{k_2}} = \sum_{\nu=1}^{n} (-1)^{\nu+1} \sum_{1 \leq \ell_1 < \cdots < \ell_\nu \leq n} \overline{T_{\ell_\nu}} \cdots \overline{T_{\ell_1}} f_{k_1 \ell_1 \zeta_{\ell_2} \cdots \zeta_{\ell_\nu} \overline{\zeta_{k_2}}}\,, \quad 1 \leq k_1, k_2 \leq n\,,$$

can be shown, see [2], 5.3.

It remains to solve the Dirichlet problem for pluriharmonic functions u in $\mathbb{D}$. Representing u as before by two analytic functions φ and ψ as $u = \varphi + \overline{\psi}$ then $\varphi + \overline{\psi} = \gamma$ on $\partial_0 \mathbb{D}$ means

$$\operatorname{Re}\{\varphi + \psi\} = \operatorname{Re}\gamma, \quad \operatorname{Im}\{\varphi - \psi\} = \operatorname{Re} i\{\psi - \varphi\} = \operatorname{Im}\gamma \text{ on } \partial_0 \mathbb{D}$$

for the analytic functions $\varphi + \psi$, $i(\psi - \varphi)$ in $\mathbb{D}$.

Asking $\operatorname{Re}\gamma$ and $\operatorname{Im}\gamma$ to satisfy (11.27) and observing that $K(z,\zeta)$ is a real kernel the solvability condition for γ is just (11.27). The solutions to our both problems are then via (11.28) with real constants c_1 and c_2

$$(\varphi + \psi)(z) = \frac{1}{(2\pi i)^n}\int_{\partial_0 \mathbb{D}} \operatorname{Re}\gamma(\zeta)\Big[\frac{2\zeta}{\zeta - z} - 1\Big]\frac{d\zeta}{\zeta} + ic_1 ,$$

$$(\varphi - \psi)(z) = \frac{1}{(2\pi i)^n}\int_{\partial_0 \mathbb{D}} i\operatorname{Im}\gamma(\zeta)\Big[\frac{2\zeta}{\zeta - z} - 1\Big]\frac{d\zeta}{\zeta} - c_2 .$$

Hence

$$\varphi(z) = \frac{1}{(2\pi i)^n}\int_{\partial_0 \mathbb{D}} \gamma(\zeta)\Big[\frac{\zeta}{\zeta - z} - \frac{1}{2}\Big]\frac{d\zeta}{\zeta} + i(c_1 + ic_2),$$

$$\psi(z) = \frac{1}{(2\pi i)^n}\int_{\partial_0 \mathbb{D}} \overline{\gamma(\zeta)}\Big[\frac{\zeta}{\zeta - z} - \frac{1}{2}\Big]\frac{d\zeta}{\zeta} + i(c_1 - ic_2),$$

so that

$$u(z) = \varphi(z) + \overline{\psi(z)} = \frac{1}{(2\pi i)^n}\int_{\partial_0 \mathbb{D}} \gamma(\zeta)\Big[\frac{\zeta}{\zeta - z} + \frac{\overline{\zeta}}{\overline{\zeta - z}} - 1\Big]\frac{d\zeta}{\zeta}, \quad z \in \mathbb{D}. \tag{11.36}$$

Applying this to the pluriharmonic function $u - u_0$ in $\mathbb{D}$ satisfying

$$u - u_0 = \gamma - u_0 \quad \text{on } \partial_0 \mathbb{D}$$

shows that if

$$\frac{1}{(2\pi i)^n}\int_{\partial_0 \mathbb{D}} [\gamma(\zeta) - u_0(\zeta)]\, K(z,\zeta)\,\frac{d\zeta}{\zeta} = 0$$

this problem is solvable and the unique solution is

$$u(z) = \frac{1}{(2\pi i)^n}\int_{\partial_0 \mathbb{D}} [\gamma(\zeta) - u_0(\zeta)]\,[\frac{\zeta}{\zeta - z} + \frac{\overline{\zeta}}{\overline{\zeta - z}} - 1]\,\frac{d\zeta}{\zeta}, \quad z \in \mathbb{D}.$$

What remains is to evaluate the integrals involving u_0, see [1,2]. The following result is obtained.

Theorem 9 *Let the $f_{k\ell}$'s satisfy the compatibility conditions in (11.33)*

$$\frac{1}{(2\pi i)^n}\int_{\partial_0 \mathbb{D}} \overline{f_{k\ell\overline{\zeta_j}}(\zeta)} \prod_{\substack{\mu=1\\ \mu\neq k}}^{n} \frac{d\zeta_\mu}{\zeta_\mu - z_\mu}\,\frac{d\overline{\zeta_k}}{\overline{\zeta_k - z_k}} = \frac{1}{(2\pi i)^n}\int_{\partial_0 \mathbb{D}} \overline{f_{\ell k\zeta_j}(\zeta)} \prod_{\substack{\mu=1\\ \mu\neq \ell}}^{n} \frac{d\zeta_\mu}{\zeta_\mu - z_\mu}\,\frac{d\overline{\zeta_\ell}}{\overline{\zeta_\ell - z_\ell}},$$

$$k \neq \ell, \quad 1 \le k,\ell \le n,$$

and together with γ

$$\sum_{k=1}^{n} \frac{1}{(2\pi i)^n} \int_{\partial_0 \boldsymbol{D}} \gamma(\zeta)\, K(z,\zeta)\, \frac{d\zeta}{\zeta}$$

$$= \sum_{\mu=2}^{n} \sum_{\nu=1}^{\mu-1} \sum_{\substack{1 \le \ell_1 < \cdots < \ell_{\mu-\nu} \le n \\ 1 \le k_1 < \cdots < k_\nu \le n \\ cd\{\ell_1, \ldots, \ell_{\mu-\nu}, k_1, \ldots, k_\nu\} = \mu}} \frac{(-1)^\mu}{\mu} \int_{\boldsymbol{D}_{k_1}} \cdots \int_{\boldsymbol{D}_{k_\nu}} \int_{\boldsymbol{D}_{\ell_1}} \cdots \int_{\boldsymbol{D}_{\ell_{\mu-\nu}}} f_{k_1 \ell_1 \overline{\zeta_{k_2}} \cdots \overline{\zeta_{k_\nu}} \zeta_{\ell_2} \cdots \zeta_{\ell_{\mu-\nu}}}$$

$$\times \prod_{\tau=1}^{\nu} \frac{\overline{z_{k_\tau}}}{1 - \overline{z_{k_\tau}} \zeta_{k_\tau}} \prod_{\tau=1}^{\mu-\nu} \frac{z_{\ell_\tau}}{1 - z_{\ell_\tau} \overline{\zeta_{\ell_\tau}}} \prod_{\tau=1}^{\nu} d\xi_{k_\tau} d\eta_{k_\tau} \prod_{\tau=1}^{\mu-\nu} d\xi_{\ell_\tau} d\eta_{\ell_\tau} \,.$$

Then the Dirichlet problem for system (11.33) is uniquely solvable. The solution is given by

$$u(z) = \frac{1}{(2\pi i)^n} \int_{\partial_0 \boldsymbol{D}} \gamma(\zeta) \prod_{k=1}^{n} \frac{1 - |z_k|^2}{|\zeta_k - z_k|^2} \frac{d\zeta_k}{\zeta_k}$$

$$+ \Big[\sum_{\substack{\mu,\nu=1 \\ 2 \le \mu+\nu \le n}}^{n} \sum_{\rho=0}^{\min\{\mu,\nu\}} + \sum_{\substack{\mu,\nu=1 \\ n+1 \le \mu+\nu \le 2n}}^{n} \sum_{\rho=\mu+\nu-n}^{\min\{\mu,\nu\}} \Big]$$

$$\times \sum_{\substack{1 \le \sigma_1 < \cdots < \sigma_\rho \le n \\ 1 \le k_1 < \cdots < k_{\nu-\rho} \le n \\ 1 \le \ell_1 < \cdots < \ell_{\mu-\rho} \le n}} \frac{1}{\pi^{\mu+\nu-\rho}} \int_{\boldsymbol{D}_{\sigma_1}} \cdots \int_{\boldsymbol{D}_{\sigma_\rho}} \int_{\boldsymbol{D}_{k_1}} \cdots \int_{\boldsymbol{D}_{k_{\nu-\rho}}} \int_{\boldsymbol{D}_{\ell_1}} \cdots \int_{\boldsymbol{D}_{\ell_{\mu-\rho}}}$$

$$\times f_{\sigma_1 \sigma_1 \overline{\zeta_{\sigma_2}} \cdots \overline{\zeta_{\sigma_\rho}}\, \overline{\zeta_{k_1}} \cdots \overline{\zeta_{k_{\nu-\rho}}} \zeta_{\sigma_2} \cdots \zeta_{\sigma_\rho} \zeta_{\ell_1} \cdots \zeta_{\ell_{\mu-\rho}}} \prod_{\lambda=1}^{\rho} \log \left| \frac{\zeta_{\sigma_\lambda} - z_{\sigma_\lambda}}{1 - \overline{z_{\sigma_\lambda}} \zeta_{\sigma_\lambda}} \right|^2 d\xi_{\sigma_\lambda} d\eta_{\sigma_\lambda}$$

$$\times \prod_{\lambda=1}^{\nu-\rho} \Big(\frac{1}{\zeta_{k_\lambda} - z_{k_\lambda}} + \frac{\overline{z_{k_\lambda}}}{1 - \overline{z_{k_\lambda}} \zeta_{k_\lambda}} \Big) d\xi_{k_\lambda} d\eta_{k_\lambda} \prod_{\lambda=1}^{\nu-\rho} \frac{d\xi_{\ell_\lambda} d\eta_{\ell_\lambda}}{\zeta_{\ell_\lambda} - z_{\ell_\lambda}}, \quad z \in \boldsymbol{D}\,.$$

Here for $\rho = 0$ the term $f_{\sigma_1 \sigma_2 \overline{\zeta_{\sigma_2}} \cdots \overline{\zeta_{\sigma_\rho}}\, \overline{\zeta_{k_1}} \cdots \overline{\zeta_{k_{\nu-\rho}}} \zeta_{\sigma_2} \cdots \zeta_{\sigma_\rho} \zeta_{\ell_1} \cdots \zeta_{\ell_{\mu-\rho}}}$ has to be interpreted as $f_{k_1 \ell_1 \overline{\zeta_{k_2}} \cdots \overline{\zeta_{k_\nu}} \zeta_{\ell_2} \cdots \zeta_{\ell_\mu}}$ and the product $\prod_{\lambda=1}^{\rho}$ has to be understood as 1. For details see [2], 5.3.

References

[1] H. BEGEHR, A. DZHURAEV: *Schwarz problem for Cauchy Riemann systems in several complex variables.* S. Stoilov memory volume, World Scientific, Singapore, to appear.

[2] H. BEGEHR, A. DZHURAEV: *Introduction to several complex variables and partial differential equations.* Addison Wesley Longman, Harlow, to appear.

[3] H. BEGEHR: *Boundary value problems in $\mathbb{C}$ and $\mathbb{C}^n$.* Acta Math. Vietnamica, to appear.

[4] H. BEGEHR: *Complex analytic methods for partial differential equations.* ZAMM 76 (1996), S2, 21–24.

[5] H. BEGEHR: *Elliptic second order equations.* Functional analytic methods in complex analysis and applications to partial differential equations. Eds. A.S. Mshimba, W. Tutschke. World Sci., Singapore, 1995, 115–152.

[6] H. BEGEHR, R.P. GILBERT: *Transformations, transmutations, and kernel functions. I, II.* Longman, Harlow, 1992, 1993.

[7] A. BITSADZE: *On the uniqueness of solutions of the Dirichlet problem for elliptic partial differential equations.* Uspechi Math. Nauk 3 (1948), no. 6, 211–212 (Russian).

[8] A. DZHURAEV: *On boundary value problems in the unit ball in $\mathbb{C}^n$.* Ross. Akad. Nauk Dokl. 343 (1995), no. 4.

[9] A. DZHURAEV: *Degenerate and other problems.* Longman, Harlow, 1992.

[10] A. DZHURAEV: *On Riemann–Hilbert boundary value problems in several complex variables.* Complex Variables, Theory Appl. 24 (1996), 287–303.

[11] A. DZHURAEV: *The Dirichlet, Neumann and other overdetermined boundary value problems in the unit ball of $\mathbb{C}^n$.* Preprint, FU Berlin, 1996.

[12] N. KERZMAN: *Singular integrals in complex analysis.* Proc. Symp. Pure Math. AMS 35 (1979), 2, 3–41.

[13] W. RUDIN: *Function theory in the unit ball of $\mathbb{C}^n$.* Springer Verlag, New York etc., 1980.

12 ON THE BOUNDARY BEHAVIOUR OF POLYANALYTIC FUNCTIONS OF TWO VARIABLES IN JORDAN BICYLINDERS

M. Balk, W. Tutschke and V. Vassilenkov

Dedicated to Robert P. Gilbert on the occasion of his 65th birthday

Abstract:

The idea of splitting a non-holomorphic function into two summands one of which is holomorphic in some region and the other has a limit in some points of its boundary is used here in the study of the boundary behaviour of polyanalytic functions of several complex variables. For simplicity we restrict us to the case of two complex variables. Here two Fatou type theorems and one boundary uniqueness theorem for polyanalytic functions of two variables are proved.

12.1 NOTATIONS AND TERMINOLOGY

Hereafter the bold letters and bold figures denote bi-indices: $\mathbf{n} = (n_1, n_2)$, $\mathbf{k} = (k_1, k_2)$, $\mathbf{0} = (0,0)$, $\mathbf{1} = (1,1)$, $\mathbf{n} - \mathbf{1} = (n_1 - 1, n_2 - 1)$, etc. The symbol $\mathbf{z}$ denotes a point in the complex space C^2, $\mathbf{z} = (z_1, z_2)$, and the symbol $\mathbf{A}(\mathbf{z})$ denotes an ordered pair of two analytic functions: $\mathbf{A}(\mathbf{z}) = (A_1(z_1), A_2(z_2))$. We use also the following abbreviations: $[\mathbf{k}!] = k_1! k_2!$,

$$[(\overline{\mathbf{z}} - \mathbf{A}(\mathbf{z}))^{\mathbf{k}}] = (\overline{z}_1 - A_1(z_1))^{k_1} (\overline{z}_2 - A_2(z_2))^{k_2},$$

$[\mathbf{n}] = n_1 \cdot n_2$, $|\mathbf{n}| = n_1 + n_2$; $\mathbf{0} \le \mathbf{k} \le \mathbf{n} - \mathbf{1}$ means the pair of inequalities:

$$0 \le k_1 \le n_1 - 1, 0 \le k_2 \le n_2 - 1.$$

H. Florian et al. (eds.), Generalized Analytic Functions, 111–126.

Let **G** be some simply connected region in the space C^2. A function $f(\mathbf{z})$ $(\mathbf{z} = (z_1, z_2))$ which is infinitely differentiable in **G** is called *polyanalytic* (in short: p.a.) of bi-order $\mathbf{n} = (n_1,\ n_2)$ if it satisfies in **G** the system of differential equations:

$$\partial^{n_j} f/\partial \overline{z}_j^{\,n_j} = 0 \qquad (j = 1,2; \partial/\partial \overline{z}_j = (\partial/\partial x_j + i \cdot \partial/\partial y_j)/2, z_j = x_j + i \cdot y_j),$$

or in short:

$$\partial^{\mathbf{n}} f/\partial \overline{\mathbf{z}}^{\mathbf{n}} = \mathbf{0}$$

(see [1, 3, 4, 7]). It is known that every p.a. function f in **G** of bi-order $\mathbf{n} = (n_1, n_2)$ is representable in the form

$$f(\mathbf{z}) = \sum_{\mathbf{k}=\mathbf{0}}^{\mathbf{n}-\mathbf{1}} h_{\mathbf{k}}(\mathbf{z})\overline{\mathbf{z}}^{\mathbf{k}}. \tag{12.1}$$

(Here $\mathbf{0} = (0,0), \mathbf{n} - \mathbf{1} = (n_1 - 1, n_2 - 1), \mathbf{k} = (k_1, k_2)$, $h_{\mathbf{k}}(\mathbf{z})$ is a function holomorpic in **G**,

$$\overline{\mathbf{z}}^{\mathbf{k}} = \overline{z}_1{}^{k_1} \cdot \overline{z}_2{}^{k_2};$$

the summation from $\mathbf{k} = \mathbf{0}$ to $\mathbf{k} = \mathbf{n} - \mathbf{1}$ means the summation over all bi-indices $\mathbf{k} = (k_1, k_2)$, for which

$$0 \leq k_i \leq n_i - 1 \qquad (i = 1,2)).$$

Now let G_1 and G_2 be two Jordan regions respectively in the complex planes C_1 and C_2 of the variables z_1 and z_2; let L_i be the boundary of G_i in the plane C_i: $L_i = \partial G_i$ $(i = 1,2)$. The set $\mathbf{L} = L_1 \times L_2$ is the distinguished (or Bergmann-Shilov) boundary (in short: d-boundary) of the Jordan bicylinder $\mathbf{G} = G_1 \times G_2$. Let $L_i (i = 1,2)$ contain an analytic arc l_i (this will mean hereafter that l_i is the homeomorphic image of some *interval* (a_i, b_i) given by some holomorphic function $z_i = h_i(t), a_i < t < b_i, h_i'(t) \neq 0)$. Denote by $A_i(z_i)$ the Schwarz function of the arc l_i (i.e. l_i can be given in some its neighbourhood D_i by the "bianalytic"(in short: b.a.) equation $\overline{z}_i = A_i(z_i)$ where A_i is a holomorphic function in D_i). [E.g.the Schwarz function of an open (without its ends) arc of the unit circumference is $1/z$]. Two analytic arcs are called *non-kindred* if their Schwarz functions are not analytic continuations from each other. The symbol $\mathbf{A}(\mathbf{z})$ will denote the pair of functions $(A_1(z_1), A_2(z_2))$ and will be called *the Schwarz function of the analytic bi-arc* $\mathbf{l} = l_1 \times l_2$. This bi-arc will be called *coherent with the region* **G** if $\mathbf{A}(\mathbf{z})$ is a holomorphic vector-function in the entire region **G**.

Let $f(\mathbf{z})$ be a p.a. function of bi-order $\mathbf{n} = (n_1, n_2)$ (see (12.1)) in the bicylinder **G**. The auxiliary holomorphic function

$$h(\mathbf{z}) = \sum_{\mathbf{k}=\mathbf{0}}^{\mathbf{n}-\mathbf{1}} h_{k}(\mathbf{z})[(\mathbf{A}(\mathbf{z}))^{\mathbf{k}}]$$

(here $[(\mathbf{A}(\mathbf{z}))^{\mathbf{k}}]$ means: $(A_1(z_1))^{k_1} \cdot (A_2(z_2))^{k_2}$) is called *coherent with the p.a. function* f *on the bi-arc* $\mathbf{l} = l_1 \times l_2$.

Let c_i be some point on $L_i (i = 1,2)$,and let $S_i(c_i)$ be some Stolz sector in G_i with the vertex c_i ($S_i(c_i)$ is understood to be an open set and its angle is supposed to be less than π). The set $\mathbf{S}(\mathbf{c}) = S_1(c_1) \times S_2(c_2)$ is called a *Stolz bi-sector* in $\mathbf{G} = G_1 \times G_2$. In an obvious way the notions of the ***bi-radius*** and the ***bi-angle*** of a given bi-sector can be introduced. The bi-sector $\mathbf{S}(\mathbf{c}) = S_1(c_1) \times S_2(c_2)$ is said to be an inner bi-sector for the bi-sector $\mathbf{S}'(\mathbf{c}) = S_1'(c_1) \times S_2'(c_2)$ if $[\mathbf{S}] \subset (\mathbf{S}' \cup \{\mathbf{c}\}$. The least of the angles between the boundary radii of S_i and S_i' is called the slope of S_i to S_i'; denote it $\alpha_i (i = 1, 2)$; call the least of these two angles the slope of the bi-sector $\mathbf{S}$ to the bi-sector $\mathbf{S}'$.

The number B is called *the angular limit* of the function f in the point $\mathbf{c} = (c_1, c_2)$ of the d-boundary $\mathbf{L}$ if f tends to B while the point $\mathbf{z}$ tends to $\mathbf{c}$ along any Stolz bi-sector with the vertex $\mathbf{c}$.

Let G_1 be some region in the complex plane and let p be some positive number. As it is known the function $\phi(z_1)$ is said to be L_p-bounded in G_1 (or to belong to the space $L_p(G_1)$) if the following condition is fulfilled:

$$\iint_{G_1} |\phi(z_1)|^p dx_1 dy_1 < +\infty;$$

the left side of this inequality in the degree $1/p$ is called the L_p-norm of the function ϕ and is denoted as

$$\|\phi(\cdot)\|_{L_p(G_1)}.$$

Quite similar notions and notations are used in the case of functions $g(z_1, z_2)$ of two complex variables defined in a region $\mathbf{G}$ of the complex space C^2.

Let a function $g = g(z_1, z_2)$ be defined in some Jordan bicylinder $\mathbf{G} = G_1 \times G_2$ and be - as a function of z_1 - L_p-bounded in G_1 for each fixed z_2 from G_2. This function is said to be *L_p-uniformly continuous (or uniformly continuous in the mean of order p) in $\mathbf{G}$ with regard to the variable z_2* if for each fixedly chosen number $\varepsilon > 0$ there exists such number $\delta = \delta(\varepsilon)$ that for any numbers z_2', z_2'' from G_2 the inequality

$$| z_2'' - z_2' | < \delta$$

implies - for each z_1 from G_1 -

$$\|g(\cdot, z_2') - g(\cdot, z_2'')\|_{L_p(G_1)} < \varepsilon.$$

Note that the demand of L_p-uniform continuity of some function $g(z_1, z_2)$ is always fulfilled if the function g is in the region $\mathbf{G}$ Hölder-continuous with an arbitrary exponent $\alpha > 0$. Indeed,

$$| g(z_1, z_2') - g(z_1, z_2'') | \leq H \, | z_2' - z_2'' |^\alpha$$

implies

$$\|g(\cdot, z_2') - g(\cdot, z_2'')\|_{L_p(G_1)} \leq H \, | z_2' - z_2'' |^\alpha \cdot (\text{mes } G_1)^{1/p}.$$

Following A. Zygmund [14] we shall say that a function $f(\mathbf{z})$ defined in the bicylinder $\mathbf{G}$ *behaves restrictedly* (or *behaves boundedly*) in the boundary point $\mathbf{c} = (c_1, c_2)$ if there exists at least one such Stolz bi-sector $\mathbf{S}(\mathbf{c})$ with the vertex $\mathbf{c}$ for which the set $f(\mathbf{S}(\mathbf{c}))$ is not dense in the complex plane C (i.e. its complement to the plane C contains at least one disk). Clearly every function bounded in the bicylinder $\mathbf{G}$ behaves restrictedly (in the Zygmund sense) in every such boundary point for which there exists a Stolz bi-sector. The opposite statement is obviously wrong.

12.2 THE MAIN RESULTS

A.Zygmund proved such Fatou type theorem: Every function f holomorphic in the unit bi-disk and behaving restrictedly in almost every point of the d-boundary $\mathbf{L}$ of this region has a finite angular limit in almost every point of $\mathbf{L}$ (see [14], Chapter 17, Theorem 4.2.2). For p.a. functions this statement is wrong. We intend to give here the additional demands which permit to secure the conclusion of this Fatou type theorem in the case of p.a. functions.

Theorem 1 *Suppose that:*
a) G_k (k=1,2) is the unit disk in the complex plane, $\mathbf{G} = G_1 \times G_2$, and $\mathbf{L} = L_1 \times L_2$ is the d-boundary of $\mathbf{G}$;
b) E_k (k=1,2) is a set of positive measure on L_k,and $\mathbf{E} = E_1 \times E_2$;
c) the function $f(\mathbf{z})$ is polyanalytic (of some bi-order $\mathbf{n} = (n_1, n_2)$) in the bi-disk $\mathbf{G}$. Then the function f has a finite angular limit in almost every point of the set $\mathbf{E}$ if and only if the following conditions are fulfilled:
1) f behaves restrictedly in almost every point of $\mathbf{E}$;
2)for almost every point $\mathbf{c}$ belonging to $\mathbf{E}$ the two auxiliary p.a. functions

$$f_i = (z_i - c_i)\partial f/\partial \bar{z}_i (i = 1, 2)$$

have in the point $\mathbf{c}$ the angular limit zero.

Remark 1. In particular, f has a finite angular limit in almost every point of $\mathbf{L}$ if all the three functions $f, \partial f/\partial \bar{z}_1, \partial f/\partial \bar{z}_2$ are bounded in $\mathbf{G}$.

Zygmund's theorem can be used for obtaining a Fatou type theorem for a wide class of non-holomorphic (not necessarily polyanalytic) functions:

Theorem 2 *Suppose that:*
1) the function f is defined and twice continuously differentiable in the bi-disk $\mathbf{G} = G_1 \times G_2$;
2) there exists such number $p > 2$ that the first order areolar derivative $g_j(z_1, z_2) = \partial f/\partial \bar{z}_j$ (j=1,2) of f belongs to $L_p(G_j)$ for each z_{3-j} from G_{3-j}, and is L_p-uniformly continuous (uniformly continuous in the mean of order p) with respect to the "parametric variable" (z_{3-j});
3)the second order derivative

$$g_{1,2} = \partial^2 f/\partial \bar{z}_1 \partial \bar{z}_2$$

belongs to $L_p(\mathbf{G})$;
4)f behaves restrictedly (in the Zygmund sense) in almost every point of the distinguished boundary $\mathbf{L}$ of $\mathbf{G}$.
Then f has a finite angular limit in almost every point of the d-boundary $\mathbf{L}$.

Hereafter for p.a.functions of two variables a boundary uniqueness theorem of the Luzin-Privalov type will be proved:

Theorem 3 *Suppose that*
1) $\mathbf{G}$ *is a Jordan bi-cylinder (not necessarily a bi-disk),*

$$\mathbf{G} = G_1 \times G_2,$$

and $\mathbf{L}$ *is its d-boundary,* $\mathbf{L} = L_1 \times L_2$*;*
2) the curve $L_i (i = 1, 2)$ *contains* n_i *analytic arcs* l_{i,j_i} $(j_i = 1, ..., n_i)$ *pairwise non-kindred;*
3) every bi-arc

$$l_{1,p} \times l_{2,q} \quad (1 \leq p \leq n_1, \quad 1 \leq q \leq n_2)$$

is coherent with the region $\mathbf{G}$ *and contains a set* $E_{p,q}$ *of positive measure (on this bi-arc);*
4) f *is a p.a. function of bi-order* $\mathbf{n} = (n_1, n_2)$ *in* $\mathbf{G}$*;*
5) f *has the angular limit zero in every point of every boundary set* $E_{p,q}$ *mentioned above.*
Then $f(\mathbf{z}) \equiv 0$ *in* $\mathbf{G}$.

12.3 PROOFS

For proving Theorems 1-3 we shall need some Lemmas.

Lemma 1 . *Suppose that*
1) $f(\mathbf{z})$ *is a p.a.function of order* $\mathbf{n} = (n_1, n_2)$ *in the Jordan bi-cylinder* $\mathbf{G}$*;*
2) the d-boundary $\mathbf{L}$ *of* $\mathbf{G}$ *contains an analytic bi-arc* $\mathbf{l}$ *with the b.a. equation* $\overline{\mathbf{z}} = \mathbf{A}(\mathbf{z})$.
Then the following identity is true in $\mathbf{G}$ *near this bi-arc:*

$$f(\mathbf{z}) = h(\mathbf{z}) + p(\mathbf{z})$$

where $h(\mathbf{z})$ *is the holomorphic function coherent with* f *on the bi-arc* $\mathbf{l}$*, and*

$$p(\mathbf{z}) = (\overline{z}_1 - A_1(z_1))\partial f/\partial \overline{z}_1 + (\overline{z}_2 - A_2(z_2))\partial f/\partial \overline{z}_2 +$$

$$\sum_{\mathbf{k}=1}^{\mathbf{n}-1} [1/\mathbf{k}!](\overline{\mathbf{z}} - \mathbf{A}(\mathbf{z}))^{\mathbf{k}} \cdot \partial^{|\mathbf{k}|} f/\partial \overline{\mathbf{z}}^{\mathbf{k}}.$$

Proof. Put

$$\mathbf{w} = (w_1, w_2), \quad \mathbf{w_0} = (w_{0,1}, w_{0,2}).$$

Consider the auxiliary holomorphic function

$$F(\mathbf{z}, \mathbf{w}) = \sum_{\mathbf{k}=0}^{\mathbf{n}-1} \mathbf{h_k}(\mathbf{z})\mathbf{w}^{\mathbf{k}}.$$

For fixed $\mathbf{z}, \mathbf{w_0} = (w_{0,1}, w_{0,2})$ we have, by the Taylor formula,

$$F(\mathbf{z}, \mathbf{w}) = F(\mathbf{z}, \mathbf{w_0}) + (w_1 - w_{0,1})\partial F/\partial w_1|_{\mathbf{w}=\mathbf{w_0}} +$$

$$(w_2 - w_{2,0})\partial F/\partial w_2|_{\mathbf{w}=\mathbf{w_0}} +$$
$$\sum_{\mathbf{k}=1}^{\mathbf{n}-1}[1/\mathbf{k}!](\mathbf{w}-\mathbf{w_0})^{\mathbf{k}}\partial^{|\mathbf{k}|}F/\partial\mathbf{w}^{\mathbf{k}}.$$

Put $\mathbf{w} = \overline{\mathbf{z}}, \mathbf{w_0} = \mathbf{A}(\mathbf{z})$. Then it is clear that Lemma 1 is true.

Lemma 2 . *Suppose that*
1) $\mathbf{G}$ *is some bicylinder having on its d-boundary* $\mathbf{L}$ *an analytic bi-arc* $\mathbf{l}$ *with the b.a. equation* $\overline{\mathbf{z}} = \mathbf{A}(\mathbf{z})$ *(here* $\mathbf{A}(\mathbf{z}) = (A_1(z_1), A_2(z_2))$*);*
2) $\mathbf{c} = (c_1, c_2)$ *is an arbitrary point on* $\mathbf{l}$.
Then there exists such (sufficiently small) bi-disk $\mathbf{d}(\mathbf{c}, \mathbf{r})$ *and such positive constants* $M_i (i = 1, 2)$ *that for every point* $\mathbf{z} = (z_1, z_2)$ *from* $\mathbf{d}(\mathbf{c}, \mathbf{r})$ *the following inequalities are true:*

$$|\overline{z}_i - A_i(z_i)| \leq M_i|z_i - c_i|.$$

Proof. Let $D_i (i = 1, 2)$ be some region containing the arc l_i. Consider an arbitrary closed disk $d_i = d(c_i, r_i)$ belonging to D_i. Take into account that $\overline{\mathbf{c}} = \mathbf{A}(\mathbf{c})$. We have for every $z_i \in d_i$:

$$\begin{aligned} |\overline{z}_i - A_i(z_i)| &\leq |\overline{z}_i - \overline{c}_i| + |A_i(c_i) - A_i(z_i)| \\ &\leq |z_i - c_i| + |\int_{c_i}^{z_i} A_i'(t)dt| \leq M_i|z_i - c_i| \end{aligned}$$

where $M_i = 1 + m_i, \quad m_i = \max|A_i'(z_i)|$ for $z_i \in d_i, \quad i = 1, 2$.

Lemma 3 . *Suppose that:*
1) $\mathbf{S}(\mathbf{c})$ *is some bi-sector with the vertex* $\mathbf{c} = (c_1, c_2)$*;*
2) $f(\mathbf{z})$ *is some p.a. function in* $\mathbf{S}(\mathbf{c})$ *satisfying there the inequality*

$$|f(\mathbf{z})| < M.$$

Then for every choice of a bi-sector $\mathbf{S^0}(\mathbf{c})$ *inner to* $\mathbf{S}(\mathbf{c})$ *and being of sufficiently small bi-radius there exists for every bi-index* $\mathbf{k}$ *such constant* $T_{\mathbf{k}}^0$ *not depending on* f *and* M *that in* $\mathbf{S^0}(\mathbf{c})$ *the following inequality is true:*

$$|\mathbf{z} - \mathbf{c}|^{\mathbf{k}}|\partial^{|\mathbf{k}|}f/\partial\overline{\mathbf{z}}^{\mathbf{k}}| \leq T_{\mathbf{k}}^0 \cdot M. \tag{12.2}$$

Proof. We shall make use of the following estimate of the Cauchy-Nicolescu type (see [1]): If some function $f(z_1, z_2)$ is polyanalytic in a bi-disk $\mathbf{D} = D_1 \times D_2$ of bi-radius $\mathbf{R} = (R_1, R_2)$ and if $|f| < M$ in $\mathbf{D}$, then for each bi-index $\mathbf{k} = (k_1, k_2)$ there exists such constant $T_{\mathbf{k}}^{00}$ (not depending on f, M, R) that in the centre of $\mathbf{D}$ the following inequality is true:

$$R_1^{k_1} \cdot R_2^{k_2}|\partial^{|\mathbf{k}|}f/\partial\overline{\mathbf{z}}^{\mathbf{k}}| \leq T_{\mathbf{k}}^{00}. \tag{12.3}$$

Let α be the slope of $\mathbf{S^0}$ to $\mathbf{S}$. Consider in S_i $(i = 1, 2)$ a disk D_i with the centre in some point z_i of S_i^0 and maximal possible radius R_i. If the radius of S_i^0 is chosen sufficiently small the disk D_i will touch at least one boundary radius of S_i. Therefore the radius R_i of D_i satisfies the condition $R_i \geq |z_i - c_i| \sin\alpha$. From here and (12.3) we get (12.2).

Proof of Theorem 1.

Sufficiency of conditions 1, 2: Denote by $\mathbf{E}^0$ the set of all those points of $\mathbf{E}$ for which both conditions 1 and 2 are fulfilled. Let $\mathbf{c} = (c_1, c_2)$ be an arbitrary point from $\mathbf{E}^0$, and let $\mathbf{S}^0(\mathbf{c})$ be such Stolz bi-sector in which f "behaves restrictedly" (i.e. the complement of the set $f(\mathbf{S}^0(\mathbf{c}))$ with respect to the plane C contains some disk); clearly in every Stolz bi-sector $\mathbf{S}^{00}(\mathbf{c})$ inner to $\mathbf{S}^0(\mathbf{c})$ the function f "behaves restrictedly" as well. Denote by $\mathbf{S}(\mathbf{c})$ and $\mathbf{S}'(\mathbf{c})$ two Stolz bi-sectors with the vertex $\mathbf{c}$ possessing the following property: the bi-angle of the bi-sector $\mathbf{S}^0(\mathbf{c})$ is an inner bi-angle for the bi-angle of the bi-sector $\mathbf{S}(\mathbf{c})$, and the last is an inner bi- angle for the bi-angle of $\mathbf{S}'(\mathbf{c})$. Clearly we can always choose the bi-radii of these bi-sectors so small that $\mathbf{S}^0(\mathbf{c})$ is an inner bi-sector for $\mathbf{S}(\mathbf{c})$ and the later is an inner one for $\mathbf{S}'(\mathbf{c})$. Chose arbitrarily a number $\varepsilon > 0$. Take into account that the p.a. function $f_1(\mathbf{z})$ has in the point $\mathbf{c}$ the angular limit zero; therefore $| f_1(\mathbf{z}) |< \varepsilon$ in $\mathbf{S}'(\mathbf{c})$ if its bi-radius is chosen sufficiently small; similarly we may assume that $| f_2(\mathbf{z}) |< \varepsilon$ in $\mathbf{S}'(\mathbf{c})$.

Let $\mathbf{k} = (k_1, k_2)$ be an arbitrary bi-index,

$$\mathbf{k} \geq \mathbf{1}; \mathbf{k}' = (k_1 - 1, k_2), \mathbf{k}'' = (k_1, k_2 - 1).$$

For each index $\mathbf{k}'$ there exists - in accordance with Lemma 3 - such number $C_{\mathbf{k}'}$ (not depending on ε) that the inequality

$$| \mathbf{z} - \mathbf{c} |^{\mathbf{k}'} | \partial^{|\mathbf{k}'|} f / \partial \overline{\mathbf{z}}^{\mathbf{k}'} |< C_{\mathbf{k}'} \cdot \varepsilon$$

is true in $\mathbf{S}(\mathbf{c})$ if its bi-radius is sufficiently small; in other words,

$$| \mathbf{z} - \mathbf{c} |^{\mathbf{k}} | \partial^{|\mathbf{k}|} f / \partial \overline{\mathbf{z}}^{\mathbf{k}} |< C_{\mathbf{k}'} \cdot \varepsilon.$$

Similarly one can prove that in the case of a sufficiently small bi-radius of $\mathbf{S}(\mathbf{c})$ we have in $\mathbf{S}(\mathbf{c})$:

$$| \mathbf{z} - \mathbf{c} |^{\mathbf{k}} | \partial^{|\mathbf{k}|} f / \partial \overline{\mathbf{z}}^{\mathbf{k}} |< C_{\mathbf{k}''} \cdot \varepsilon$$
$$| \mathbf{z} - \mathbf{c} |^{\mathbf{k}} | \partial^{|\mathbf{k}|} f / \partial \overline{z}^{\mathbf{k}} |< C_{\mathbf{k}''} \cdot \varepsilon$$

(here $C_{\mathbf{k}''}$ is some constant). Therefore (see Lemma 1) we have in $\mathbf{S}(\mathbf{c})$:

$$| p(\mathbf{z}) |< C \cdot \varepsilon,$$

where $C > 0$ is some constant.

Now making use of Lemma 1 we can conclude that the holomorphic function $h(\mathbf{z})$ behaves boundedly in the boundary point $\mathbf{c}$. According to condition 1 the p.a. function $f(\mathbf{z})$ behaves restrictedly in $\mathbf{c}$, i.e. $f(\mathbf{S}(\mathbf{c}))$ does not contain some disk $\mathbf{D}$. Clearly we can always assume that $\mathbf{D}$ does not contain the point $\mathbf{w} = \mathbf{0}$. Since ε can be chosen arbitrarily small we may assume that some disk $\mathbf{D}'$ (with the same centre as $\mathbf{D}$, but of lesser radius) does not contain -for any $\mathbf{z}$ belonging to $\mathbf{S}^0(\mathbf{c})$ - the point

$w = f(\mathbf{z}) - p(\mathbf{z})$. Hence the holomorphic function $h(\mathbf{z}) = f(\mathbf{z}) - p(\mathbf{z})$ is of restricted behaviour in $\mathbf{c}$. The same reasoning may be repeated for any point $\mathbf{c} \in \mathbf{E^0}$. By the Zygmund theorem the function h has a finite angular limit in almost every point of $\mathbf{E}$. But since $p(\mathbf{z})$ has in almost every point of $\mathbf{E}$ the angular limit zero, it follows from Lemma 1 that f possesses the same property. Thus the sufficiency of conditions 1,2 is proved.

Necessity of conditions 1, 2: Let f have an angular limit in almost every point $\mathbf{c}$ of some set $\mathbf{E} = E_1 \times E_2$, which is a subset of $\mathbf{L}$.Then f has a finite limit B in $\mathbf{c}$ along any Stolz bi-sector $\mathbf{S}(\mathbf{c})$ - and therefore "behaves boundedly" in $\mathbf{c}$. Thus condition 1 is fulfilled.

Choose an arbitrary Stolz bi-sector $\mathbf{S}(\mathbf{c})$ and choose a Stolz bi-sector $\mathbf{S}'(\mathbf{c})$ for which $\mathbf{S}(\mathbf{c})$ is an inner one. We may assume that the bi-radius of $\mathbf{S}(\mathbf{c})$ is not more than a half of the bi-radius of $\mathbf{S}'(\mathbf{c})$. By condition, f has the limit B while $\mathbf{z}$ tends to $\mathbf{c}$ along the bi-sector $\mathbf{S}'(\mathbf{c})$; therefore for every $\varepsilon > 0$ the inequality

$$| f(\mathbf{z}) - B | < \varepsilon$$

is true in $\mathbf{S}'(\mathbf{c})$ if its bi-radius is sufficiently small. Therefore - making use of the Cauchy-Nicolescu type inqualities (see Lemma 3)- we have

$$| z_k - c_k | \cdot | \partial(f(\mathbf{z}) - B)/\partial \overline{z}_k | < C \cdot \varepsilon (k = 1, 2)$$

(here C is some constant). This just means that the condition 2 is fulfilled in the point $\mathbf{c}$.

Lemma 4 . *If the function $g(z_1, z_2)$ is defined and continuous in some bicylinder $\mathbf{G} = G_1 \times G_2$ and is - for some fixed $p > 2$ - uniformly L_p-continuous with respect to the variable z_2, then the Pompeiu integral*

$$(T_1 g)(z_1, z_2) = -\frac{1}{\pi} \iint\limits_{G_1} \frac{g(\zeta_1, z_2)}{\zeta_1 - z_1} d\xi_1 d\eta_1 .$$

has a finite limit in every boundary point of the bicylinder $\mathbf{G}$.

Proof. We shall make use of three known inequalities: 1) If mes G denotes the measure of G where G is a domain in the complex plane, then one has

$$\iint\limits_{G} \frac{d\xi d\eta}{|\zeta - z|^q} \leq \frac{2\pi}{2 - q} \left(\frac{\text{mes } G}{\pi} \right)^{1 - \frac{q}{2}}$$

(E. Schmidt's inequality);

2) For any positive p, q with $1/p + 1/q = 1$ we have

$$\left| \iint\limits_{G} \phi(z) \psi(z) dx dy \right| \leq \left(\iint\limits_{G} |\phi(z)|^p \, dx dy \right)^{\frac{1}{p}} \left(\iint\limits_{G} |\psi(z)|^q \, dx dy \right)^{\frac{1}{q}}$$

(Hölder's inequality);

3) For any fixed domain G in the complex plane (with finite measure) and for any fixed number q, $1 < q < 2$, there exist such constants C_1 and C_2 that for any numbers z' and z'' belonging to G the following relation is true:

$$\iint\limits_G \frac{d\xi d\eta}{|\zeta - z'|^q |\zeta - z''|^q} \leq C_1 + C_2 |z' - z''|^{2-2q}$$

(J. Hadamard; see, for instance, [12] or [11]).

Hölder's inequality implies

$$|(T_1 g)(z_1, z_2') - (T_1 g)(z_1, z_2'')|$$

$$\leq \left\| g(z_1, z_2') - g(z_1, z_2'') \right\|_{L_p(G_1)} \cdot \frac{2\pi}{2-q} \left(\frac{\text{mes } G_1}{\pi} \right)^{1-\frac{q}{2}}.$$

Since g is L_p-uniformly continuous with respect to z_2, the function $T_1 g$ turns out to be uniformly continuous in G as a function of z_2.

Now we shall prove that the same integral $T_1 g$ is uniformly continuous in G as a function of z_1. Indeed, we have obviously

$$(T_1 g)(z_1', z_2) - (T_1 g)(z_1'', z_2) =$$

$$-\frac{z_1' - z_1''}{\pi} \iint\limits_{G_1} g(\zeta_1, z_2) \frac{d\xi_1 d\eta_1}{(\zeta_1 - z_1')(\zeta_1 - z_1'')}. \tag{12.4}$$

The absolute value of the last integral can be estimated by

$$\|g(\cdot, z_2)\|_{L_p(G_1)} \left(\iint\limits_{G_1} \frac{d\xi_1 d\eta_1}{c|\zeta_1 - z_1'|^q |\zeta_1 - z_1''|^q} \right)^{\frac{1}{q}}.$$

Take into consideration that the above introduced uniform continuity in the mean (L_p-uniform continuity) with respect to z_2 implies uniform boundedness of the L_p-norms of the functions $g(\cdot, z_2)$ (with respect to the parametric variable z_2).

Consequently, up to a constant factor the absolute value of the expression (12.4) can be estimated by

$$(C_1 |z_1' - z_1''|^q + C_2 |z_1' - z_1''|^{2-q})^{\frac{1}{q}}$$

This proves that $T_1 g$ is continuous in G as a function of z_1, and the continuity is uniform with respect to z_1.

Thus $T_1 g$ is *separately* uniformly continuous in G with regard to z_2 and with regard to z_1.

It follows from the triangle inequality that a function of two variables in G is necessarily uniformly continuous in G if it is separately uniformly continuous (note that a function can have discontinuities if the separate continuity is not uniform). Hence $T_1 g$ is uniformly continuous in G.

This implies - as we shall now show - that the function $l(\mathbf{z}) = (T_1 g)(\mathbf{z})$ has a finite limit in every point $\mathbf{z}^0$ belonging to the boundary $\partial\mathbf{G}$ of the bicylinder $\mathbf{G}$. Indeed, choose an arbitrary (small) number $\varepsilon > 0$. Consider in $\mathbf{G}$ an arbitrary sequence of points $\mathbf{z}^{(n)}$ converging to $\mathbf{z}^0$. Since it is a Cauchy sequence and $l(\mathbf{z})$ is uniformly continuous in $\mathbf{G}$, the inequality

$$| \, l(\mathbf{z}^{(m)}) - l(\mathbf{z}^{(n)}) \, | < \varepsilon$$

is true for all sufficiently large indices m and n. This implies - according to the Cauchy criterion - the convergence of $\mathbf{z}^{(n)}$ to some number - say, B. It is clear, that the limit does not depend on the choice of the sequence $l(\mathbf{z}^{(n)})$ converging to $\mathbf{z}^0$.

Remark 2. Under the conditions of Lemma 4 the Pompeiu integral $T_1 g$ coincides in the bicylinder $\mathbf{G}$ with a function continuous in the closure $[\mathbf{G}]$ of this bicylinder.

Lemma 5 . *If the function $g = g(z_1, z_2)$ is continuous in the bicylinder $\mathbf{G}$ and is L_p-bounded in $\mathbf{G}$ for some $p > 2$ then the function $T_1 T_2 g$ is continuous in the closure of $\mathbf{G}$.*

Proof. First we shall prove the uniform continuity of $T_1 T_2 g$ with respect to z_1. Choose the number q satisfying the equality $1/p + 1/q = 1$. Using Hölder's inequality, one gets

$$|(T_1 T_2 g)(z_1', z_2) - (T_1 T_2 g)(z_1'', z_2)| \leq$$

$$|z_1' - z_1''| \cdot \|g\|_{L_p(G_1 \times G_2)} \cdot c \left(\iint\limits_{G_1} \frac{d\xi_1 d\eta_1}{c|\zeta_1 - z_1'|^q |\zeta_1 - z_1''|^q} \right)^{\frac{1}{q}} \cdot \left(\iint\limits_{G_2} \frac{d\xi_2 d\eta_2}{|\zeta_2 - z_2|^q} \right)^{\frac{1}{q}} .$$

It is known that the second integral in this expression is bounded by some constant (see [12], Chapter 1, Section 6, formula (6.7)). Applying to the first integral in the same expression the Hadamard inequality and reasoning as in the proof of Lemma 5 we conclude that $T_1 T_2 g$ is uniformly continuous in $\mathbf{G}$ with respect to z_1.

Similar considerations lead to the conclusion that the function $T_1 T_2 g$ is uniformly continuous in $\mathbf{G}$ with respect to the variable z_2. Reasoning as in the proof of Lemma 4 we conclude that $T_1 T_2 g$ has a finite limit in each boundary point of $\mathbf{G}$ and can be extended to a function continuous in the closure $[\mathbf{G}]$).

Lemma 6 . *Suppose that*
1) the function $f(\mathbf{z}) = f(z_1, z_2)$ is defined and twice continuously differentiable in the bicylinder $\mathbf{G}$;
2) there exists such number $p > 2$ that the derivatives
$g_k(z_1, z_2) = \partial f / \partial \bar{z}_k (k = 1, 2)$ and $g_{1,2}(z_1, z_2) = \partial^2 f / \partial \bar{z}_1 \partial \bar{z}_2$
are L_p-bounded functions in $\mathbf{G}$;
3) the function $g_1(z_1, z_2)$ (resp. $g_2(z_1, z_2)$) is L_p-uniformly continuous in $\mathbf{G}$ with regard to z_2 (resp. z_1).
Then there exists such function $\Phi(\mathbf{z})$ holomorphic in $\mathbf{G}$ and such function $\Psi(\mathbf{z})$ continuous in $[\mathbf{G}]$ that in $\mathbf{G}$ the following identity is true:

$$f(\mathbf{z}) = \Phi(\mathbf{z}) + \Psi(\mathbf{z}).$$

Proof. The function

$$\Phi = f - T_1 g_1 - T_2 g_2 + T_1 T_2 g_{1,2} \tag{12.5}$$

turns out to be *holomorphic*.

In order to prove it consider the weak (or generalized) areolar derivative - namely in the sense of Sobolev-Vekua (see [12], Chapter 1, Section 5, formula (5.7)); here we shall use for it the notation $\tilde{\partial}/\partial\overline{\mathbf{z}}$. Taking into account the identity

$$\tilde{\partial}/\partial\overline{\mathbf{z}}(Tg) = g,$$

one can detect that for k=1,2

$$\tilde{\partial}/\partial\overline{z}_k \Phi = 0.$$

According to the Weyl lemma (see [12], Chapter 1, Section 5, theorem 1.15) the function Φ is separately holomorphic with regard to each of the variables $z_k (k = 1,2)$. By the Hartogs theorem Φ is a holomorphic function with regard to the pair of variables z_1, z_2 in $\mathbf{G}$. Consequently, the function f under consideration can be presented in the form (12.5) where

$$\Psi(\mathbf{z}) = T_1 g_1 + T_2 g_2 - T_1 T_2 g_{1,2}.$$

By Lemmas 4 - 6 the function Ψ is continuous in $[\mathbf{G}]$.

Proof of Theorem 2. According to Lemma 6 the function f is representable in the form $f = \Phi + \Psi$ where Φ is holomorphic in $\mathbf{G}$ and Ψ is continuous in the closure $[\mathbf{G}]$. By Zygmund's theorem the function Φ has an angular limit in almost each boundary poimt $\mathbf{z}^0$; since Ψ is continuous in this point (see Lemmas 4-6) Ψ has a finite limit in $\mathbf{z}^0$. Hence f has a finite limit in this point.

Lemma 7 . *Suppose that*

$$1) \quad (A_{i,k_i}, \quad 1 \leq k_i \leq n_i), \quad (i = 1,2)$$

are two cortèges (finite sequences) of complex numbers, and

$$V_{n_i}(A_{i,k_i}, 1 \leq k_i \leq n_i), \quad (i = 1,2)$$

(in short: V_{n_i}) are their corresponding Vandermondians (Vandermonde determinants);

$$2) \quad P(\mathbf{Z}) = \sum_{\mathbf{k}=\mathbf{0}}^{\mathbf{n}-\mathbf{1}} a_{\mathbf{k}} \mathbf{Z}^{\mathbf{k}}$$

is a polynomial in two complex variables, $\mathbf{Z} = (Z_1, Z_2)$, of bi-degree $\mathbf{n} - \mathbf{1} = (n_1 - 1, n_2 - 1)$, with complex coefficients $a_{\mathbf{k}}$,

$$\mathbf{k} = (k_1, k_2), \quad \mathbf{0} \leq \mathbf{k} \leq \mathbf{n} - \mathbf{1};$$

3) the polynomial $P(\mathbf{Z})$ turns into zero in every point

$$\mathbf{A_k} = (A_{1,k_1}, A_{2,k_2}), \quad \mathbf{1} \leq \mathbf{k} \leq \mathbf{n}.$$

Then the system of [n] *linear (with regard to the coefficients* $a_{\mathbf{k}}$ *of the polynomial* P*) equations*

$$P(\mathbf{A_k}) = 0, \qquad \mathbf{1 \leq k \leq n},$$

has the determinant

$$W(\mathbf{A_k}, \qquad \mathbf{1 \leq k \leq n})$$

(in short: $W_{\mathbf{n}}$*) which satisfies the equality*

$$W_{\mathbf{n}} = C \cdot (V_{n_1})^{n_2} \cdot (V_{n_2})^{n_1},$$

where C *is a constant (independent of the* A_k *and* B_k, $C \neq 0$).

(In reality $C = \pm 1$ - but this doesn't matter for the proof of Theorem 3).

Proof. The proof of Lemma 7 is based on the Laplace theorem for determinants: Suppose that
1) D is a determinant of some order m;
2) $\{M_i\}$ is the set of all its minors of order k which can be composed from the columns belonging to the first k lines of D, ($1 \leq i \leq p = C_m^k$) (in every minor the columns are arranged in the order of growth of their numbers in D);
3)the $\{N_i\}$ are the corresponding cofactors (algebraic complements) to the minors $\{M_i\}$.
Then $D = \sum_{i=1}^{p} M_i \cdot N_i$. (see e.g. [9] , Chapter 2).

In order to avoid cumbersome calculations we shall confine us only to clarifying the idea of the proof of Lemma 7 in some concrete specific case: $n_1 = 3, n_2 = 2$. In this case we can present W (up to a factor ± 1) in the form of such determinant:

$$\begin{vmatrix} 1 & A_1 & A_1^2 & B_1 & A_1B_1 & A_1^2B_1 \\ 1 & A_2 & A_2^2 & B_1 & A_2B_1 & A_2^2B_1 \\ 1 & A_3 & A_3^2 & B_1 & A_3B_1 & A_3^2B_1 \\ 1 & A_1 & A_1^2 & B_2 & A_1B_2 & A_1^2B_2 \\ 1 & A_2 & A_2^2 & B_2 & A_2B_2 & A_2^2B_2 \\ 1 & A_3 & A_3^2 & B_2 & A_3B_2 & A_3^2B_2 \end{vmatrix}$$

Decomposing this determinant over the minors of order 3 belonging to the first three lines we detect that all these minors are divisible by the determinant (denote it V_1)

$$\begin{vmatrix} 1 & A_1 & A_1^2 \\ 1 & A_2 & A_2^2 \\ 1 & A_3 & A_3^2 \end{vmatrix}$$

As to their algebraic complements we can also note that they all are divisible by V_1. Hence the polynomial $W = W(A_1, A_2, A_3, B_1, B_2)$ is divisible by V_1^2. Quite similarly one can notice that W is divisible by V_2^3 where V_2 is the determinant (denote it V_2)

$$\begin{vmatrix} 1 & B_1 \\ 1 & B_2 \end{vmatrix}$$

Hence W is divisible by $V_1^2V_2^3$. But W and $V_1^2 \times V_2^3$ are polynomials (in A_1, A_2, A_3, B_1, B_2) of the same degree. Therefore $W = C \cdot V_1^2V_2^3$, where $C = const \neq 0$. (Putting e.g.$A_1 = 0, A_2 = 1, A_3, B_1 = 0, B_2 = 1$, one can easily find the constant C and show that $C = \pm 1$).

Proof of Theorem 3. Let

$$f(\mathbf{z}) = \sum_{\mathbf{k}=0}^{\mathbf{n}-1} \mathbf{h_k}(\mathbf{z})\overline{\mathbf{z}}^{\mathbf{k}}.$$

Let the bi-arc $l_{p,q}$ have the b.a. equation

$$\overline{\mathbf{z}} = \mathbf{A}(\mathbf{z}) = (A_{1,p}(z_1), A_{2,q}(z_2)),$$

Consider the holomorphic function h coherent with f on $l_{p,q}$:

$$h(\mathbf{z}) = \sum_{\mathbf{k}=0}^{\mathbf{n}-1} h_{\mathbf{k}}(\mathbf{z})[(\mathbf{A}(\mathbf{z}))^{\mathbf{k}}].$$

Since f has the angular limit zero in every point of the arc $l_{p,q}$, the same is true for h; according to a Luzin type boundary uniqueness theorem found by Bergmann and Marcinkiewicz [8] for functions holomorphic in a bicylinder) this implies that h vanishes identically in $\mathbf{G}$. Hence

$$\sum_{\mathbf{k}=0}^{\mathbf{n}-1} h_{\mathbf{k}}(\mathbf{z})[(\mathbf{A}(\mathbf{z}))^{\mathbf{k}}] \equiv 0 \tag{12.6}$$

(here $\mathbf{A}(\mathbf{z}) = (A_{1,p}(z_1), A_{2,q}(z_2))$. This identity is true for every pair of indices (p, q) , $1 \le p \le n_1, 1 \le q \le n_2$. For the system of $[\mathbf{n}](i.e. n_1 \cdot n_2)$ equations (linear with regard to the functions $h_{\mathbf{k}}(\mathbf{z})$) we can find its determinant $W_{\mathbf{n}}$. According to Lemma 7

$$W_{\mathbf{n}} = \psi_1(z_1) \cdot \psi_2(z_2),$$

where $\psi_i(z_i)(i = 1, 2)$ are holomorphic functions (each of only one complex variable). Choose in $G_i(i = 1, 2)$ a subregion G_i^0 compactly belonging to G_i. Since for every fixed $i(i = 1, 2)$ the analytic functions $A_{i,j_i}(1 \le j_i \le n_i)$ are pairwise distinct, the function $\psi_i(i = 1, 2)$ has in G_i^0 only a finite number of zeros. Therefore there exists in G_i^0 such subregion G_i^{00} in which ψ_i has no zeros at all. This means that in some subregion $\mathbf{G}^{00}$ of $\mathbf{G}$ the function $W_{\mathbf{n}}$ has no zero points. Therefore (see the system of equations (12.6)) in every point of $\mathbf{G}^{00}$ all functions $h_{\mathbf{k}}$ $(\mathbf{0} \le \mathbf{k} \le \mathbf{n} - \mathbf{1})$ are identically equal to zero. This means that all $h_{\mathbf{k}} \equiv 0$ in $\mathbf{G}$ and hence $f(\mathbf{z}) \equiv 0$. Theorem 3 is proved.

12.4 GENERALIZATIONS

The crucial role in our considerations was played by the two representations of non-analytic functions exposed in Lemmas 1 and 6. Similar representations can be given for non-holomorphic functions in the n-dimensional complex space where n is an arbitrary positive integer. Fix our attention to the possibility of generalization of Lemma 6. Consider the polycylinder

$$\Omega = \Omega_1 \times \cdots \times \Omega_n$$

where $\Omega_j \quad (j = 1, ..., n)$ is a bounded region in the z_j-plane. Suppose $w = f(\mathbf{z}) = f(z_1, ..., z_n)$ is defined and n times continuously differentiable in Ω. Denote the derivative

$$\frac{\partial^k f}{\partial \overline{z}_{i_1} \cdots \partial \overline{z}_{i_k}}$$

with respect to the variables $\overline{z}_{i_1}, ..., \overline{z}_{i_k}$ each different from each other) as

$$h_{i_1 \cdots i_k}.$$

Denote, further, the Pompeiu T-operator in Ω_j as T_j. Then the function

$$\Phi = f - \sum_{\lambda=1}^{n} (-1)^{\lambda+1} \sum\nolimits^{*} T_{j_1} \cdots T_{j_\lambda} h_{j_1 \cdots j_\lambda} \tag{12.7}$$

turns out to be holomorphic (here Σ^* means summation over all combinations of indices $j_1, ..., j_\lambda$ which are different from each other (see [10]).Consequently, the function f under consideration can be presented in the form

$$f = \Phi + \sum_{\lambda=1}^{n} (-1)^{\lambda+1} \sum\nolimits^{*} T_{j_1} \cdots T_{j_\lambda} h_{j_1 \cdots j_\lambda}, \tag{12.8}$$

and boundary properties of f are determined by the boundary properties of the holomorphic function Φ and of the further terms on the right-hand side of (12.8). The notion of L_p-uniform continuity (now - with regard to a group of variables) can obviously be extended to the case of functions in n-dimensional polycylinders. This permits to find conditions of representability of a function f non-holomorphic in some n-dimensional polycylinder Ω as a sum of a function holomorphic in Ω and a function continuous in the closure $[\Omega]$. And this is just a possible approach in the search for theorems of the Fatou-Plessner type for non-analytic (specifically - polyanalytic) functions. Consider two examples (for the sake of simplicity we restrict us to the case $n = 2$).

Example 1. It follows from one theorem due to L.A. Ayzenberg (see [2]) that a function holomorphic and bounded in some bi-disk **G** has in almost each point of its full (**3-dimensional**) boundary a finite limit along arbitrary paths not tangential to this boundary. Consequently *each non-analytic function f bounded in the bi-disk* **G** *and satisfying there the conditions 1-3 of Theorem 2 (given above) has a finite limit along non-tangential paths in almost each point of its full (3-dimensional) boundary.*

Example 2. Suppose that: Ω is a Jordan bicylinder in C^2; M is a complexly one-dimensional manifold in C^2 given by a pair of analytic functions $z_k = g_k(t)(k = 1, 2)$; $\Gamma = M \cap \Omega$ is a rectifiable curve; $f(z_1, z_2)$ is a non-analytic (in particular - polyanalytic) function in Ω; $\Phi(t) = f(g_1(t), g_2(t))$; the curve Γ corresponds to some curve Γ_t in the plane of the complex variable t. Therefore we can speak about points on Γ_t which are Fatou points or Plessner points for the function $\Phi(t)$; the corresponding points on Γ will be called the Fatou points and (resp.) Plessner points for the function $f(z_1, z_2)$ with regard to the manifold M. Using the considerations described above one can prove the following statement:

Suppose that: the function $f(z_1, z_2)$ is defined and continuous in some bicylinder Ω; the first areolar derivatives $\partial f/\partial\overline{z_k}(k = 1, 2)$ are L_p-uniformly continuous (resp. with regard to the "complementary variable" z_{3-k}) in Ω; the mixed second areolar derivative $\partial^2 f/\partial\overline{z_1}\partial\overline{z_2}$ is L_p-bounded in Ω ;M is a complexly one-dimensional analytic manifold in C^2. Then almost every point of the curve $\Gamma = M \cap \Omega$ is either a Fatou point or a Plessner point for the function f with regard to the manifold M.

References

[1] V. Avanissian and A. Traore, Sur le fonctions polyanalytic de plusieurs variables. C.R. Acad. Sci. Paris, Ser. AB, vol. 286, N 17, 1978, pp. 7743-7746.

[2] L.A. Ayzenberg , An extention of P. Fatou's theorem. "Uchenye Zapiski", Moscow Reg. Ped. Inst., vol. 57, pp. 11-18. 1957.

[3] M. B. Balk , Polyanalytic Functions. Berlin, Akademie-Verlag, 1991.

[4] M. B. Balk, On a uniqueness theorem for polyanalytic functions of two complex variables. In "Smolenskii Matematicheskii Sbornik", vol.2, pp. 8-12, Smolensk, 1969 (in Russian).

[5] M.B. Balk and W. Tutschke, Boundary theorems of the Gehring-Lohwater and Plessner types for polyanalytic functions. "Zeitschrift für Analysis und ihre Anwendungen", 1992, vol. 11, No. 4, pp. 553-558.

[6] M.B. Balk and V. P. Vassilenkov, On boundary properties of polyanalytic functions in Jordan regions. "Izv. Vysssh. Uch. Zav., Matematika", 1992, no. 8, pp. 3-12, Kazan (in Russian).

[7] M.B. Balk and M.F. Zuev, On polyanalytic functions. "Uspekhi Mat. Nauk", 1970, vol.25, no. 5, pp. 203-226.

[8] S. Bergmann and J. Marcinkiewicz, Sur les fonctions analytiques de deux variables complexes. Fund. Math., 33, pp. 75-94, 1945.

[9] A.K. Sushkevich, Higher Algebra. Moscow 1937.

[10] W. Tutschke, Partielle komplexe Differentialgleichungen in einer und in mehreren komplexen Variablen. Berlin 1977.

[11] W. Tutschke, Partielle Differentialgleichungen. Klassische, funktionalanalytische und komplexe Methoden. Teubner-Texte zur Mathematik, Bd. 27., Leipzig 1983.

[12] I. N. Vekua, Generalized analytic functions. Oxford 1962.

[13] A. Zygmund , On the boundary values of functions of several complex variables, I., Fund. Math., vol. 36 ,1949, pp. 207-235.

[14] A. Zygmund , Trigonometric Series, vol.2. Cambridge 1988.

13 GENERALIZED ANALYTIC FUNCTIONS IN DIFFERENTIAL GEOMETRY

Z. D. Usmanov†

Dedicated to Professor Robert P. Gilbert on the occasion of his 65th birthday

Let S be a regular surface, and let

$$\mathbf{r} = \mathbf{r}(x, y)$$

be any smooth parametrization of S. We suppose that S is defined in a domain D so that $(x, y) \in D$ and $\mathbf{r}(x, y) \in C^p(D)$, $p \geq 3$.

13.1 TWO QUADRATIC DIFFERENTIAL FORMS OF S.

The first fundamental form is given by the formula

$$I = ds^2 = (d\mathbf{r}d\mathbf{r}) = Edx^2 + 2Fdxdy + Gdy^2,$$

where $E = \mathbf{r}_x^2$, $F = (\mathbf{r}_x\mathbf{r}_y)$, $G = \mathbf{r}_y^2$. This form defines the line element (or the matrix) of the surface S. All geometric concepts, which can be computed through the coefficients E, F, G, belong to the *intrinsic geometry* of S. The concepts are the length of a curve situated on S, the angle between two intersecting curves, the area of figures on S, the geodesic curvature of a curve and geodesic lines, the Christoffel symbols of the first and the second kind.

Note that by definition ds^2 is positive and $E, G > 0$, $a = EG - F^2 > 0$.

Let n be the unit vector of the normal to the surface S. Obviously

†Partially supported by a grant from the NATO Science Committee, no.OUTR.CRG 960930

H. Florian et al. (eds.), Generalized Analytic Functions, 127–138.

$$\mathbf{n} = \frac{\mathbf{r}_x \times \mathbf{r}_y}{\sqrt{a}}.$$

The following scalar product

$$II = -(d\mathbf{r} d\mathbf{r}) = Ldx^2 + 2Mdxdy + Ndy^2$$

is called the *second quadratic differential form* of S and L, M, N are called its coefficients.

The second form jointly with the first one defines concepts belonging to the extrinsic geometry of S. We mention only some of them: the normal curvature of a curve on S, Euler's difference, the mean curvature and certainly the principal (gaussian) curvature of the surface expressed by the formula

$$K = \frac{LN - M^2}{EG - F^2} \quad (aK = LN - M^2). \tag{13.1}$$

Depending on the sign of K, we realize the classification of points on the surface S. Namely, a point is of elliptic, parabolic (in particular, a flat point) or hyperbolic type if $K > 0$, $K = 0$ or $K < 0$, respectively.

13.2 DEFORMATIONS OF S.

Let t be a numerical parameter and

$$\mathbf{r}_t = \mathbf{r}(x, y, t)$$

be the radius–vector of a one–parameter family of surfaces S_t, $t \in [0, 1]$. We suppose that $\mathbf{r}_t$ is continuous with respect to t and continuously differentiable with respect to $(x, y) \in D$. Moreover, $\mathbf{r}_0 \equiv \mathbf{r}$. The last means that the family of S_t contains the given surface S. In that case a surface S_t (t is fixed in [0,1]) is said to be a continuous *deformation* of S. Now in mathematical papers a great number of different deformations are analysed. Here we restrict ourselves to discussing classical deformations that are known as *bendings* (finite, continuous, precise or infinitesimal) of the surface S.

13.2.1

Let us consider a deformation of the surface S satisfying the condition

$$ds_t^2 = ds^2.$$

This means the surfaces S and S_t have equal first fundamental forms. As was said earlier, in this case S and S_t have the same intrinsic geometry. The non–trivial example of the discussed deformation is a cylinder obtained from a piece of a plane, and vice versa.

The considered deformation is called *the bending* of the surface S, and the surfaces S and S_t are then said to be *isometric*.

13.2.2

Consider the situation when

$$E_t = E,\ F_t = F,\ G_t = G$$
$$L_t = L,\ M_t = M,\ N_t = N$$

i.e. we examine a deformation of the surface S that preserves the coefficients of its first and second quadratic forms. What can we say about the surface S_t? The answer to this question is extracted from the Bonnet theorem, [1]: $S_t = S$, i.e. S_t coincides with S in the sense that, although they can be situated in different positions in space, the surface S_t is obtained from S by motion of the last one as a rigid body.

13.2.3

Let two sets of functions $E(x,y)$, $F(x,y)$, $G(x,y)$, and $L(x,y)$, $M(x,y)$, $N(x,y)$, be given in a some domain D, $(x,y) \in D$. We solve the problem of searching a surface (defined in D) that has the first and second forms with the above coefficients. It also follows from the mentioned Bonnet theorem that for this aim it is necessary and sufficient that

- the first form must be positive ($E, G > 0,\ EG - G^2 > 0$);
- E, F, G and L, M, N must jointly satisfy the Gauss equation presented by formula (13.1) and the following system of the equations discovered by Peterson and Codazzi:

$$\frac{\partial L}{\partial y} - \frac{\partial M}{\partial x} - \Gamma^1_{12} L + (\Gamma^1_{11} - \Gamma^2_{12}) M + \Gamma^2_{11} N = 0, \quad (13.2)$$
$$(13.3)$$
$$\frac{\partial N}{\partial x} - \frac{\partial M}{\partial y} + \Gamma^1_{22} L + (\Gamma^2_{22} - \Gamma^1_{12}) M - \Gamma^2_{12} N = 0, \quad (13.4)$$

where $\Gamma^{\alpha}_{\beta,\gamma}$ are the Christoffel symbols of the second kind, [2].

The given information about formula (13.1) needs some more explanation. In the present situation we should write it in the form

$$LN - M^2 = f.$$

The function f is sufficiently complicated, but we do not need its explicit expression. For our aim it is necessary to know that f depends only on the coefficients E, F, G of the first fundamental form and their derivatives of first and second orders. This is the essence of the famous Gauss theorem which establishes that the principal curvature K is an object of the intrinsic geometry.

13.3 CONJUGATE ISOMETRIC PARAMETRIZATION ON S.

It is known that some geometric problems are simplified under utilization of a conveniently selected coordinate system. In this paper we shall deal only with surfaces of positive gaussian curvature and for them shall widely use *the so called conjugate isometric coordinates*, [2]. We shall denote them by (u, v).

By definition, in those coordinates the second quadratic form takes the canonical expression:

$$II = \Lambda(u,v)(du^2 + dv^2), \quad \Lambda \neq 0.$$

Let $z = x + iy$ and $w = u + iv$, $i^2 = 1$. The existence proof of conjugate isometric coordinates on a surface is equivalent to finding so called *homeomorphic solutions* $w = w(z)$ of the equation

$$\partial_{\bar{z}} w - q(z)\partial_z w = 0, \tag{13.5}$$

where $2\partial_{\bar{z}} = \partial_x + i\partial_y$, $2\partial_z = \partial_x - i\partial_y$, and

$$q(z) = \frac{L - N + 2iM}{L + N + 2\sqrt{LN - M^2}}.$$

The above solutions establish a one–to–one mutually continuous correspondence of the G–domain of the z–plane onto a Δ–domain of the w–plane.

It is important to note that $|q(z)| < 1$ for $K \geq k_0 > 0$.

It is clear that we can also transform the first fundamental form to the canonical expression. With respect to the first form the coordinate system is called *isometric*.

13.4 EQUATIONS FOR BENDINGS OF A SURFACE

It follows from section 13.2 that two quadratic differential forms (one of them being positively definite) satisfying the Gauss and Peterson–Codazzi equations, determine precisely, up to motion of the space, the surface, for which the above forms are basic. Therefore, the problem of searching the surfaces, that are bendings of a given surface, can be reduced to investigating solutions of the Gauss and Peterson–Codazzi equations where the coefficients of the second quadratic form are unknown functions and the coefficients of the first fundamental form are considered to be given.

As before, we use the notation L, M, N and L_t, M_t, N_t for the coefficients of the second quadratic form of a given surface S and for an unknown surface S_t, obtained from S by a bending. But in future instead of the unknown L_t, M_t, N_t, it is convenient to examine the differences

$$\Delta L = L_t - L,\ \Delta M = M_t - M,\ \Delta N = N_t - N$$

as the new unknown functions. In this case the condition $\Delta L = \Delta M = \Delta N = 0$ will characterizes only the trivial bending of S.

Introducing the new unknown functions w (complex–valued) and Π (real) by the formula

$$w = \Delta M + \frac{i}{2}(\Delta L - \Delta N),$$
$$2\Pi = \Delta L + \Delta N,$$

and then adapting equations (13.4) for the functions ΔL, ΔM, ΔN, we get

$$\partial_{\bar{z}}w + Aw + B\overline{w} - i\partial_z\Pi + \frac{1}{2}[(\Gamma_{12}^1 - \Gamma_{11}^2) - i(\Gamma_{22}^1 - \Gamma_{12}^2)]\Pi = 0,$$

where

$$A = \partial_{\bar{z}}\ln K^{1/4} = -\frac{1}{4}[(\Gamma_{11}^1 + \Gamma_{22}^1) + i(\Gamma_{11}^2 + \Gamma_{22}^2)],$$

$$B = -\frac{1}{4}[(\Gamma_{11}^1 - 2\Gamma_{12}^2 - \Gamma_{22}^1) + i(\Gamma_{22}^2 - 2\Gamma_{12}^1 - \Gamma_{11}^2)].$$

We shall not write the Gauss equation out, since for surfaces with an arbitrary sign of the principal curvature it has a complicated form. As to a surface of positive curvature we can consider its given interior coordinates (x, y) as conjugate isometric and in accordance with section 3 we can suppose that $L(x, y) = N(x, y)$ and $M(x, y) = 0$. In this case from (13.4) we get

$$\frac{\partial L}{\partial y} = (\Gamma_{12}^1 - \Gamma_{11}^2)L, \quad \frac{\partial L}{\partial x} = (\Gamma_{12}^2 - \Gamma_{22}^1)L$$

and then

$$(\Gamma_{12}^1 - \Gamma_{11}^2)\Pi - \Pi_y = -L(\Pi/L)_y,$$
$$(\Gamma_{22}^1 - \Gamma_{12}^2)\Pi + \Pi_x = L(\Pi/L)_x.$$

Using these formulas in (13.4) we derive

$$\partial_{\bar{z}}w + Aw + B\overline{w} = iL\partial_z(\Pi/L) \tag{13.6}$$

It remains to attach the Gauss equation

$$\Pi^2 + 2L\Pi - |w|^2 = 0 \tag{13.7}$$

which is obtained by combining the formulas

$$LN - M^2 = aK,$$

$$(L + \Delta L)(N + \Delta N) - (M + \Delta M)^2 = aK,$$

expressing the gaussian curvature for the given surface and its bending.

Complex equation (13.6) and real equation (13.7) compose a full system of equations for the description of finite bendings of the given surface by the functions $w(z)$ and $\Pi(z)$.

This approach was utilized in [3] for investigating continuous bendings of surfaces with positive curvature and certain boundary conditions along the surface edge.

13.5 CONTINUATION

In the previous section finite bendings of surfaces were characterized by three real equations. Here we display that the considerable problem can be described by an unknown complex-valued function satisfying a complex equation.

For this purpose we introduce an isometric parametrization (x, y) on the surface S. In that case the first fundamental form of S takes the canonical form:

$$I = ds^2 = \Lambda(x, y)(dx^2 + dy^2).$$

Then we compute

$$\Gamma^1_{12} = \Gamma^2_{22} = \frac{1}{2}\partial_y \ln \Lambda,$$
$$\Gamma^2_{12} = \Gamma^1_{11} = \frac{1}{2}\partial_x \ln \Lambda,$$
$$\Gamma^2_{11} = -\frac{1}{2}\partial_y \ln \Lambda,$$
$$\Gamma^1_{22} = -\frac{1}{2}\partial_x \ln \Lambda.$$

Substituting these expressions for Christoffel's symbols in equations (13.1) and (13.4) adapted for the unknown surface we obtain

$$\Lambda^2 K = L_t N_t - {M_t}^2,$$
$$\frac{\partial L_t}{\partial y} - \frac{\partial M_t}{\partial x} = \frac{1}{2}(L_t + N_t)\partial_y \ln \Lambda,$$
$$\frac{\partial M_t}{\partial y} - \frac{\partial N_t}{\partial x} = -\frac{1}{2}(L_t + N_t)\partial_x \ln \Lambda.$$

Obviously, this system has the trivial solution: $L_t = L$, $M_t = M$, $N_t = N$.

It remains to introduce the new complex-valued function $w(z)$, $z = x + iy$, by the formula

$$\Lambda w = \frac{N - L}{2} + iM,$$

and then the Peterson–Codazzi equations take the following form:

$$\partial_{\overline{z}} w + \partial_z \sqrt{K + |w|^2} + w\partial_{\overline{z}} \ln \Lambda = 0.$$

This equation can be presented in equivalent forms, for example:

$$\partial_{\overline{z}} w + q_1(z, w)\partial_z w + q_2(z, w)\partial_z \overline{w} + w\partial_{\overline{z}} \ln \Lambda + l(z, w) = 0,$$

where

$$q_2 = \overline{q}_1 = \frac{w}{2\sqrt{K + |w|^2}},$$
$$l = (\partial_z K)/2\sqrt{K + |w|^2}.$$

Now we take the sign of gaussian curvature K into account. If $K > 0$, then $|q_1| + |q_2| < 1$, and some results from the theory of elliptic systems can be used for investigating the obtained equation. Namely, in this way some theorems on monotypy and rigidity for regular convex surfaces with an edge were obtained in [4, 5].

If the function $w(z)$ is defined then attracting the Gauss equation, all the coefficients L_t, M_t, N_t can be calculated. Consequently, a surface S_t, being a finite bending of S, will be determined precisely, up to motions in space, through its coefficients of the two quadratic forms.

We point out that this modern state of the theory of surface deformations was recently observed in [6].

13.6 INFINITESIMAL BENDINGS OF S

For this type of bendings a one–parameter family of surfaces S_t is described by the equation

$$S_t : \mathbf{r}_t = \mathbf{r}(x, y) + t\tau(x, y),$$

where t is a sufficiently small parameter and τ is a smooth vector field. By definition, the surfaces S_t are said to be *infinitesimal bendings* of the surface S if the difference between the squares of their linear elements is a quantity of order t^2, i.e.

$$ds_t^2 - ds^2 = O(t^2).$$

Since

$$ds_t^2 = d\mathbf{r}^2 + 2td\mathbf{r}d\tau + t^2 d\tau d\tau,$$

and $d\mathbf{r}^2 = ds^2$, we can subordinate the vector τ to the condition

$$d\mathbf{r}d\tau = 0,$$

which is called the equation of infinitesimal bendings of S. But further, we do not apply this equation and shall in the following use the Gauss and Peterson–Codazzi equations.

13.7 EQUATIONS FOR INFINITESIMAL BENDINGS OF REGULAR CONVEX SURFACES

Following our tradition in studying deformations of surfaces by the Gauss and Peterson–Codazzi equations, we again return to them. Under infinitesimal deformations a deformed surface S_t will have $L_t = L + t\delta L$, $M_t = M + t\delta M$, $N_t = N + t\delta N$ as coefficients of its second quadratic form. As was told earlier, it is convenient to examine variations $\delta L, \delta M, \delta N$ as unknown functions. The corresponding equations for them are obtained from the above equations, see also [2]:

$$L\delta N + N\delta L - 2M\delta M = 0,$$

$$\frac{\partial}{\partial x}(\delta M) - \frac{\partial}{\partial y}(\delta L) + \Gamma^1_{12}\delta L - (\Gamma^1_{11} - \Gamma^2_{12})\delta M - \Gamma^2_{11}\delta N = 0,$$

$$\frac{\partial}{\partial y}(\delta M) - \frac{\partial}{\partial x}(\delta N) - \Gamma^1_{22}\delta L - (\Gamma^2_{22} - \Gamma^1_{12})\delta M - \Gamma^2_{12}\delta N = 0.$$

If we suppose that the surface S has positive curvature and the coordinates (x, y) on S are conjugate isometric ($L = N$, $M = 0$) then the above system takes the following form:

$$\frac{\partial}{\partial x}(\delta M) - \frac{\partial}{\partial y}(\delta L) - (\Gamma^1_{11} - \Gamma^2_{12})\delta M + (\Gamma^1_{12} + \Gamma^2_{11})\delta L = 0,$$

$$\frac{\partial}{\partial y}(\delta M) + \frac{\partial}{\partial x}(\delta L) - (\Gamma^2_{22} - \Gamma^1_{12})\delta M - (\Gamma^1_{22} + \Gamma^2_{12})\delta L = 0.$$

Using the notation $z = x + iy$, $i_2 = -1$, and introducing the complex function

$$U(z) = K^{1/4}(z)(\delta M + i\delta L),$$

where $K(z)$ is the gaussian curvature of the surface S, we rewrite the preceding system in the form:

$$\partial_{\bar{z}} U + \overline{B(z)U} = 0. \tag{13.8}$$

Here

$$B(z) = -e^{i\psi(z)}\partial_{\bar{z}}(Arch\frac{H}{\sqrt{K}}) - \frac{\partial_z K}{4K}e^{2i\psi(z)}, \tag{13.9}$$

and

$$e^{i\psi(z)} = \frac{a^-(z)\sqrt{K(z)}}{2\sqrt{a(z)\Upsilon(z)}}.$$

In these formulas $H(z)$ is the mean curvature of S, $\Upsilon(z)$ is the Euler difference, $a(z) = EG - F^2$ is the discriminant of the first fundamental form (E, F, G are its coefficients), $a^-(z) = E - G + 2iF$, see [2]. Note also that by the function $U(z)$ the variations of the normal curvature k_s and the geodesic torsion τ_s of some curve Γ on the surface S can be expressed by, [2]:

$$\delta k_s = Re[\gamma_1 U] \quad \delta\tau_s = Re[\gamma_2 U].$$

If Γ is the boundary curve of a simply–connected piece of S then the increment in the arguments of γ_1 and γ_2 are equal to zero under a single circuit of Γ. Composing a linear combination of δk_s and $\delta\tau_s$, we get

$$\lambda(s)\delta k_s + \mu(s)\delta\tau_s \equiv Re[(g(z)U(z)].$$

Here $\lambda(s)$, $\mu(s)$ are functions given on Γ by its arc length s and $g(z)$ is a function, determined through $\lambda(s), \mu(s), \gamma_1, \gamma_2$ and computed in the conjugate isometric coordinates. This expression shows that one admissible geometric interpretation of the Riemann–Hilbert boundary problem is the search of infinitesimal bendings of a surface of positive curvature with an in advance given linear combination of δk_s and $\delta\tau_s$ along the surface edge.

Now it is necessary to note that in the case when a surface is sufficiently regular and has positive curvature in all its points, solutions of geometric problems are not differed from the solutions of the corresponding analytic ones. This means that changing, where it is necessary, the words *solution, number of independent solutions and so on* to the words *infinitesimal bending, family of infinitesimal bendings and so on*, we get geometric interpretations for analytic results. In this connection we do not list numerous results in this field referring to the well–known book of I.Vekua,[2].

13.8 EQUATIONS FOR INFINITESIMAL BENDINGS OF SURFACES WITH A FLAT POINT,[7]

Let S be a sufficiently regular surface with positive curvature everywhere except some point O being a flat point. We introduce a cartesian system of coordinates $Oxyz$ with the origin at the flat point and axes x, y situated in the tangent plane of S. Then the surface for a neighbourhood of O can be given by the equation

$$z_1 = \sum_{k=0}^{n} a_{k,n-k} x_1^k y_1^{n-k} + R(x_1, y_1),$$

where n ($n \geq 3$) is an integer and $a_{k,n-k}$ are constants. We suppose that $R(x_1, y_1)$ is a sufficiently regular function and $R = O[(x_1^2 + y_1^2)^{(n+1)/2}]$ as $x_1, y_1 \mapsto 0$. In polar coordinates ($x_1 = r_1 \cos\varphi$, $y_1 = r_1 \sin\varphi$) the surface equation has the form

$$z_1 = r_1^n f(\varphi) + R(x_1, y_1), \tag{13.10}$$

where $f(\varphi) = \sum_{k=0}^{n} a_{k,n-k} (\cos\varphi)^k (\sin\varphi)^{n-k}$. The requirement of a positive curvature for a neighbourhood of the point $(0,0)$ imposes on $f(\varphi)$ the restriction

$$-(n-1)\left(\frac{df}{d\varphi}\right)^2 + nf\frac{d^2 f}{d\varphi^2} + n^2 f^2 > 0. \tag{13.11}$$

Besides we account $f(\varphi) > 0$.

The first summand in the right–hand side of (13.10) defines the structure of the surface for a neighbourhood of the flat point. The model surface

$$z_1 = r_1^n f(\varphi) \tag{13.12}$$

corresponds to it.

Further we consider surfaces (13.10) and (13.12) under wider assumptions on n and $f(\varphi)$. We assume that n ($n > 2$) is a real number, $f(\varphi)$ is a 2π–periodic function from the class $C^3[0, 2\pi]$ satisfying inequality (13.11).

To get the equations of infinitesimal bendings of surface (13.10) we construct the conjugate isometric parametrization on it using the way pointed out in section 3. We prove its existence and establish its asymptotic behaviour for a neighbourhood of the flat point. It helps us to compute all geometric concepts in formula (13.9) and to get the equation of infinitesimal bendings on the basis of (13.8):

$$\partial_{\overline{z}} w - \frac{\lambda(\varphi) + b_0(z)}{2\overline{z}} \overline{w} = 0, \quad z \in G, \tag{13.13}$$

where

$$w(z) = z^2 K^{1/4}(z)(\delta M + i\delta L), \tag{13.14}$$

and the point $z = 0$ is inside the domain G, $\lambda(\varphi)$ is a 2π–periodic continuous function, b_0 is continuous in G, moreover $b_0 = O(|z|^\alpha)$, $\alpha > 0$, $z \to 0$ (more detailed information on $\lambda(\varphi)$ is contained in [7]).

The equation

$$\partial_{\overline{z}} \Phi - \frac{\lambda(\varphi)}{2\overline{z}} \overline{\Phi} = 0, \quad z \in G, \tag{13.15}$$

is a model equation for (13.13). It characterizes infinitesimal bendings of (13.12).

For $\lambda(\varphi) = \lambda_0$ and $\lambda(\varphi) = \lambda_0 e^{im\varphi}$ (λ_0 is a complex number and m is an integer) the investigation of equation (13.13) is based on the two–dimensional integral equation

$$w(z) = \Phi(z) + P_G \overline{w} = 0, \quad z \in G, \tag{13.16}$$

which establishes the contact between continuous solutions of equations (13.13) and (13.15) and is the main apparatus for constructing the theory of equation (13.13). Uniquely solvability of equation (13.16) is proved only in the above cases. The first of them finds the realisation for the surface in the following form:

$$z_1 = r_1^n, \quad n > 2.$$

For that object the influence of the flat point on infinitesimal bendings was analysed in [7]. As was stated, the available isolated flat point strengthens the resistance of the surface with respect to infinitesimal bendings. This fact can be explained by formula (13.14), in which the function $w(z)$ must have a zero of order $(n - 2 + 2p)/2\sqrt{n-1}$ as $z \to 0$, where p is an integer and denotes the order of regularity for infinitesimal bendings of the surface. This indicates the difference between the geometric and analytic problems. Any solution of equation (13.13) (for $\lambda(\varphi) = \lambda_0$) must satisfy that additional condition. It generates additional restrictions to infinitesimal bendings, see for details in [7].

As to surfaces (13.10) there is no progress in studying its infinitesimal bendings.

13.9 EQUATIONS FOR INFINITESIMAL BENDINGS OF SURFACES WITH A CONIC POINT,[7]

This equation has the following form:

$$\partial_{\bar{z}} w - \frac{e^{i\varphi}}{2} \frac{b(z)}{r - r_0} \overline{w} = 0, \quad z \in G, \tag{13.17}$$

where $z = x + iy$ is a conjugate isometric parametrisation on the surface, $z = re^{i\varphi}$, r_0 is a positive number, G is a doubly–connected domain, one of its edges being the circle $L_0 = \{z : |z| = r_0\}$ and the other is a simple closed contour L_1, enveloping L_0. Besides, $b(z)$ is continuous in $\overline{G}$ and $b(r_0 e^{i\varphi}) = \lambda$, λ is a complex parameter. We should note that this equation was obtained only for surfaces of revolution,[8]. Those surfaces are sufficiently regular everywhere except a conic point situated in the surface pole and have positive curvature.

As for surfaces with a flat point so for surfaces with a conic one the connection between the analytic and geometric problems is stated by the formula (13.14).

It remains to mention that although the problem of infinitesimal bendings of surfaces with conic point was formulated in the 60s and a result for equation (13.17) was obtained in the 70s, no principal progress in this field of investigation is observed until now.

References

[1] V.F.KAGAN, Basis of surface theory, part 2, Moscow, GosTehIzd., 1948, pp.407 (in Russian).

[2] I.N.VEKUA, Generalized analytic functions, International series of monographs on pure and applied mathematics, v.25, Pergamon Press, Oxford-London-NewYork-Paris, 1962, pp.668.

[3] G.S.BARKHIN, V.T.FOMENKO, On bendings of positive curvature surfaces with an edge, Sibirsk. Math. Z., 1963, v.4, no.1, 32 - 47 (in Russian).

[4] K.M.BELOV, On monotypy of regular convex surfaces with an edge, Dokl. Akad. Nauk SSSR, 1959, v.127, no.2

[5] K.M.BELOV, Notations to the problem of rigidity for surfaces with positive curvature and with an edge, Dokl. Akad. Nauk SSSR, 1960, v.131, no.3.

[6] I.H.SABITOV, Local theory of surfaces bendings, Itogi nauki i tekhniki, Modern problems of mathematics, Moscow, VINITI, 1989, 196 - 270.

[7] Z.D.USMANOV, Generalized Chauchy–Riemann systems with a singular point, Addison Wesley Longman, Harlow, 1997.

[8] L.G.MIKHAILOV, Z.D.USMANOV, Infinitesimal bendings of a surface of revolution with positive curvature and with a conical or parabolic point at the pole, Doklady Akad. Nauk SSSR, tom 166, no.4, 1966, 174 - 177.

14 THE BROTHERS RIESZ THEOREM: COMPLEX AND REAL VERSIONS

A. Cialdea

Dedicated to Prof. R. P. Gilbert on his 65th birthday.

14.1 INTRODUCTION.

In their only joint paper [26] F. and M. Riesz proved this famous result:

Theorem 1 *If a trigonometric series and its conjugate series*

$$\frac{a_0}{2}+\sum_{k=1}^{\infty}(a_k\cos k\vartheta+b_k\sin k\vartheta)\,,\qquad \sum_{k=1}^{\infty}(a_k\sin k\vartheta-b_k\cos k\vartheta)$$

are both Fourier-Stjelties series, then they are ordinary Fourier series.

In other words, if we have two real measures α, β defined on the Borel sets of $[0,2\pi]$ such that

$$\int_0^{2\pi}\cos k\vartheta\,d\alpha=\int_0^{2\pi}\sin k\vartheta\,d\beta,\quad \int_0^{2\pi}\sin k\vartheta\,d\alpha=-\int_0^{2\pi}\cos k\vartheta\,d\beta \qquad (k=1,2,\dots) \qquad (14.1)$$

then these measures have to be absolutely continuous, i.e. there exist two L^1 functions f and g such that

$$\alpha(E)=\int_E f(\vartheta)\,d\vartheta,\qquad \beta(E)=\int_E g(\vartheta)\,d\vartheta$$

H. Florian et al. (eds.), Generalized Analytic Functions, 139–149.

for any Borel set $E \subset [0, 2\pi]$. The interest of this result in the theory of Fourier series is evident. On the other hand theorem 1 can be easily rewritten in a "complex" form:

Theorem 2 *If μ is a complex measure defined on the Borel sets of the unit circle $C = \{z \in \mathbf{C} \mid |z| = 1\}$ such that*

$$\int_C e^{ik\vartheta} d\mu = 0 \qquad k = 1, 2, \ldots \tag{14.2}$$

then μ is absolutely continuous, i.e. there exists a function $f \in L^1(C)$ such that

$$\mu(E) = \int_E f(\vartheta) d\vartheta$$

for any Borel set E of C.

For different proofs of this result, see [13], [14], [18], [19], [29], [32], [33], [37]. This beautiful theorem gave rise to a long series of papers and "*in its direct applications as well as the generalizations it has inspired, this has proved to be one of the more important theorems of the century*" (R. B. Burckel, *Math. Rev.*, 96k:43009). Recently also a mechanical interpretation was given in [6].

For variants of this theorem on the real line instead of the circle, see [17] and the papers quoted there. Abstract versions of it have been obtained in the theory of function algebras (see e.g. [2], [35]) and in the theory of ordered groups (see e.g. [20], [21], [28]). It should be impossible to mention all the papers written so far on this topic. In the present paper we want to make a survey of the more "concrete" generalizations, especially as far multidimensional extensions are concerned.

14.2 COMPLEX VERSIONS.

Before considering higher dimensional versions of the Brothers Riesz theorem, we have to observe that theorem 2 says also that if a measure μ is the trace on C of a function which is holomorphic in D, then μ has to be absolutely continuous. This statement is still true in more general domains. In fact let $D \subset \mathbf{C}$ be a bounded domain with a Lyapunov boundary S; let us denote by D_ρ a family of domains contained in D obtained in the following way: let $\lambda(z)$ be a unit vector defined on S and continuously differentiable such that $\lambda(z) \cdot \nu(z) \geq \rho_0 > 0$, $\forall\, z \in S$. By S_ρ we denote the hypersurface $z_\rho = z + \rho\lambda(z)$, $z \in S$, where $|\rho| \leq \rho_0$ (ρ_0 small enough). D_ρ is the bounded domain whose boundary is S_ρ. The domains D_ρ with $0 < \rho \leq \rho_0$ are contained in D, while D is contained in D_ρ with $-\rho_0 \leq \rho < 0$.

It is natural to say that the measure $\alpha \in M(S)$ is the trace on S of the function f holomorphic in D if

$$\lim_{\rho \to 0+} \int_{S_\rho} \omega f dz = \int_S \omega\, \dot{z}\, d\alpha \qquad \forall\, \omega \in C^\infty(\mathbf{C}) \, .$$

Theorem 3 *Under the previous hypothesis on D, if the function f is holomorphic in D and it admits a trace $\alpha \in M(S)$, then α has to be absolutely continuous.*

As remarked before, if D is the unit disk, this theorem is perfectly equivalent to 2. For the proof of 3 see [15].

We have also to mention that the Brothers Riesz theorem is closely related to two other important statements in the theory of holomorphic functions of one complex variable; the first one can be found in [26]:

Theorem 4 *Let $f \in L^1(C)$ be such that*

$$\int_C f(\vartheta)\, e^{ik\vartheta} d\vartheta = 0 \qquad k = 1, 2, \dots .$$

If $f = 0$ on a set of positive measure on C, then $f = 0$ a.e..

The other one was proved independently by Carleson [5] and Rudin [27].

Theorem 5 *If W is a closed set of Lebesgue zero measure on the unit circle C and if f is a continuous function on W, then there exists a continuous function F on the closure of the unit disk D which is holomorphic in the interior of the disk such that $F(z) = f(z)$, $\forall\ z \in W$.*

For these two results see also [3], [13], [25].

Let us consider now some generalization of theorem 2 in the theory of holomorphic functions of several complex variables.

One generalization is due to Cole and Range (see [11], [30]). In order to describe their result, we have to recall some definitions. Let B be the unit ball in $\mathbf{C}^n$ and S its boundary. By $M(S)$ we denote the set of complex Borel measures defined on S and by $M_0(S)$ the set of all the probability measures $\rho \in M(S)$ such that

$$f(0) = \int_S f\, d\rho$$

for any f holomorphic in B and continuous on $\overline{B}$. A measure $\mu \in M(S)$ is said to be a Henkin measure (or A-measure) if

$$\lim_{i \to \infty} \int_S f_i\, d\mu = 0$$

for every Montel sequence $\{f_i\}$, i.e. for any sequence such that i) f_i are holomorphic in B and continuous on $\overline{B}$; ii) the sequence $\{f_i\}$ is uniformly bounded on $\overline{B}$; iii) $\lim_{i\to\infty} f_i(z) = 0$ for any $z \in B$. The Cole-Range theorem is the following:

Theorem 6 *If $\mu \in M(S)$ is a Henkin measure then it is absolutely continuous with respect to some $\rho \in M_0(S)$.*

Now if μ is a measure orthogonal to holomorphic functions, i.e. if:

$$\int_S f\, d\mu = 0$$

for any f holomorphic in B and continuous up to S, it is obvious that μ is a Henkin measure. Therefore theorem 6 implies immediately that if $\mu \in M(S)$ is orthogonal to holomorphic functions, then it must be absolutely continuous with respect to some measure in $M_0(S)$. Since in the case $n = 1$ there is only the one-dimensional Lebesgue measure in $M_0(S)$, we have that the Cole-Range theorem generalizes the original Brothers Riesz theorem.

In [1] a different point of view is considered. At first let us remark that theorem 2 can be stated in a slight more precise way:

Theorem 7 *Let μ be a measure on C; (14.2) holds if and only if there exists a function f belonging to the Hardy space H^1 such that*

$$\mu(E) = \int_E f\, d\mu$$

for any Borel set $E \subset C$.

Let D now be a bounded domain with piecewise smooth boundary in $\mathbf{C}^n$. The Hardy space H^1 can be introduced also for such domains; for the details we refer to [1]. Let us denote by ω the following form

$$\omega(w, z) = \sum_{k=1}^{n} (-1)^{k-1} w_k\, dw[k] \wedge dz$$

where $w = (w_1, \dots, w_n)$, w_k being C^1 functions such that $\sum_{k=1}^{n} w_k(z) z_k = 1$ on ∂D and $[k]$ means that the term dw_k is omitted. Let $O(\partial D)$ be the space of continuous functions φ on ∂D such that

$$\int_{\partial D} f(z)\, \varphi(z)\, \omega(w, z) = 0$$

for any f holomorphic in D and continuous on $\overline{D}$.

Theorem 8 *Let $\mu \in M(\partial D)$;*

$$\int_{\partial D} \varphi\, d\mu = 0 \qquad \forall\, \varphi \in O(\partial D)$$

if and only if there exists $f \in H^1$ such that

$$\mu(E) = \int_E f(z) \omega$$

for any Borel set $E \subset \partial D$.

Since for $n = 1$ the space of traces on the unit circle of functions f holomorphic in the unit disk and continuous up to the boundary with $f(0) = 0$ is nothing but $O(\partial D)$, it is clear that theorem 8 is another generalization of the classical Brothers Riesz Theorem (for the proof, see [1], where other related results can be found).

A version of the theorem in which the measure is defined only on the Shilov boundary of a bounded circular strongly starlike domain of $\mathbf{C}^n$ is given in [36].

14.3 REAL VERSIONS.

Theorem 2 has an interesting meaning also in the theory of singular integral operators. In fact we have

Theorem 9 *Let μ be a finite measure on $\mathbf{R}$. If its Hilbert transform is also a finite measure β, i.e. if*

$$\int_{\mathbf{R}} d\mu_y \int_{\mathbf{R}} \frac{g(x)}{y-x} dx = -\int_{\mathbf{R}} g\, d\beta \qquad \forall\, g \in \mathring{C}^{\infty}(\mathbf{R})$$

then μ and β are absolutely continuous.

Let us consider the Riesz transforms in $\mathbf{R}^{n-1}$:

$$R_k(f) = \int_{\mathbf{R}^{n-1}} \frac{x_k - y_k}{|x-y|^n} f(y)\, dy \qquad (k = 1, 2, \ldots, n-1)$$

where the integrals must be understood as singular integrals. In [34] the following theorem, which clearly generalizes 9, is proved

Theorem 10 *Let μ be a finite measure on $\mathbf{R}^{n-1}$. If all its Riesz transforms are finite measures β_k, i.e. if*

$$\int_{\mathbf{R}^{n-1}} R_k(g)\, d\mu = -\int_{\mathbf{R}^{n-1}} g\, d\beta_k \qquad \forall\, g \in \mathring{C}^{\infty}(\mathbf{R}^n)$$

$(k = 1, 2, \ldots, n-1)$ then $\mu, \beta_1, \ldots, \beta_{n-1}$ are absolutely continuous.

The version 3 of the Brothers Riesz Theorem can be immediately stated also in the following "real" way:

Theorem 11 *If u and v are two harmonic conjugate functions , i.e. if they are solution of the Cauchy-Riemann system: $u_x = v_y$, $u_y = -v_x$ in a bounded domain $\Omega \subset \mathbf{R}^2$ with a Lyapunov boundary and both their traces on $\partial\Omega$ are measures, then these measures are absolutely continuous.*

This result is interesting, because it is very easy to give an example of a harmonic function whose trace on $\partial\Omega$ is a *non* absolutely continuous measure. If we look for a generalization of this theorem in $\mathbf{R}^n$ at first we have to consider generalizations of Cauchy-Riemann system. This can be done in several ways. In [8] the following definition is proposed.

Let U be a non homogeneous differential form, i.e.

$$U = \sum_{k=0}^{n} u_k$$

u_k being (real) differential forms of degree k defined in a domain $\Omega \subset \mathbf{R}^n$. We say that U is *self-conjugate* if $dU = \delta U$, i.e. if $\delta u_1 = 0$, $du_{k-1} = \delta u_{k+1}$, $(k = 1, \ldots, n-1)$, $du_{n-1} = 0$, where d is the differential and $\delta = (-1)^{n(k+1)+1} * d*$ is the codifferential, "$*$" denoting the Hodge star operator with respect to the canonical metric in $\mathbf{R}^n$. Holomorphic functions of one complex variable, solutions of Moisil-Theodorescu system, quaternionic hyperholomorphic functions, harmonic vectors (i.e. solutions of the system $\{\operatorname{div}\mathbf{v}=0,\ \operatorname{curl}\mathbf{v}=0\}$ can be identified with particular self-conjugate differential forms. Moreover several properties of holomorphic functions of one complex variable hold true for such forms (for the details see [8]). The theorem 11 generalizes in the following way:

Theorem 12 *If the self-conjugate form $U \in C_0^1(\Omega)\oplus\ldots\oplus C_n^1(\Omega)$ is such that U and $*U$ admit traces on $\partial\Omega$ in the sense of k-measures then these traces are absolutely continuous.*

For the details and the proof we refer to [7]. For the concept of k-measure, which roughly speaking is a differential form of degree k whose coefficients are measures, we refer to [16]. We can also write result 12 by means of some "orthogonality" conditions. In order to do that, let us denote by $\{\omega_h\}$ the system of homogeneous harmonic polynomials, which can be obtained by ordering in one sequence the polynomials $|x|^k Y_{ks}(\frac{x}{|x|})$ $(k=0,1,2,\ldots;\ s=1,\ldots,p_{nk};\ p_{nk}=(2k+n-2)\frac{(k+n-3)!}{(n-2)!k!})$, where $Y_{k1},\ldots,Y_{kp_{nk}}$ is a complete system of (surface) spherical harmonics of degree k. Denoting by $M_k(\partial\Omega)$ the space of the k-measures on $\partial\Omega$ and by $\omega_h^{i_1\ldots i_k}$ the k-form $\omega_h(x)\,dx^{i_1}\ldots dx^{i_k}$, we have (see [7]):

Theorem 13 *Let Ω be such that $\mathbf{R}^n-\overline{\Omega}$ is connected. If $\alpha=(\alpha^0,\ldots,\alpha^{n-1})$, $\widetilde{\alpha}=(\widetilde{\alpha}^n,\ldots,\widetilde{\alpha}^1)\in M_0(\partial\Omega)\oplus\ldots\oplus M_{n-1}(\partial\Omega)$ are such that*

$$\int_{+\partial\Omega}[\alpha^k\wedge *d\omega_h^{i_1\ldots i_k}-\delta\omega_h^{i_1\ldots i_k}\wedge\widetilde{\alpha}^k+d\omega_h^{i_1\ldots i_k}\wedge\widetilde{\alpha}^{k+2}-\alpha^{k-2}\wedge *\delta\omega_h^{i_1\ldots i_k}]=0$$

(where $\alpha^k=\widetilde{\alpha}^{n-k}=0,\ k=-2,-1,n$) for any $1\le i_1<\cdots<i_k\le n$, $h=1,2,\ldots$, $k=0,1,\ldots,n$ then $\alpha,\widetilde{\alpha}$ are absolutely continuous.

For different generalizations of the Brothers Riesz theorem in $\mathbf{R}^n$ see [10], [12].

In Section 1 we have seen that the Brothers Riesz Theorem is interesting in the theory of Fourier series. We have to mention that Muckenhoupt and Stein [23] generalized this theorem for ultraspherical expansions. In fact they consider the series

$$\sum_{n=0}^{\infty} a_n P_n^{\lambda}(\cos\vartheta) \qquad (\lambda>0) \tag{14.3}$$

where P_n^λ are defined by the expansion: $(1-2t\,r+r^2)^{-\lambda}=\sum_{k=0}^{\infty} r^k P_k^\lambda(t)$ (note that $\lambda^{-1}P_k^\lambda(\cos\vartheta)\to 2k^{-1}\cos k\vartheta$ as $\lambda\to 0$ and that $P_k^{\frac{1}{2}}(t)$ are the classical Legendre Polynomials). They introduce the following series

$$2\lambda\sum_{n=1}^{\infty}\frac{a_n}{n+2\lambda}\sin\vartheta\,P_{n-1}^{\lambda+1}(\cos\vartheta)$$

as the series conjugate to (14.3) and they deeply study them. Among many results they prove the following one

Theorem 14 *Let μ_1 and μ_2 be two measures such that*

$$\int_0^\pi \sin^{2\lambda}\vartheta\, d|\mu_1| < \infty \qquad \int_0^\pi \sin^{2\lambda}\vartheta\, d|\mu_2| < \infty .$$

If

$$\mu_1 \sim \sum_{n=0}^{\infty} a_n P_n^\lambda(\cos\vartheta), \qquad \mu_2 \sim 2\lambda \sum_{n=1}^{\infty} \frac{a_n}{n+2\lambda} \sin\vartheta\, P_{n-1}^{\lambda+1}(\cos\vartheta)$$

then μ_1 and μ_2 are absolutely continuous.

This theorem is evidently analogous to 1. A similar result for Jacobi expansions can be found in [22].

As far as multidimensional expansions are concerned, Bochner [4] proved the following theorem for multiple Fourier series:

Theorem 15 *If*

$$\sum_{m_1=-\infty}^{+\infty} \cdots \sum_{m_n=-\infty}^{+\infty} a_{m_1\ldots m_n} e^{-i(m_1t_1+\ldots+m_nt_n)} \tag{14.4}$$

is the Fourier series of a measure defined on the torus $0 \le t_j < 2\pi$ $(j = 1, \ldots, n)$, and if the Fourier coefficients are $\ne 0$ only for the indices $m_1, \ldots, m_n$ such that $m_1\sigma_1 + \ldots + m_n\sigma_n \ge 0$ for all $(\sigma_1, \ldots, \sigma_n)$ of some n-dimensional neighborhood, then the measure is absolutely continuous, i.e. (14.4) is the Fourier series of a L^1 function.

For different proofs of Bochner's theorem by means of more general results see [28], [31], [33].

Let us consider now a Laplace series of a measure μ defined on the Borel sets of the $(n-1)$-dimensional sphere S:

$$\sum_{h=0}^{\infty}\sum_{k=1}^{p_{nh}} a_{hk} Y_{hk}, \qquad a_{hk} = \int_S Y_{hk}\, d\mu . \tag{14.5}$$

In [9] the following expansion is introduced as the series conjugate to (14.5):

$$\sum_{h=1}^{\infty}\sum_{k=0}^{p_{nh}} \sqrt{\frac{h}{n+h-2}}\, a_{hk}\, \psi_{hk} \tag{14.6}$$

where

$$\psi_{hk} = \frac{1}{\sqrt{h(n+h-2)}} \sum_{j=1}^{n-1} (-1)^{n-1-j} \sqrt{g}\, g^{ij} \frac{\partial Y_{hk}}{\partial \varphi_j}\, d\varphi^1 \ldots \widehat{j} \ldots d\varphi^{n-1}$$

and $\varphi_1, \ldots, \varphi_{n-1}$ are the usual polar coordinates, $\{g^{ij}\}$ is the inverse matrix of the relevant metric tensor $\{g_{ij}\}$ on S and $\widehat{j}$ means that $d\varphi^j$ is omitted. We remark that

(14.6) is not a series of scalar functions but a series of differential forms of degree $n-2$ on S. If $n=3$, e.g., this series is

$$\sum_{h=1}^{\infty}\sum_{k=0}^{2h}\frac{a_{hk}}{h+1}\left[\frac{1}{\sin\varphi}\frac{\partial Y_{hk}}{\partial\vartheta}d\varphi-\sin\varphi\frac{\partial Y_{hk}}{\partial\varphi}d\vartheta\right]$$

while for $n=2$ we obtain the classical conjugate trigonometric series. Series (14.6) has a precise meaning; if u is a harmonic function in the unit ball whose trace on the sphere is given by (14.5), then (14.6) represents the restriction on the sphere of the differential form $*v$, where v is conjugate to u, i.e. v is such that $du=\delta v$, $dv=0$ (in other words the non homogeneous form $u+v$ is self-conjugate), and the restriction of v on the sphere is equal to zero. The next theorem (see [9]) provides a Brothers Riesz Theorem for Laplace series. Here $\underset{S}{*}\psi$ denotes the adjoint of ψ on S with respect to the usual metric on S.

Theorem 16 *Let (14.5) be a Laplace series of a measure $\mu\in M(S)$; if its conjugate series (14.6) is a "Fourier series" of a $(n-2)$-measure, i.e. if there exists $\beta\in M_{n-2}(S)$ such that*

$$a_{hk}=\sqrt{\frac{n+h-2}{h}}\int_S\beta\wedge\underset{S}{*}\psi_{hk}\qquad(h=1,2,\ldots;\ k=1,\ldots,p_{nh})\tag{14.7}$$

and moreover if

$$\int_S\beta\wedge\underset{S}{*}\gamma=0\qquad\forall\,\gamma\in C^{\infty}_{n-2}(\mathbf{R}^n)\ :\ d\gamma=0\text{ on }S\,,\tag{14.8}$$

then μ and β are absolutely continuous.

Note that, if $n=2$, conditions (14.7) are nothing but (14.1), while conditions (14.8) can be always satisfied, because in this case the only 0-forms closed on S are the functions which are constant on S. But if $n>2$ without conditions (14.8) the result is false, as easy examples can show.

References

[1] Aizenberg, A., Dautov, Sh. A.: *Differential Forms Orthogonal to Holomorphic Functions or Forms, and their Properties*, Trans. of Math. Monogr. 56, AMS, 1983.

[2] Barbey, K., König, H.: *Abstract analytic function theory and Hardy algebras*, Lecture Notes in Math. 593, Springer Verlag, Berlin Heidelberg New York, 1977.

[3] Bishop, E.: A general Rudin-Carleson theorem, *Proc. Amer. Math. Soc.*, 13, 1962, 140–143.

[4] Bochner, B.: Boundary Values of Analytic Functions in Several Variables and of Almost Periodic Functions, *Ann. of Math.*, 45, 1944, 708–722.

[5] Carleson, L.: Representations of continuous functions, *Math. Z.*, 66, 1957, 447–451.

[6] Cassisa, C., Fichera, G.: A mechanical interpretation of the Brothers Riesz Theorem, in *Elasticity: Mathematical Models and Applications*, edited by Eason & Ogden, Ellis Horwood, 1990, 35–48.

[7] Cialdea, A.: The Brothers Riesz Theorem for conjugate differential forms in $\mathbf{R}^n$, to appear on *Appl. Anal.*.

[8] Cialdea, A.: On the theory of self-conjugate differential forms, to appear on *Atti Sem. Mat. Fis. Univ. Modena.*

[9] Cialdea, A.: The Brothers Riesz Theorem in $\mathbf{R}^n$ and Laplace series, to appear on *Proc. of DEMPh*, Tbilisi, 1997.

[10] Coifman, R. R., Dahlberg, B.: Singular integral characterizations on non-isotropic H^p-spaces and the F. and M. Riesz Theorem, *Proc. Symp. Pure Math.*, 35, 1979, 231–234.

[11] Cole, B., Range, R. M.: A-measures on complex manifolds and some applications, *J. Functional Anal.*, 11, 1972, 393–400.

[12] Davis, K. M., Gilbert, J. E., Kunze,-R.-A.: Elliptic differential operators in harmonic analysis. I. Generalized Cauchy-Riemann systems, *Amer. J. Math.*, 113, 1991, 75–116.

[13] Doss, R.: Elementary proof of the Rudin-Carleson theorem and the F. and M. Riesz theorems, *Proc. Amer. Math. Soc.*, 82, 1981, 599–602.

[14] Douglas, R. G.: *Banach algebras techniques in operator theory*, Academic Press, New York, 1972.

[15] Fichera, G.: Approssimazione uniforme delle funzioni olomorfe mediante funzioni razionali aventi poli semplici prefissati, Nota I & II, *Rend. Acc. Naz. Lincei*, 27, 1959, 193–201, 317–323.

[16] Fichera, G.: Spazi lineari di k-misure e di forme differenziali, in *Proc. of Intern. Symposium on Linear Spaces*, Jerusalem 1960, Israel Ac. of Sciences and Humanities, Pergamon Press, 1961, 175–226.

[17] Forelli, F.: The theorem of F. and M. Riesz for unbounded measures, in *The Madison symposium on complex analysis*, edited by Nagel & Stout, Contemp. Math. 137, Amer. Math. Soc., Providence, RI, 1992, 221–234.

[18] Hoffman, K.: *Banach spaces of analytic functions*, Prentice-Hall, New Jersey, 1962.

[19] Katznelson, Y.: *An introduction to harmonic analysis*, Dover, New York, 1976.

[20] Koshi, S.: On F. and M. Riesz theorem, in *Functions, series, operators*, Vol. II, edited by Sz.-Nagy & Szabados, Colloq. Math. Soc. Janos Bolyai 35, North-Holland Publ. Co., Amsterdam Oxford New York, 1983, 739–743.

[21] Koshi, S.: Recent developments on the F. and M. Riesz theorem, in *Topics in complex analysis*, edited by Jakobczak & Plesniak, Banach Center Publications, 31, 1995, 227–231.

[22] Li, Z.: Hardy spaces for Jacobi expansions. I. The basic theory, *Analysis*, 16, 1996, 27–49.

[23] Muckenhoupt, B., Stein, E. M.: Classical expansions and their relation to conjugate harmonic functions, *Trans. Amer. Math. Soc.*, 118, 1965, 17–92.

[24] Øksendal, B.: A short proof of the F. and M. Riesz theorem, *Proc. Amer. Math. Soc.*, 30, 1971, 204.

[25] Pigno, L., Saeki, S.: On the theorem of F. and M. Riesz, *Colloq. Math.*, 60/61, 1990, 315–320.

[26] Riesz, F. and M.: Über Randwerte einer analytischen Funktion, *Quatrième Congrès des mathématiciens scandinaves*, Stockolm, 1916, 27–44.

[27] Rudin, W.: Boundary values of continuous analytic functions, *Proc. Amer. Math. Soc.*, 7, 1956, 808–811.

[28] Rudin, W.: *Fourier analysis on groups*, Interscience Publ., New York London, 1962.

[29] Rudin, W.: *Real and complex analysis*, McGraw-Hill, New York, 1974.

[30] Rudin, W.: *Function Theory in the Unit Ball of* $\mathbf{C}^n$, Springer Verlag, Berlin Heidelberg New York, 1980.

[31] Saeki, S.: The F. and M. Riesz theorem and singular measures, *Proc. Amer. Math. Soc.*, 90, 1984, 391–196.

[32] Sarason, D.: Invariant subspaces, in *Topics in operator theory*, edited by Pearcy, Math. Surveys 13, Amer. Math. Soc., Providence, R.I., 1974, 1–47.

[33] Shapiro, J. H.: Subspaces of $L^p(G)$ spanned by characters: $0 < p < 1$, *Israel J. Math.*, 29, 1978, 248–264.

[34] Stein, E. M., Weiss, G.: On the Theory of Harmonic Functions of Several Variables, *Acta Math.*, 103, 1960, 25-62.

[35] Stout, E. L.: *The Theory of Uniform Algebras*, Bogden and Quigley, Tarrytown on Hudson, New York, 1971.

[36] Znamenskaya, L. N.: Generalization of the theorem of F. and M. Riesz and the existence of the multidimensional Carleman formula, *Siberian Math. J.*, 29, 1988, 573–577.

[37] Zygmund, A.: *Trigonometric Series*, Cambridge University Press, 1979.

15 ON THE THEORY OF SERIES OF EXPONENTIALS

A. Ju. Timofeev

Series of the form

$$\sum_{n=1}^{\infty} a_n e^{-\lambda_n \cdot z}, \quad z = x + iy, \tag{15.1}$$

where $a_1, a_2, ...$ are constants, and $\lambda_1, \lambda_2, ...$ are arbitrary real numbers tending monotonically to $+\infty$, are called Dirichlet series. For every Dirichlet series (15.1) there exists a real number α, such that the series is absolutely convergent for $Rez > \alpha$ and absolutely divergent for $Rez < \alpha$. (In the extreme cases we may have $\alpha = \pm\infty$). The number α is called the abscissa of absolute convergence, the half-plane $Rez > \alpha$ the half-plane of absolute convergence. We may analogically determine the abscissa of convergence β and the half-plane of convergence. If in the series (15.1) the exponents λ_n satisfy the condition $\lambda_n^{-1} \cdot \ln n \to 0$ then

$$\alpha = \beta = \overline{\lim}_{n\to\infty} \frac{\ln |a_n|}{\lambda_n}.$$

This formula is a generalization of the Cauchy-Hadamard formula for the radius of convergence of a power series.

Let $\lambda_1, \lambda_2, ...$ be an arbitrary sequence of complex numbers (not necessarily tending to ∞). Then set D of points of absolute convergence of the series of exponentials

$$\sum_{n=1}^{\infty} a_n e^{\lambda_n \cdot z}, \quad z = x + iy, \tag{15.2}$$

H. Florian et al. (eds.), Generalized Analytic Functions, 151–153.

is convex. There exists series of exponentials (15.2) with coefficients a_n, not all equal to 0, to the sum $F(z) \equiv 0$ convergent.

While every function analytic in a circle is expansible in a power series in this circle, the problem of determining necessary and sufficient conditions which must be satisfied by a function analytic in a convex domain D in order that it be expansible in a series of exponentials, is very difficult.

The theory of the representation of analytic functions in convex domain D by exponential series is account in the book [1].

Let $L(\lambda)$ be an entire functions of exponential type, $\bar{D}$ its conjugate diagram, $\lambda_1, \lambda_2, \ldots$ the simple zeros of $L(\lambda)$. Let $\gamma(t)$ be the Borel function associated with the function $L(\lambda), h(\theta)$ is the indicator function of $L(\lambda)$. The functions

$$\psi_k(t) = \frac{1}{L'(\lambda_k)} \cdot \int_0^\infty \frac{L(\lambda)}{\lambda - \lambda_k} \cdot e^{-t\cdot\lambda} \cdot d\lambda \ (k \geq 1)$$

are analytic in domain $\mathbb{C} \setminus \bar{D}$ and form a system dual to the system $\{\exp(\lambda_m z)\}$:

$$\frac{1}{2\pi i} \int_C e^{\lambda_m t} \cdot \psi_k(t) dt = \delta_{mk} = \begin{cases} 0, & m \neq k, \\ 1, & m = k. \end{cases}$$

Here C is a closed contour enveloping $\bar{D}$.

With a function $F(z)$ analytic in $\bar{D}$ we associate the series

$$F(z) \sim \sum_{n=1}^\infty a_n e^{\lambda_n \cdot z}, \text{ where } a_n = \frac{1}{2\pi i} \int_C F(t) \cdot \psi_n(t) dt. \tag{15.3}$$

A. F. Leont'ev [1] established necessary and sufficient conditions of $L(\lambda)$ for the convergence of the series (15.3) to $F \in \kappa(\bar{D})$ in domain D. The last limitation on the convergence in D arises in the nature of the matter. If $F(z) \in \kappa(\bar{D})$ is expandable in the series (15.3), converging in $G \supset \bar{D}$, then $F(z)$ must automatically satisfy the equation

$$\frac{1}{2\pi i} \int_C \gamma(t) F(t+z) dt = 0.$$

But this means that $F(z)$ cannot be an arbitrary functions from $\kappa(\bar{D})$; for example, it cannot be the function $e^{\lambda \cdot z}$, $\lambda \neq \lambda_k$.

Theorem 1 *In order that the series (15.3) constructed for an arbitrary $F \in \kappa(\bar{D})$ convergence in D it is necessary and sufficient that for arbitrary $\varepsilon > 0$*

$$\ln | L'(\lambda_k) | > [h(\varphi_k) - \varepsilon] \cdot | \lambda_k |, \lambda_k = | \lambda_k | \cdot e^{i} \varphi_k, k > k_0(\varepsilon). \tag{15.4}$$

Theorem 2 *In order that the series (15.3) converges everywhere to $F \in \kappa(\bar{D})$ in the topology of $\kappa(D)$, it is necessary and sufficient that the following conditions be satisfied:*

1. *(15.4) is satisfied;*
2. *$L(\lambda)$ of completely regular growth, that is, when there exists an increasing sequence r_k of real tending to infinity for which*

$$r_k^{-1} \cdot \ln \left| L\left(r_k e^{i\theta}\right) \right| \to h(\theta)$$

uniformly in θ.

Theorem 3 *There exists entire functions $L(\kappa)$ of completely regular growth such that (15.4) for an arbitrary bounded convex region D.*

The construction principle of the series differs in the case of unbounded domains (plane, half-plane etc.) from one of bounded domains D.

Let $\{\exp(\lambda_k \cdot z)\}_{k=1}^{\infty}$ is incomplete system of functions in $\kappa(D)$. The class $\kappa(\lambda_k, D)$ of functions $F(z)$ of the form

$$F(z) = \lim_{n\to\infty} \sum_{k=1}^{Pn} a_{k,n} \cdot \exp(\lambda_k \cdot z), z \in D,$$

(the convergence is uniform on compact subsets of the domain D) is study for example in the papers [2]-[3]. It is evident that $\kappa(\{\lambda_n\}, D) \subset \kappa(D)$. Let $\{\lambda_n\} = \lambda_n' \bigcup \{\lambda_n''\}$. In [3] necessary and sufficient conditions for the representation $\kappa(\{\lambda_n\}, D) = \kappa(\{\lambda_n'\}, D) + \kappa(\{\lambda_n''\}, D)$ were obtained.

References

[1] A. F. Leont'ev. Series of exponentials. *Nauka*, Moskow, 1976 (Russian).

[2] A. F. Leont'ev. Sequences of polynomials of exponentials. *Nauka*, Moskow, 1980 (Russian).

[3] A. Ju. Timofeev. Splitting a sequence of exponential polynomials. Dokl. Akad. Nauk SSSR 260 (1981), 43-46 (Russian); English transl. in Soviet Math. Dokl. Vol. 24 (1981), No. 2, 236-238.

II Applications of Partial Differential Equations to Mechanics

16 A BOUNDARY-FINITE ELEMENT METHOD FOR POTENTIAL FLOW PAST AN AIRFOIL*

George C. Hsiao

and Michael D. Marcozzi

This paper is dedicated to Professor Robert P. Gilbert on the occasion of his 65th birthday

Abstract: A finite element–boundary element coupling procedure is applied to the computation of the potential flow past an airfoil. By utilizing a representation of the flow exterior to a circular auxiliary boundary, the reduced variational problem on an annular region is solved by the finite element method alone. Galerkin approximate solutions are constructed and error estimates obtained. [1]

16.1 INTRODUCTION

A steady uniform irrotational flow of inviscid incompressible fluid past a thin airfoil is usually modeled by the exterior boundary problem in terms of the velocity field $\mathbf{q} := (q_1, q_2)$ (cf. [2], [17]):

*This paper contains part of the lecture delivered at the Conference on " Generalized Analytic Functions – They and Applications" under the title *Boundary Element Methods for Problems in Mechanics.*

H. Florian et al. (eds.), Generalized Analytic Functions, 157–181.

$$
\begin{aligned}
\nabla \cdot \mathbf{q} = \frac{\partial q_1}{\partial x_1} + \frac{\partial q_2}{\partial x_2} &= 0 \text{ in } \Omega' := I\!R^2 \setminus \Omega \cup \Gamma \\
\nabla \times \mathbf{q} = \frac{\partial q_2}{\partial x_2} - \frac{\partial q_2}{\partial x_2} &= 0 \text{ in } \Omega' , \\
\mathbf{q} \cdot \mathbf{n} &= 0 \text{ on } \Gamma_0 , \\
\mathbf{q} - \mathbf{q}_\infty &= o(1) \text{ as } |\mathbf{x}| \to \infty , \\
\lim_{\mathbf{x} \to \mathbf{T}} |\mathbf{q}(\mathbf{x})| &= |\mathbf{q}(\mathbf{T})| < \infty
\end{aligned}
\tag{16.1}
$$

Here Γ is the profile of the airfoil which has interior Ω and one corner point $\mathbf{T}$, the trailing edge (see Figure 1). By $\mathbf{q}_\infty = (q_{\infty,1}, q_{\infty,2})$, we denote the given free-stream velocity and by $\mathbf{n}$ the unit normal vector which will face away from any bounded interior region. The last condition in the formulation is widely known as the *Kutta-Joukowski condition.* Alternatively, one may introduce a stream function Ψ such that $\mathbf{q} = (\nabla\Psi)^\perp := \left(\frac{\partial \Psi}{\partial x_2}, \frac{\partial \Psi}{\partial x_1}\right)$, and reformulate (16.1) as an exterior boundary value problem. More precisely, denoting by $\psi(\mathbf{x})$ the disturbance stream function due to the presence of the airfoil and setting $\Psi(\mathbf{x}) = \Psi_\infty(\mathbf{x}) + \psi(\mathbf{x})$, where $\Psi_\infty(\mathbf{x}) = q_{\infty,2}x_1 - q_{\infty,1}x_2$, we may reformulate the problem (16.1) as an exterior boundary value problem for ψ:

$$
\begin{aligned}
\Delta\psi &= 0 \ \text{ in } \ \Omega' , \\
\psi &= \psi_0 \ \text{ on } \ \Gamma_0 , \\
\psi(\mathbf{x}) + \frac{\kappa}{2\pi} \log|\mathbf{x}| &= \varpi + O(|\mathbf{x}|^{-1}) \ \text{ as } \ |\mathbf{x}| \to \infty .
\end{aligned}
\tag{16.2}
$$

We remark that in (16.2), both constants ϖ and κ are unknown. Physically, κ represents the circulation of the flow about the airfoil such that $\kappa = \int_{\Gamma_0} \mathbf{q} \cdot d\mathbf{x}$ (cf. [32]), which will be determined by the Kutta–Joukowski condition of (16.1) (see also, [13], [19]).

In the recent papers by the authors [23],[24], a boundary-field equation method has been introduced for treating the problem defined by (16.2). This method is a combined boundary and field equation method which reduces the problem to a finite domain through the use of an auxiliary boundary. The present paper is a continuation of [24] and is devoted to the numerical implementation of the method and the asymptotic error estimates for the approximate solutions.

We organize the paper as follows. In §16.2, we discuss the variational formulation introduced in [24] and summarize some basic results established therein. Section 16.3 deals with the descretization of (16.2) and asymptotic error estimates. The finite element approximation will be discussed in §16.4. Section 16.5 is devoted to the treatment of a modified finite element method. In the final section, §16.6, we include some numerical experiments.

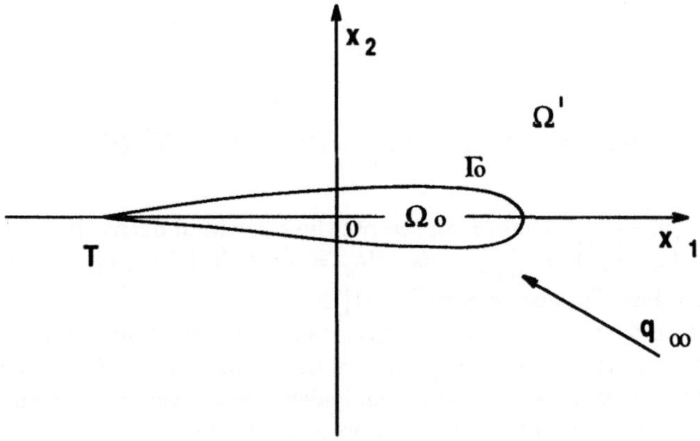

Figure 1. *Profile in a uniform stream.*

16.2 VARIATIONAL FORMULATION

To resolve the problem (16.2) onto a domain of finite extent, we introduce a circular auxiliary boundary Γ into the region exterior to the profile, dividing Ω' into a bounded annular region Ω_F and an exterior region Ω_B. Letting then ψ_F and ψ_B be restrictions of ψ to Ω_F and Ω_B, respectively, the coupling of the two parts of ψ follows from the so-called *transmission* condition on Γ:

$$\frac{\partial \psi_F^-}{\partial \mathbf{n}} = \frac{\partial \psi_B^+}{\partial \mathbf{n}} =: \sigma \text{ and } \psi_F^- = \psi_B^+$$

(cf. [22]). Representing ψ_B in the unbounded exterior region Ω_B such that

$$\psi_B(\mathbf{x}) = \int_\Gamma \mu(y) \frac{\partial}{\mathbf{n_y}} \gamma(\mathbf{x}, \mathbf{y})\, ds_\mathbf{y} - \int_\Gamma \sigma(\mathbf{y}) \gamma(\mathbf{x}, \mathbf{y})\, ds_\mathbf{y} + \varpi \ , \ \mathbf{x} \in \Omega_B \ ,$$

we obtain the coupled differential–boundary integral system:

$$\begin{gathered} \Delta \psi_F = 0 \text{ in } \Omega_F \ ; \ \psi_F = \psi_0 \text{ on } \Gamma_0 \ , \\ \frac{\partial \psi_F^-}{\partial \mathbf{n}} = \sigma \text{ on } \Gamma \ , \\ \left(\tfrac{1}{2} I - K\right) \psi_F^- + V\sigma - \omega = 0 \text{ on } \Gamma \ , \end{gathered} \tag{16.3}$$

and

$$\int_\Gamma \sigma\, ds = -\kappa \ ,$$

subject to the Kutta–Joukowski condition (cf. 16.1). Here $\gamma(\mathbf{x}, \mathbf{y}) = -\frac{1}{2\pi} \log |\mathbf{x} - \mathbf{y}|$ is the fundamental solution and V and K are the single–layer and double–layer potential operators on Γ, respectively (e.g., [22]). As a consequence of the special nature of the auxiliary boundary (i.e., circular), the integral equation in (16.3) may be *inverted* at the continuous level. We obtain in place of (16.3) then the following problem: *Find* $\psi_F \in H^1_{-\Psi_\infty}(\Omega_F) := \{v \in H^1(\Omega_F) \mid v|_{\Gamma_0} = -\Psi|_{\Gamma_0}\}$ *and* $\kappa \in \mathbb{R}$ *such that*

$$a(\psi_F, v) + \left\langle \dot{v}, (2V)\dot{\psi}_F \right\rangle = \ell(v) \ ; \ \forall v \in H^1_{\Gamma_0}(\Omega_F) , \tag{16.4}$$

where $\ell(v) = \langle v, -\frac{\kappa}{2\pi a}\rangle$. Here, $\dot{v}$ represents the tangential derivative (on Γ), $a(u,v) = \int_{\Omega_F} \nabla u \cdot \nabla v \, dx$, $\langle \chi, v\rangle = \int_{\Gamma_0} \chi v \, ds$, $H^1_{\Gamma_0} = \{v \in H^1(\Omega_F) \mid v|_{\Gamma_0} = 0\}$, and $H^m = W^{m,2}$, the standard Sobolev spaces (cf. [1]).

The Kutta-Joukowski condition ensures that the correct solution to (16.4) is sufficiently regular at the trailing edge **T**. By supposing $\psi_F \in H^2(\Omega_F)$, the problems (16.3) and (16.4) may be shown to be equivalent. Moreover, the formulation is well–posed. In particular, the following theorem is established:

Theorem 1 *For given* $\mathbf{q}_\infty \in \mathbb{R}^2$, *there exists a unique solution pair* $(\kappa, \psi_F) \in \mathbb{R} \times H^2(\Omega_F)$ *satisfying (16.4) such that*

$$|\kappa| + \|\psi_F\|_{H^2(\Omega_F)} \leq \Lambda \|\Psi_\infty|_{\Omega_F}\|_{H^1(\Omega_F)} \tag{16.5}$$

holds, for some positive constant Λ, *independent of* $\mathbf{q}_\infty$.

In order to realize the unique value of the circulation κ for which $\psi_F \in H^2(\Omega_F)$, we replace (16.4) by the following two basic problems: *Find* $\psi^{(i)} \in H^1_{\Gamma_0}(\Omega_F)$ *for* $i = 0, 1$, *such that*

$$a(\psi^{(i)}, v) + \left\langle \dot{v}, (2V)\dot{\psi}^{(i)} \right\rangle = \begin{cases} a(v, \psi_\infty) + \left\langle \dot{v}, (2V)\dot{\Psi}_\infty \right\rangle, & if i=0; \\ \langle v, -\frac{1}{2\pi a}\rangle, & if i=1, \end{cases} \tag{16.6}$$

for all $v \in H^1_{\Gamma_0}(\Omega_F)$, *where* $\psi_F = \left(\psi^{(0)} - \Psi_\infty\right) + \kappa\psi^{(1)}$. In the following, let $u_1 := r^{\pi/\omega} \sin \frac{\pi}{\omega_0}\omega$, where $r := \text{dist}(\mathbf{x}, \mathbf{T})$, and $\phi : K \to \overline{K}$, the mapping from the Cartesian coordinates (x_1, x_2) in the nonlinear corner region K to the polar coordinates (r, ω) with respect to the linear corner region $\overline{K}$ (see Figure 2). The solutions to (16.6) admit then expansions

$$\psi^{(i)} = c_0^{(i)} \eta u_1 \circ \phi + w^{(i)} , \tag{16.7}$$

where η is a C^∞ cut–off function of *small* support about **T**, $c_0^{(i)} \in \mathbb{R}$ and $w \in H^1_{\Gamma_0}(\Omega_F) \cap H^2(\Omega_F)$ is the regular remainder (cf. [9], [11] [12]). It follows then that $\psi_F = \psi^{(0)} + \kappa\psi^{(1)}$ would have no singular component provided $\kappa \equiv -c_0^{(0)}/c_0^{(1)}$. That this is the case is a consequence of the following (see also [10], [8], [29]):

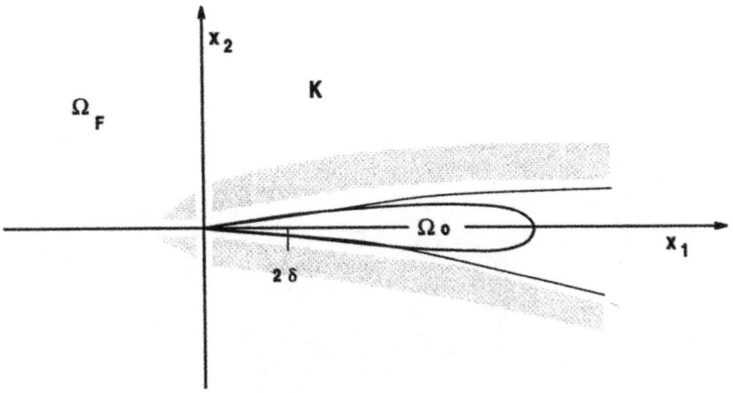

Figure 2. *The corner region K.*

Theorem 2 *Let $(\kappa, \mathbf{q}_\infty) \in I\!R \times I\!R^2$ be given. Then the problem (16.4) admits one and only one solution $\psi^{(i)}$ $(i = 0, 1)$, which is of the form (16.7). Moreover, there exists a positive constant Λ, independent of κ and $\mathbf{q}_\infty$ such that the estimate*

$$\|\psi^{(i)}\|_{H^1(\Omega_F)} \leq \Lambda \left(\|\Psi_\infty|_{\Omega_F}\|_{H^1(\Omega_F)} + |\kappa|\right)$$

holds. The constant $c_0^{(1)}$ is, in addition, non-zero.

16.3 ASYMPTOTIC ERROR ESTIMATES

Towards obtaining a discretization of (16.6), the domain Ω_F is approximated by an (indexed) polygonal region Ω_h whose boundaries Γ_h and $\Gamma_{0,h}$ are supposed piecewise linear with nodes on Γ and Γ_0, respectively. The trailing edge $\mathbf{T}$ is taken as a node of the approximate boundary. As the frontiers of Ω_h will be assumed *close* to those of Ω_F for h sufficiently small, the existence of a bounded open set $\tilde{\Omega}$ may be assumed such that $\Omega_F \cup \Omega_h \subset \tilde{\Omega}$, for all such h. The interior boundary of $\tilde{\Omega}$ will be labeled $\tilde{\Gamma}_0$ (see Figure 3). The space $H^1_{\Gamma_0}(\Omega_F)$ is isomorphic to the subspace $\tilde{V} := \left\{\tilde{v} \in H^1_{\tilde{\Gamma}_0}(\tilde{\Omega}) \mid \tilde{v}|_{\tilde{\Omega}\setminus\Omega_F} = 0\right\}$. Denoting again by A and $\ell^{(i)}$ the respective trivial extensions to $H^1_{\tilde{\Gamma}_0}(\tilde{\Omega})$, it follows that $\psi^{(i)} \in H^1_{\Gamma_0}(\Omega_F)$ is a solution of (16.6) if and only if the associate element $\tilde{\psi}^{(i)} \in \tilde{V}$ satisfies the *extended* variational problem

$$A(\tilde{\psi}^{(i)}, \tilde{v}) = \ell^{(i)}(\tilde{v}) \,, \tag{16.8}$$

for any $\tilde{v} \in \tilde{V}$, where integration is again performed over Ω_F and Γ.

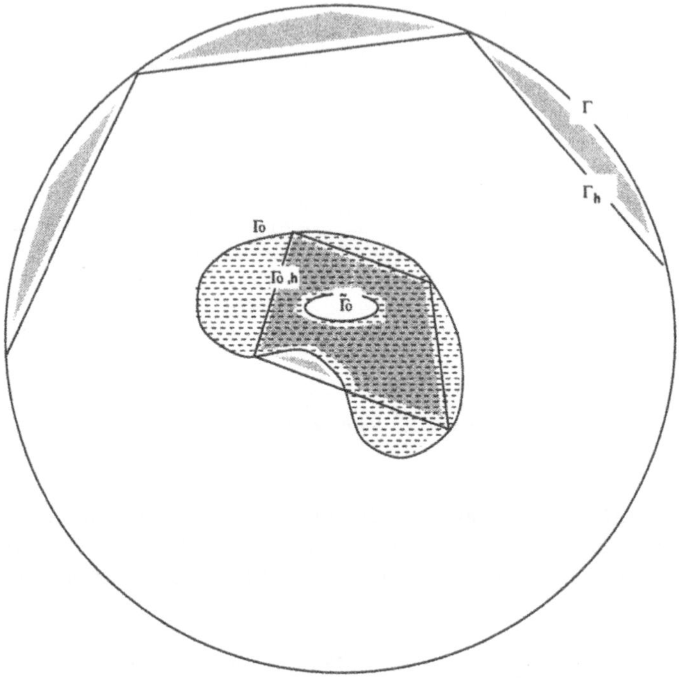

Figure 3. *The extended domain* $\tilde{\Omega}$.

To formulate the Galerkin procedure, let $\tilde{V}_h$ be a finite dimensional subspace of $H^1_{\tilde{\Gamma}_0}(\tilde{\Omega})$ for which restrictions to the regions bounded by $\Gamma_{0,h}$ and $\tilde{\Gamma}_0$ vanish and Γ_h and Γ are linear. More specifically, let $\chi_h : \Gamma_h \to \Gamma$ denote a mapping such that for any $\mathbf{y} \notin \Gamma_h \cap \Gamma$, $\chi_h(\mathbf{y})$ equals the intersection of the line perpendicular to Γ_h at $\mathbf{y}$ and the curve Γ. For sufficiently small h, it follows that χ_h is a bijection (cf. [28]). For any $\tilde{v}_h \in \tilde{V}_h$ then, $\tilde{v}_h(\mathbf{y}) = \tilde{v}_h(\mathbf{x})$, where $\mathbf{x}$ lies in the region bounded by Γ and Γ_h on the line segment connecting the points $\mathbf{y} \in \Gamma_h$ and $\chi_h(\mathbf{y}) \in \Gamma$. Moreover,

$$\int_{\Gamma_h} \tilde{v}_h(s)ds = \int_{\Gamma} \tilde{v}_h \circ \chi_h^{-1}(s)\, J\Big(\chi_h^{-1}(s)\Big)\, ds\,,$$

for any $\tilde{v}_h \in \tilde{V}_h$, where $J(\chi_h^{-1}) = \left|\frac{\partial \chi_h^{-1}}{\partial s}\right|$. The Galerkin solution of (16.8) is the element $\tilde{\psi}_h^{(i)} \in \tilde{V}_h$ for which

$$A_h(\tilde{\psi}_h^{(i)}, \tilde{v}_h) = \ell_h^{(i)}(\tilde{v}_h)\,, \tag{16.9a}$$

where

$$A_h(\tilde{u}_h, \tilde{v}_h) := a_h(\tilde{u}_h, \tilde{v}_h) + \left\langle \frac{d}{ds}\gamma_{\Gamma_h}\tilde{v}_h, (2V_h)\frac{d}{ds}\gamma_{\Gamma_h}\tilde{u}_h \right\rangle_h \tag{16.9b}$$

and

$$\begin{aligned}\ell_h^{(i)}(\tilde{v}_h) \quad &:= \quad \delta_{i1}\left\langle \gamma_{\Gamma_h}\tilde{v}_h, -\tfrac{1}{2\pi a}\right\rangle_h \\ &+ \quad \delta_{i0}\left[a_h(\Psi_\infty, \tilde{v}_h) + \left\langle \tfrac{d}{ds}\gamma_{\Gamma_h}\tilde{v}_h, (2V_h)\tfrac{d}{ds}\gamma_{\Gamma_h}\Psi_\infty\right\rangle_h\right] ,\end{aligned} \tag{16.9c}$$

such that

$$a_h(\tilde{u}_h, \tilde{v}_h) := \int_{\Omega_h} \nabla\tilde{u}_h \cdot \nabla\tilde{v}_h \, d\mathbf{x} , \tag{16.9d}$$

for any $\tilde{u}_h, \tilde{v}_h \in \tilde{V}_h$. Here $\langle\cdot,\cdot\rangle_h$ denotes the duality pairing on $H^{-1/2}(\Gamma_h) \times H^{1/2}(\Gamma_h)$ and (16.9a-16.9d) the *conforming* discrete problem on Ω_h.

Isomorphic with $\tilde{V}_h$ is the space, noted V_h, of restrictions to the computational domain Ω_h. The *discrete* problem is then to find $\psi_h^{(i)} \in V_h$ satisfying

$$A_h(\psi_h^{(i)}, v_h) = \ell_h^{(i)}(v_h) \tag{16.10}$$

for all $v_h \in V_h$. In particular, $\psi_h^{(i)} \in V_h$ is a solution of (16.9a-16.9d) if and only if $\tilde{\psi}_h^{(i)} \in \tilde{V}_h$ is a solution of (16.10). Moreover, letting μ_j, $j = 1, \cdots, N$ form a basis for V_h, then

$$\psi_h^{(i)} = \sum_{j=1}^{N} \nu_j^{(i)} \mu_j , \tag{16.11}$$

and the Galerkin equations (16.10) are equivalent to the linear system

$$\sum_{j=1}^{N} A_h(\mu_j, \mu_k)\nu_j^{(i)} = \ell^{(i)}(\mu_k) \quad , k = 1, 2, \cdots, N , \tag{16.12}$$

for the unknown coefficients $\nu_j^{(i)}$.

We now compare formulations over $\tilde{\Omega}$, for which the following coerciveness property and abstract error estimate of *Strang* type hold.

Lemma 3 *Let $\tilde{V}_h$ approximate $H^1_{\tilde{\Gamma}_0}(\tilde{\Omega})$ such that for any $\tilde{v} \in H^1_{\tilde{\Gamma}_0}(\tilde{\Omega})$, there exists a sequence $\tilde{v}_h \in \tilde{V}_h$ for which $\|\tilde{v}_h - \tilde{v}\|_{H^1(\tilde{\Omega})} \to 0$ as $N = \dim \tilde{V}_h \to \infty$ and let $\Pi_h : H^1_{\tilde{\Gamma}_0}(\tilde{\Omega}) \to \tilde{V}_h$ denote the associate interpolation operator. Moreover, suppose that*

$$|A(\tilde{v}_h, \tilde{v}_h) - A_h(v_h, v_h)| \le c\, h\, \|v_h\|^2_{H^1(\Omega_h)} , \tag{16.13}$$

for all $v_h \in V_h$, then

(i) there exists a constant $\alpha \in I\!R$, $\alpha > 0$, such that

$$A_h(v_h, v_h) \ge \alpha \|v_h\|^2_{H^1(\Omega_h)} ,$$

for all $v_h \in V_h$;

(ii) *there exists a constant $c \in \mathbb{R}$, $c > 0$, such that (i=0,1)*

$$
\begin{aligned}
c\,\|\tilde{\psi}^{(i)} - \psi_h^{(i)}\|_{H^1(\Omega_h)} &\leq \inf_{v_h \in V_h} \Big[\|\tilde{\psi}^{(i)} - v_h\|_{H^1(\Omega_h)} \\
&+ \sup_{w_h \in V_h} \frac{|A(\tilde{v}_h, \tilde{w}_h) - A_h(v_h, w_h)|}{\|w_h\|_{H^1(\Omega_h)}}\Big] \\
&+ \delta_{i0} \sup_{\tilde{w}_h \in V_h} \frac{|\ell^{(i)}(\tilde{v}) - \ell_h^{(i)}(w_h)|}{\|w_h\|_{H^1(\Omega_h)}} ,
\end{aligned}
$$

where $\tilde{\psi}^{(i)}$ is the solution of the extended variational problem (16.8); for all h *sufficiently small.*

Proof. Let v_h denote an arbitrary element in the space V_h.
(i) *A_h is $H^1(\Omega_h)$–elliptic.* With the assumption (16.13), we may write:

$$
\begin{aligned}
A_h(v_h, v_h) &= -[A(\tilde{v}_h, \tilde{v}_h) - A_h(v_h, v_h)] + A(\tilde{v}_h, \tilde{v}_h) \\
&\geq A(\tilde{v}_h, \tilde{v}_h) - [A(\tilde{v}_h, \tilde{v}_h) - A_h(v_h, v_h)] \\
&\geq A(\tilde{v}_h, \tilde{v}_h) - ch\|v_h\|^2_{H^1(\Omega_h)} \\
&\geq \alpha'\|v_h\|^2_{H^1(\Omega_h)} - ch\|v_h\|^2_{H^1(\Omega_h)} ,
\end{aligned}
$$

as

$$
A(\tilde{v}_h, \tilde{v}_h) \geq \alpha'\|\tilde{v}_h\|^2_{H^1(\tilde{\Omega})} \geq \alpha'\|v_h\|^2_{H^1(\Omega_h)} ,
$$

where $\tilde{v}_h|_{\Omega_h} = v_h$. It then follows that

$$
A_h(v_h, v_h) \geq (\alpha' - ch)\|v_h\|^2_{H^1(\Omega_h)} \geq \alpha\|v_h\|^2_{H^1(\Omega_h)} ,
$$

for all h sufficiently small (see also [21]).
(ii) *An Estimate of the Strang Type.* From the result (i), we have that (i= 0,1)

$$
\begin{aligned}
\alpha\|\psi_h^{(i)} - v_h\|^2_{H^1(\Omega_h)} &\leq A_h\left(\psi_h^{(i)} - v_h, \psi_h^{(i)} - v_h\right) \\
&= A(\psi^{(i)} - \tilde{v}_h, \tilde{\psi}_h^{(i)} - \tilde{v}_h) \\
&+ A\left(\tilde{v}_h, \tilde{\psi}_h^{(i)} - \tilde{v}_h\right) - A_h\left(v_h, \psi_h^{(i)} - v_h\right) \\
&- \ell^{(i)}(\tilde{\psi}_h^{(i)} - \tilde{v}_h) + \ell^{(i)}(\psi_h^{(i)} - v_h) ,
\end{aligned}
$$

or

$$
\begin{aligned}
\alpha\|\psi_h^{(i)} - v_h\|_{H^1(\Omega_h)} &\leq \frac{A(\tilde{\psi}^{(i)} - \tilde{v}_h, \tilde{\psi}_h^{(i)} - \tilde{v}_h)}{\|\psi_h^{(i)} - v_h\|_{H^1(\Omega_h)}} \\
&+ \frac{A(\tilde{v}_h, \tilde{\psi}_h^{(i)} - \tilde{v}_h) - A_h(v_h, \psi_h^{(i)} - v_h)}{\|\psi_h^{(i)} - v_h\|_{H^1(\Omega_h)}} \\
&- \frac{\ell^{(i)}(\tilde{\psi}_h^{(i)} - \tilde{v}_h) - \ell_h^{(i)}(\psi_h^{(i)} - v_h)}{\|\psi_h^{(i)} - v_h\|_{H^1(\Omega_h)}} .
\end{aligned}
$$

Using the continuity of the bilinear form A then,

$$\begin{aligned}
\frac{|A(\tilde{\psi}^{(i)} - \tilde{v}_h, \tilde{\psi}_h^{(i)} - \tilde{v}_h)|}{\|\psi_h^{(i)} - v_h\|_{H^1(\Omega_h)}} &\leq \frac{M\|\tilde{\psi}^{(i)} - \tilde{v}_h\|_{H^1(\tilde{\Omega})}\|\tilde{\psi}_h^{(i)} - \tilde{v}_h\|_{H^1\tilde{\Omega})}}{\|\psi_h^{(i)} - v_h\|_{H^1(\Omega_h)}} \\
&\leq \frac{cM\|\tilde{\psi}^{(i)} - v_h\|_{H^1(\Omega_h)}\|\psi_h^{(i)} - v_h\|_{H^1\Omega_h)}}{\|\psi_h^{(i)} - v_h\|_{H^1(\Omega_h)}} \\
&= cM\|\tilde{\psi}^{(i)} - v_h\|_{H^1(\Omega_h)} ,
\end{aligned}$$

as $H^1(\Omega_h)$ is isomorphically embedded in $H^1(\tilde{\Omega})$. Hence,

$$\begin{aligned}
\alpha\|\psi_h^{(i)} - v_h\|_{H^1(\Omega_h)} &\leq cM\|\tilde{\psi}^{(i)} - v_h\|_{H^1(\Omega_h)} \\
&+ \sup_{0\neq w_h\in V_h} \frac{|A(\tilde{v}_h, \tilde{w}_h) - A_h(v_h, w_h)|}{\|w_h\|_{H^1(\Omega_h)}} \\
&+ \sup_{0\neq w_h\in V_h} \frac{|\ell^{(i)}(\tilde{w}_h) - \ell_h^{(i)}(w_h)|}{\|w_h\|_{H^1(\Omega_h)}} .
\end{aligned}$$

Combining the above inequalities with the triangle inequality

$$\|\tilde{\psi}^{(i)} - \psi_h^{(i)}\|_{H^1(\Omega_h)} \leq \|\tilde{\psi}^{(i)} - v_h\|_{H^1(\Omega_h)} + \|\psi_h^{(i)} - v_h\|_{H^1(\Omega_h)} ,$$

it follows that there exists a real constant $c > 0$, for which

$$\begin{aligned}
c\|\tilde{\psi}^{(i)} - \psi_h^{(i)}\|_{H^1(\Omega_h)} &\leq \|\tilde{\psi}^{(i)} - v_h\|_{H^1(\Omega_h)} \\
&+ \sup_{0\neq w_h\in V_h} \left[\frac{|A(\tilde{v}_h, \tilde{w}_h) - A_h(v_h, w_h)|}{\|w_h\|_{H^1(\Omega_h)}} \right. \\
&+ \left. \frac{|\ell^{(i)}(\tilde{w}_h) - \ell_h^{(i)}(w_h)|}{\|w_h\|_{H^1(\Omega_h)}} \right] .
\end{aligned}$$

Taking the infimum with respect to $v_h \in V_h$ completes the proof.

We will simply require the procedure converge for the case when finite elements are used as trial functions. To this end, the quality of the approximation for the $\tilde{\psi}^{(i)}|_{\Omega_h}$ by linear elements on triangular partitions is considered next. Estimates for the leading coefficients of the singular expansions and for $\tilde{\Psi}_F|_{\Omega_h}$ follow in §16.5 as well as the development and analysis of the Dual Singular Function Method. As it should be clear from the context, we identify in the sequel elements of the space and its respective restriction/extension without explicit reference in order to simplify notation (also [33], [36]).

16.4 THE FINITE ELEMENT APPROXIMATION

To apply the abstract error estimate of Lemma 3, we choose as the V_h so-called Finite Element Spaces constructed from regular 2–simplices of type 1 (and of class C^0), vanishing on the profile Γ_0; that is, linear approximating functions on triangular partitions. Here, h is a parameter taken to be the mesh size of the triangulation $\tau_h = \{\bar{\Omega}_{e,h}\}$ of $\overline{\Omega_h}$ such that $h \to 0$. The regular nature of the partitions refers to the existence of constants c and C for which each element of τ_h is contained within a circle of diameter Ch and contains a circle of diameter ch. In addition, we require that for any partition, there exists a neighborhood $U(h)$ of the trailing edge **T** for which

$$\cup_{e \in I} \overline{\Omega}_{e,h} = \overline{U(h)} \cap \overline{\Omega}_h \,, \tag{16.14}$$

such that I is an index set of cardinality at most N_0. It follows that $V_h \subset H^1_{\Gamma_0}(\Omega_h)$ (cf. [7]).

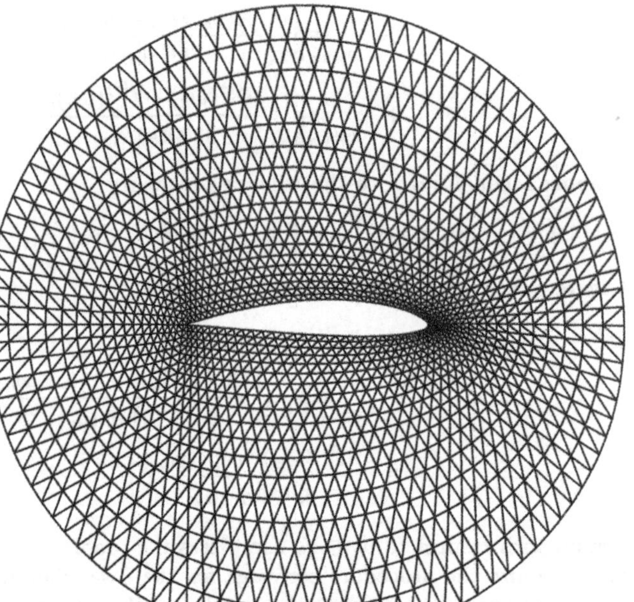

Figure 4. *Triangulation (h=0.323).*

The regularity of the solution $\psi^{(i)}$ $(i = 0, 1)$ of (16.6) afforded by the expansion $\psi^{(i)} = c_0^{(i)} \eta r^{\pi/\omega_0} \sin \frac{\pi}{\omega_0}\omega + w$ is most conveniently portrayed in the context of the Sobolev–Slobodeckiĭ spaces $H^\ell(\Omega)$ $(\ell \in I\!R_+)$ (cf. [26], [30]). To this end, we recall that $u \in H^\ell(\Omega)$ provided $u \in H^{[\ell]}(\Omega)$ and satisfies the property $I_\lambda(\partial^s u) < \infty$ for $|s| \le [\ell]$, $\ell \equiv [\ell] + \lambda$ $(0 < \lambda < 1)$. Here $[\ell]$ denotes the greatest integer less than or equal to ℓ,

$$I_\lambda(u) := \int\int_{\Omega \times \Omega} \frac{|u(\mathbf{x}) - u(\mathbf{y})|}{|\mathbf{x} - \mathbf{y}|^{2+2\lambda}} \, d\mathbf{x} \, d\mathbf{y} \,, \tag{16.15a}$$

and

$$\|u\|_{H^{\ell}(\Omega)} := \left\{ \|u\|^2_{H^{[\ell]}(\Omega)} + \sum_{|\alpha|=[\ell]} I_\lambda\left(\partial^\alpha u\right) \right\}^{1/2} , \tag{16.15b}$$

such that $\alpha = (\alpha_1, \alpha_2) \in \mathbb{N}^2$ and $\partial^\alpha u = \partial^{\alpha_1+\alpha_2} u / \partial x_1^{\alpha_1} \partial x_2^{\alpha_2}$ (cf. [35], [1]). The function $\eta r^{\pi/\omega_0} \sin \frac{\pi}{\omega_0}\omega$ may then be seen to be an element of $H^{1+r_0}(\Omega_F)$, where $r_0 := \frac{\pi}{\omega_0} - \epsilon_0$ and $\epsilon_0 > 0$ (cf. [20]). For any such ϵ_0, we define p_0 according to $\frac{1}{p_0} = \frac{1}{2} - \frac{r_0}{2}$ in which case $H^{1+r_0}(\tilde{\Omega}) \hookrightarrow W^{1,p_0}(\tilde{\Omega})$ by the Sobolev embedding theorem. Additionally, the global interpolation operator $\Pi_h : W^{1,p_0}(\Omega_F) \to V_h = \tilde{V}_h|_{\Omega_h} \subset W^{1,2}(\Omega_h) = H^1(\Omega_h)$ satisfies the relation $\Pi_h u = \sum_{j=1}^N \gamma_j \mu_j$, where the μ_j are the global basis functions of V_h (see Eq. (16.11)), for which the local interpolation is $(\Pi_h u)|_{\overline{\Omega}_{e,h}} = \Pi_{\overline{\Omega}_{e,h}} u|_{\overline{\Omega}_{e,h}}$, and $\gamma_j = \gamma_j(u)$ are interpolation coefficients (cf. [7]). The efficiency of the approximation space V_h is characterized by the following result.

Lemma 4 *For any* $\epsilon_0 > 0$, *there exists a constant* c, *independent of* h *(sufficiently small), such that* (i=0,1),

$$\|\psi^{(i)} - \Pi_h \psi^{(i)}\|_{H^1(\Omega_h)} \le c h^{r_0} \|\psi^{(i)}\|_{H^{1+r_0}(\Omega_h)} , \tag{16.16}$$

where $r_0 := \frac{\pi}{\omega_0} - \epsilon_0$.

Proof. For $p_0 > 2$, there exists embedding $W^{1,p_0}(\Omega_{e,h}) \overset{c}{\hookrightarrow} C^0(\overline{\Omega}_{e,h})$ and $W^{1,p_0}(\Omega_{e,h}) \hookrightarrow H^1(\Omega_{e,h})$. As such, it can be shown that the space V_h possesses the following local approximation properties: *there exists a constant* c, *independent of h, such that for any* $\overline{\Omega}_{e,h} \in \tau_h$ *and* $v \in W^{1,p_0}(\Omega_{e,h})$,

$$\|v - \Pi_{\overline{\Omega}_{e,h}} v\|_{H^1(\Omega_{e,h})} \le c h^{2(\frac{1}{2} - \frac{1}{p_0})} \|v\|_{W^{1,p_0}(\Omega_{e,h})} ,$$

and

$$\|v - \Pi_{\overline{\Omega}_{e,h}} v\|_{H^1(\Omega_{e,h})} \le c h^2 \|v\|_{H^2(\Omega_{e,h})} ,$$

(e.g., [7], Theorem 3.1.6).

Letting $U(h)$ denote the neighborhood of the corner defined in (16.14), it follows that

$$\begin{aligned} \|\psi^{(i)} - \Pi_h \psi^{(i)}\|^2_{H^1(\Omega_h)} &= \sum_{\overline{\Omega}_{e,h} \subset \overline{\Omega_h \setminus U(h)}} \|\psi^{(i)} - \Pi_h \psi^{(i)}\|^2_{H^1(\Omega_{e,h})} \\ &+ \|\psi^{(i)} - \Pi \psi^{(i)}\|^2_{H^1(\Omega_h \setminus U(h))} \\ &\le c h^{2(1-\frac{2}{p_0})} \sum_{\overline{\Omega}_{e,h} \subset \overline{\Omega_h \setminus U(h)}} \left\| \psi^{(i)}|_{\Omega_{e,h}} \right\|^2_{W^{1,p_0}(\Omega_{e,h})} \\ &+ c h^2 \left\| \psi^{(i)}|_{\Omega_h \setminus U(h)} \right\|^2_{H^2(\Omega_h \setminus U(h))} . \end{aligned}$$

As there are at most N_0 elements in $\Omega_h \cap U(h)$ by supposition, we have

$$\begin{aligned}
\|\psi^{(i)} - \Pi_h\psi^{(i)}\|^2_{H^1(\Omega_h)} &\leq c\,N_0\,h^{2\left(1-\frac{2}{p_0}\right)} \left\|\psi^{(i)}|_{U(h)\cap\Omega_F}\right\|^2_{W^{1,p_0}(U(h)\cap\Omega_F)} \\
&\quad + c\,h^2 \left\|\psi^{(i)}|_{\Omega_h\backslash U(h)}\right\|^2_{H^2(\Omega_h\backslash U(h))} \\
&\leq c\,N_0\,h^{2r_0} \left\|\psi^{(i)}|_{U(h)\cap\Omega_h}\right\|^2_{W^{1,p_0}(U(h)\cap\Omega_h)} + c\,h^2 \left\|\psi^{(i)}|_{\Omega_h\backslash U(h)}\right\|^2_{H^2(\Omega_h\backslash U(h))} \\
&\leq c\,N_0\,h^{2r_0} \left\|\psi^{(i)}|_{U(h)\cap\Omega_h}\right\|^2_{H^{1+r_0}(U(h)\cap\Omega_h)} + c\,h^2 \left\|\psi^{(i)}|_{\Omega_h\backslash U(h)}\right\|^2_{H^2(\Omega_h\backslash U(h))} \\
&\leq c\,h^{2r_0} \left\{ \left\|\psi^{(i)}|_{U(h)\cap\Omega_h}\right\|^2_{H^{1+r_0}(U(h)\cap\Omega_h)} + \left\|\psi^{(i)}|_{\Omega_h\backslash U(h)}\right\|^2_{H^2(\Omega_h\backslash U(h))} \right\},
\end{aligned}$$

as $H^{1+r_0} \hookrightarrow W^{1,p_0} \cap H^2$ for $0 < r_0 < 1$, from which the result follows.

With respect to the consistency errors, we have:

Lemma 5 *For any $v_h, w_h \in V_h$, there exists a constant* c *such that*

$$|A(v_h, w_h) - A_h(v_h, w_h)| \leq ch\|v_h\|_{H^1(\Omega_h)}\|w_h\|_{H^1(\Omega_h)} \tag{16.17a}$$

and

$$|\ell^{(0)}(w_h) - \ell^{(0)}_h(w_h)| \leq ch^{3/2}|\mathbf{q}_\infty|\|w_h\|_{H^1(\Omega_h)} \,. \tag{16.17b}$$

Proof. The result follows as presented in C. Johnson and J. Nedelec [25] upon noting that $\|\frac{d}{ds}\gamma\Psi_\infty - \frac{d}{ds}\gamma_{\Gamma_h}\Psi_\infty\|_{L^2(\Gamma)} \leq ch|\mathbf{q}_\infty|$.

Finally, the quality of the approximation of the discrete solution $\psi_h^{(i)}$ of (16.10) to the variational solution (16.6) is demonstrated.

Theorem 6 *For any $\epsilon_0 > 0$, there exists a constant* c, *independent of* h, *for which* (i=0,1)

$$\|\psi^{(i)} - \psi_h^{(i)}\|_{H^1(\Omega_h)} \leq ch^{r_0}\left(\|\psi^{(i)}\|_{H^1(\Omega_F)} + \delta_{i0}|\mathbf{q}_\infty|\right), \tag{16.18}$$

where $r_0 := \frac{\pi}{\omega_0} - \epsilon_0$.

Proof. From the abstract error estimate of Lemma 3 and that of (16.17a-16.17b), it follows that,

$$c\|\psi^{(i)} - \psi_h^{(i)}\|_{H^1(\Omega_h)} \leq \|\psi^{(i)} - \Pi_h\psi^{(i)}\|_{H^1(\Omega_h)} + ch\|\Pi_h\psi^{(i)}\|_{H^1(\Omega_h)} + \delta_{i0}\,ch^{3/2}|\mathbf{q}_\infty|\,,$$

where, upon employing the estimate (16.16), we find

$$\begin{aligned}
c\|\psi^{(i)} &- \psi_h^{(i)}\|_{H^1(\Omega_h)} \leq ch^{r_0}\|\psi^{(i)}\|_{H^{1+r_0}(\Omega_F)} \\
&+ ch(h^{r_0}\|\psi^{(i)}\|_{H^{1+r_0}(\Omega_f)} + \|\psi^{(i)}\|_{H^1(\Omega_h)} + \delta_{i0}\,ch^{3/2}|\mathbf{q}_\infty|) \\
&\leq ch^{r_0}(\|\psi^{(i)}\|_{H^{1+r_0}(\Omega_F)} + ch\|\psi^{(i)}\|_{H^1(\Omega_h)} + \delta_{i0}\,ch^{3/2}|\mathbf{q}_\infty|)\,.
\end{aligned}$$

The result is then obtained upon noting that $\|\psi^{(i)}\|_{H^1(\Omega_h)} = \|\psi^{(i)}\|_{H^1(\Omega_F)} + O(1)$ (cf. [28], [18]).

The proposition indicates an order of convergence of slightly more than 1/2 in the H^1-norm. This compares to a convergence rate of one were the corner not present. The estimate (16.18) is of the type obtained by M. Dobrowolski [14] for a bounded region with one significant corner singularity. See also the work by A.M. Sändig [27].

Mean squared estimates are obtained from (16.18) by employing the method of Aubin–Nitsche, providing an asymptotic error bound.

Lemma 7 *For any $v_h, w_h \in V_h$, there exists a constant* c *such that*

$$|A(v_h, w_h) - A_h(v_h, w_h)| \leq ch^{1+r_0}\|v\|_{H^{1+r_0}(\Omega_F)}\|w\|_{H^{1+r_0}(\Omega_F)}$$

and

$$|\ell^{(0)}(w_h) - \ell_h^{(0)}(w_h)| \leq ch^{3/2+r_0}|\mathbf{q}_\infty|\|w\|_{H^1(\Omega_F)} ,$$

for some $v, w \in H^{1+r_0}(\Omega_F)$.

Proof. The proof is analogous to that of Lemma 5 given the approximation of $H^{1+r_0}(\Omega_F)$ by V_h.

Theorem 8 *For any $\epsilon_0 > 0$, there exists a constant* c, *independent of* h, *for which* (i=0,1)

$$\|\psi^{(i)} - \psi_h^{(i)}\|_{L^2(\Omega_h)} \leq ch^{2r_0}(\|\psi^{(i)}\|_{H^{1+r_0}(\Omega_F)} + \delta_{i0}|\mathbf{q}_\infty|) , \tag{16.19b}$$

where $r_0 := \frac{\pi}{\omega_0} - \epsilon_0$.

Proof. For any $g \in L^2(\Omega_F)$, let $\psi_g \in H^1_{\Gamma_0}(\Omega_F)$ satisfy the auxiliary variational problem

$$A(v, \psi_g) = (g, v) ,$$

for any $v \in H^1_{\Gamma_0}(\Omega_F)$ (n.b. the form $A(\cdot,\cdot)$ is symmetric). We have then the following regularity result for ψ_g; namely,

$$\|\psi_g\|_{H^{1+r_0}(\Omega_F)} \leq c\|g\|_{L^2(\Omega_F)} , \tag{16.20b}$$

(cf. Lemma 4). Taking $v = \psi^{(i)} - \psi_h^{(i)}$ in (16.20b), we have

$$\begin{aligned}(g, \psi^{(i)} - \psi_h^{(i)}) = A(\psi^{(i)} - \psi_h^{(i)}, \psi_g) &= A(\psi^{(i)} - \psi_h^{(i)}, \psi_g - P_h\psi_g)\\ &+ A_h(\psi^{(i)}, P_h\psi_g) - A(\psi_h^{(i)}, P_h\psi_g)\\ &+ \ell^{(i)}(P_h\psi_g) - \ell_h^{(i)}(P_h\psi_g) ,\end{aligned}$$

where P_h denotes the Galerkin projection. Hence, it follows from Theorem 6 and Lemmas 5 and 7 that

$$\begin{aligned}(g,\psi^{(i)}-\psi_h^{(i)}) &\leq ch^{2r_0}\left\{\|\psi^{(i)}\|_{H^{1+r_0}(\Omega_F)}+\delta_{i0}|\mathbf{q}_\infty|\right\}\|\psi_g\|_{H^{1+r_0}(\Omega_F)}\\ &+ ch^{1+r_0}\|\psi^{(i)}\|_{H^{1+r_0}(\Omega_F)}\|\psi_g\|_{H^{1+r_0}(\Omega_F)}\\ &+ ch^{\frac{3}{2}+r_0}\delta_{i0}|\mathbf{q}_\infty|\;\|\psi_g\|_{H^{1+r_0}(\Omega_F)}\\ &\leq ch^{2r_0}(\|\psi^{(i)}\|_{H^{1+r_0}(\Omega_F)}+\delta_{i0}|\mathbf{q}_\infty|)\|\psi_g\|_{H^{1+r_0}(\Omega_F)}\,.\end{aligned}$$

We note that $r_0 := \frac{\pi}{\omega_0} - \epsilon_0 < \frac{1}{2}$. The proposition follows from the characterization

$$\|\psi^{(i)}-\psi_h^{(i)}\|_{L^2(\Omega_h)} = \sup_{g\in L^2(\Omega_F)} \frac{|(g,\psi^{(i)}-\psi_h^{(i)})|}{\|g\|_{L^2(\Omega_F)}}$$

and the estimate (16.20b).

We remark that the L^2-rate of convergence is only marginally more than one, compared with two for a domain without a corner. It is not optimal in the sense that $2r_0 \neq 1+r_0$, the difference between the indices of the spaces $H^{1+r_0}(\Omega_F)$ and $L^2(\Omega_h)$.

16.5 THE DUAL SINGULAR FUNCTION METHOD

The results of the previous section demonstrate the decreased convergence rate due to the presence of a corner singularity, the so–called *pollution* effect. As the lifting solution is dependent upon determining the circulation constant κ (cf. Theorem 1), it remains to establish the approximation rate of the coefficients of the leading singularities. In so doing, there arises a natural augmentation of the standard finite element method which restores the anticipated order of convergence. We follow here the work of Blum and Dobrowolski [6] (see also [4], [5], [15], [16], [19], [27]).

In the following, let u denote either $\psi^{(0)}$ or $\psi^{(1)}$ and $v_{-1} := -\frac{1}{i\pi}r^{-\pi/\omega_0}\sin\frac{\pi}{\omega_0}\omega$. We assume that η is a cut–off function of *small* support about the corner $\mathbf{T}$. From (16.7) then

$$u = c_0\eta u_1\circ\phi + w \tag{16.21a}$$

with $w \in H^1_{\Gamma_0}(\Omega_F)\cap H^2(\Omega_F)$ and $c_0 \in I\!R$ such that

$$c_0 = -\left(u,(\Delta i\eta v_{-1})\circ\phi\;J(\phi^{-1})\right) - \left(u, K\partial^2_\omega(i\eta v_{-1})\circ\phi\;J(\phi^{-1})\right) \tag{16.21b}$$

as per [24], where $K(|\mathbf{x}|) = \frac{K_1}{|\mathbf{x}|} + K_2 + \Theta_R(|\mathbf{x}|)$ and the constants K_1 and K_2 are real, depending on the transformation ψ, and $\Theta_R(|\mathbf{x}|) = O(|\mathbf{x}|)$. The Galerkin projection of u is found to be

$$u_h := P_h u = c_0 P_h(\eta\, u_1\circ\phi) + P_h w\,, \tag{16.22a}$$

where $P_h u_1$ satisfies (by definition)

$$A_h\left(P_h(\eta\, u_1\circ\phi), v_h\right) = \ell_h^{(i)}(v_h)\,, \tag{16.22b}$$

for all $v_h \in V_h$.

Towards obtaining estimates for the coefficient of the leading singularity and developing the Dual Singular Function Method, we define as an intermediate step an iterative procedure having as its first iterate the finite element solution and the determination of c_0. To initialize the procedure, we set $c_0^0 = 0$ and $P_h^1 u = P_h u$. The $j+1$ iteration is found such that

$$P_h^{j+1} u = c_0^j \eta \, u_1 \circ \phi + P_h^j w \, , \tag{16.23a}$$

where

$$P_h^j w := P_h(u - c_0^j \eta \, u_1 \circ \phi) = P_h u - c_0^j P_h(\eta \, u_1 \circ \phi) \tag{16.23b}$$

and

$$c_0^j := -\left(P_h^j u, \Delta(i\eta v_{-1}) \circ \phi \, J(\phi^{-1})\right) - \left(P_h^j u, K\partial_\omega^2(i\eta v_{-1}) \circ \phi \, J(\phi^{-1})\right) . \tag{16.23c}$$

We remark that the Galerkin system (16.12) is solved only once, with the iteration effecting only the coefficient c_0 and u_h in elements adjacent to the corner.

As seen in the following lemma, a post–processing of the conventional finite element produces a convergence rate of the same order as for the solution u_h.

Lemma 9 *The coefficient of the leading singularity for the first iteration converges asymptotically of order* $2r_0 := 2\left(\frac{\pi}{\omega_0} - \epsilon_0\right)$; *that is,*

$$|c_0 - c_0^1| = O(h^{2r_0}) \, , \tag{16.24}$$

for any $\epsilon_0 > 0$.

Proof. Noting that $\Delta(i\eta v_{-1}) \circ \phi \, J(\phi^{-1})$ and $K\partial_\omega^2(i\eta v_{-1}) \circ \phi \, J(\phi^{-1})$ are elements of $L^2(\Omega_F)$, it follows from (16.21b) and (16.23c) that

$$\begin{aligned} |c_0 - c_0^1| &\le \left|\left(u - P_h^1 u, \Delta(i\eta v_{-1}) \circ \phi \, J(\phi^{-1})\right)\right| + \\ &\quad + \left|\left(u - P_h^1 u, K\partial_\omega^2(i\eta v_{-1}) \circ \phi \, J(\phi^{-1})\right)\right| \\ &\le c\|u - P_h^1 u\|_{L^2(\Omega_h)}; \end{aligned}$$

by the Cauchy–Schwartz inequality. The result is then a consequence of Theorem 8.

That the iterative procedure outlined in (16.23a-16.23c) alleviates the pollution effect of the corner is seen in the next result.

Theorem 10 *The iterative procedure (16.23a-16.23c) displays the following asymptotic convergence rates:*

$$\|u - P_h^j u\|_{H^1(\Omega_F)} = O(h) \;\; ; \;\; \|u - P_h^j u\|_{L^2(\Omega_F)} = O(h^2 + H^{2jr_0}) \tag{16.25a}$$

and

$$|c_0 - c_0^j| = O(h^2 + h^{2jr_0}) , \tag{16.25b}$$

where $r_0 = \frac{\pi}{\omega_0} - \epsilon_0$, *for any* $\epsilon_0 > 0$ *and* $j \geq 2$.

Proof. From (16.22a) and (16.23a-16.23c), we see that

$$\begin{aligned} P_h^j u &= P_h u + c_0^{j-1}(\eta\, u_1 \circ \phi - P_h(\eta\, u_1 \circ \phi)) \\ &= P_h w + \left[c_0 P_h(\eta\, u_1 \circ \phi) + c_0^{j-1}(\eta\, u_1 \circ \phi - P_h(\eta\, u_1 \circ \phi))\right] , \end{aligned}$$

in which case

$$u - P_h^j u = w - P_h w + (c_0 - c_0^{j-1})(\eta\, u_1 \circ \phi - P_h(\eta\, u_1 \circ \phi))$$

and

$$\begin{aligned} \|u - P_h^j u\|_{L^2(\Omega_h)} &\leq \|w - P_h w\|_{L^2(\Omega_h)} \\ &+ |c_0 - c_0^{j-1}| \|\eta\, u_1 \circ \phi - P_h(\eta\, u_1 \circ \phi)\|_{L^2(\Omega_h)} , \end{aligned} \tag{16.26}$$

where u is provided by (16.21a). For $j = 2$, it follows from (16.26) and (16.24) that

$$\begin{aligned} \|u - P_h^2 u\|_{L^2(\Omega_h)} &\leq \|w - P_h w\|_{L^2(\Omega_h)} + ch^{2r_0} \|\eta\, u_1 \circ \phi - P_h(\eta\, u_1 \circ \phi)\|_{L^2(\Omega_h)} \\ &\leq ch^2 + ch^{2r_0} h^{2r_0} \leq c(h^2 + h^{2r_0 + 2^{r_0}}) , \end{aligned}$$

for $\epsilon_0 > 0$ and $r_0 = \left(\frac{\pi}{\omega_0} - \epsilon_0\right)$. The result for $j \geq 2$ follows by induction from (16.26). The estimate for the coefficient is an immediate consequence of the inequality $|c_0 - c_0^j| \leq \|u - P_h^j u\|_{L^2(\Omega_h)}$. The H^1–estimate follows similarly.

Remark. The proposition indicates that no improvement in convergence rate is obtained by utilizing more than two iterations. It follows that the iterative procedure (16.23a-16.23c) is most efficiently applied with one corrective step ($j = 2$) which produces convergence orders of the quality of a smooth solution.

The Dual Singular Function Method is an augmented Galerkin procedure which allows the direct determination of the leading term in the corner expansion at a rate comparable to that of a smooth domain. This is accomplished through an augmentation of the test and trial spaces such that

$$V_h^{(1)} := V_h \oplus \eta\langle u_1 \circ \phi\rangle$$

and

$$V_h^{(-1)} := V_h \oplus \eta\langle i v_{-1} \circ \phi\rangle ,$$

where $\langle\ \cdot\ \rangle$ refers to the span of the indicated element. The bilinear form A_h is extended to $V_h^{(1)} \times V_h^{(-1)}$ by the definition (16.9b) when $u \in V_h^1$, $v \in V_h$,

$$A_h(u, v) := \left(u, \Delta(v) \circ \phi\; J(\phi^{-1})\right) + \left(u, K\partial_\omega^2(v) \circ \phi\; J(\phi^{-1})\right) , \tag{16.27a}$$

for $u \in V_h$ and any v in the span of $\eta\langle iv_{-1}\rangle$, and by

$$A_h(u,v) := 1 + \big(u\circ\phi, \Delta(v)\circ\phi\, J(\phi^{-1})\big) + \big(u\circ\phi, K\partial^2_\omega(i\eta v)\circ\phi\, J(\phi^{-1})\big)\ , \tag{16.27b}$$

for any u and v in the span of $\eta\langle u_1\rangle$ and $\eta\langle iv_{-1}\rangle$, respectively. The extension (16.27b) follows from the representation of the singularity coefficient for the function ηu_1; namely,

$$1 = \int_{\overline{K}} \Delta(\eta u_1)\, i\eta v_{-1}\, r\, dr d\omega - \int_{\overline{K}} \eta u_1\, \Delta(i\eta v_{-1})\, r\, dr d\omega\ ,$$

where trivially $\eta u_1 = 1\cdot\eta u_1 + 0$ (cf. [27]).

Theorem 11 *There exists a unique solution* (i=0,1)

$$P_h^\infty\psi^{(i)} = c_0^{\infty,(i)}\eta r^{\pi/\omega_0}\sin\frac{\pi}{\omega_0}\omega\circ\phi + P_h^\infty w \in V_h^{(1)}\ , \tag{16.28}$$

satisfying, for sufficiently small h,

$$A_h(P_h^\infty\psi^{(i)}, v) = \ell_h^{(i)}(v)\ ;\ \ \forall v\in V_h^{(-1)}\ , \tag{16.29}$$

such that $\lim_{j\to\infty} P_h^j\psi^{(I)} = P_h^\infty\psi^{(i)}$*, provided the limit exists. Moreover, the following estimates hold:*

$$\|\psi^{(i)} - P_h^\infty\psi^{(i)}\|_{H^1(\Omega_h)} = O(h)\ ;\ \ \|\psi^{(i)} - P_h^\infty\psi^{(i)}\|_{L^2(\Omega_h)} = O(h^2)$$

as $h\to 0^+$.

Proof. The demonstration follows by considering, in turn, the unique solvability of the method, its characterization by the iterative scheme (16.23a-16.23c), and asymptotic estimates.

(I.) *Existence and Uniqueness.* Let $(u=\psi^{(i)})$ (i=0,1), then since $P_h^\infty u\in V_h^{(1)}$,

$$A_h(P_h^\infty u, v) = A_h(P_h u, v) = \ell_h^{(i)}(v)\ ,$$

for any $v\in V_h$, from which it follows by (16.28) and the regularity of w that

$$P_h^\infty w = P_h w = P_h u - c_0^\infty P_h(\eta\, u_1\circ\phi)\ , \tag{16.30a}$$

and so

$$P_h^\infty u = P_h u + c_0^\infty\left(\eta\, u_1\circ\phi - P_h(\eta\, u_1\circ\phi)\right)\ . \tag{16.30b}$$

From (16.30b) and (5.7) then

$$\begin{aligned} A_h(P_h^\infty u, i\eta v_{-1}\circ\phi) &= A_h(P_h u, i\eta v_{-1}\circ\phi) \\ &+ A_h\left(c_0^\infty(\eta\, u_1\circ\phi - P_h(\eta\, u_1\circ\phi)), i\eta v_{-1}\circ\phi\right) \\ &= \left(P_h u, \Delta(i\eta v_{-1})\circ\phi\, J(\phi^{-1})\right) + \left(P_h u, K\partial^2_\omega(i\eta v_{-1})\circ\phi\, J(\phi^{-1})\right) \\ &+ c_0^\infty\Big[1 - \left(P_h(\eta\, u_1)\circ\phi - \eta\, u_1\circ\phi, \Delta(i\eta v_{-1})\circ\phi\, J(\phi^{-1})\right) \\ &- \left(P_h(\eta\, u_1)\circ\phi - \eta\, u_1\circ\phi, K\partial^2_\omega(i\eta v_{-1})\circ\phi\, J(\phi^{-1})\right)\Big]\ . \end{aligned}$$

Noting that Ψ_∞ is harmonic and consequently that $\ell_h^{(i)}(i\eta v_{-1}\circ\phi)=0$, we have then

$$\begin{aligned} c_0^\infty\Big[1 &- (P_h(\eta u_1)\circ\phi-\eta u_1\circ\phi,\Delta(i\eta v_{-1})\circ\phi\, J(\phi^{-1})) \\ &- (P_h(\eta u_1)\circ\phi-\eta u_1\circ\phi,K\partial_\omega^2(i\eta v_{-1})\circ\phi\, J(\phi^{-1}))\Big] \\ &= -(P_hu,\Delta(i\eta v_{-1})\circ\phi\, J(\phi^{-1}))-(P_hu,K\partial_\omega^2(i\eta v_{-1})\circ\phi\, J(\phi^{-1}))\ , \end{aligned} \tag{16.31}$$

which is uniquely solvable for c_0^∞ as $P_h(\eta u_1)\circ\phi-\eta u_1\circ\phi$ may be made arbitrarily small depending upon the choice of h. Consequently, $P_h^\infty u$ exists and is unique.
(II.) *Characterization of $P_h^\infty u$.* From (16.30) and (16.30b), we obtain a representation for c_0^∞ comparable to (16.23c); that is,

$$\begin{aligned} c_0^\infty &= -\Big(P_h^\infty u,\Delta(i\eta v_{-1})\circ\phi\, J(\phi^{-1})\Big) \\ &- \Big(P_h^\infty u,K\partial_\omega^2(i\eta v_{-1})\circ\phi\, J(\phi^{-1})\Big)\,. \end{aligned} \tag{16.32}$$

Comparing (16.32) and (16.30a-16.30b) with (16.23a-16.23c) as $j\to\infty$ shows that $\lim_{j\to\infty}P_h^ju=P_h^\infty u$.
(III.) *Asymptotic Convergence Estimate.* We present the result for the L^2-norm. From (16.28), (16.30a-16.30b) and (16.21a) it follows that

$$P_h^\infty u=P_hw+[c_0^\infty\eta u_1\circ\phi+(c_0-c_0^\infty)P_h(\eta u_1)\circ\phi]\ ,$$

in which case

$$u-P_h^\infty u=w-P_hw+(c_0-c_0^\infty)(\eta u_1\circ\phi-P_h(\eta u_1\circ\phi))$$

and so

$$\begin{aligned} \|u-P_h^\infty u\|_{L^2(\tilde\Omega)} &\le \|w-P_hw\|_{L^2(\tilde\Omega)} \\ &+ |c_0-c_0^\infty|\|\eta u_1\circ\phi-P_h(\eta u_1\circ\phi)\|_{L^2(\tilde\Omega)}\,. \end{aligned} \tag{16.33}$$

Moreover, as seen from (16.32), (16.21b) and the Cauchy–Schwartz inequality,

$$\begin{aligned} |c_0-c_0^\infty| &= \big|(u-P_h^\infty u,\Delta(i\eta v_{-1})\circ\phi J(\phi^{-1}))\big| \\ &\le c\|u-P_h^\infty u\|_{L^2(\tilde\Omega)}\ , \end{aligned}$$

from which, for sufficiently small h, we find

$$|c_0-c_0^\infty|\ \|\eta u_1\circ\phi-P_h(\eta u_1\circ\phi)\|_{L^2(\tilde\Omega)}\le\delta\|u-P_h^\infty u\|_{L^2(\tilde\Omega)}\ ,$$

where $0<\delta<1$. Since (16.33) implies

$$\|u-P_h^\infty u\|_{L^2(\Omega_h)}\le\|w-P_hw\|_{L^2(\Omega_h)}\ ,$$

the estimate follows by the regularity of w.

Table 1. *Approximation results for $\alpha = 0^0$.*
$\kappa = 0.000\ 000$

Nodes	h	c_0	c_1	κ_h
1673	0.322 964	0.003 851	-0.142 394	0.027 042
2227	0.295 694	0.002 357	-0.148 800	0.015 837
3001	0.249 280	0.001 548	-0.148 880	0.006 299
3565	0.228 797	0.000 939	-0.149 083	0.006 299
4527	0.199 638	0.000 352	-0.148 964	0.002 364

Table 2. *Approximation results for $\alpha = 10^0$.*
$\kappa = 2.341\ 492$

Nodes	h	c_0	c_1	κ_h
1673	0.322 964	0.327 985	-0.142 394	2.303 366
2227	0.295 694	0.343 816	-0.148 800	2.310 591
3001	0.249 280	0.345 608	-0.148 880	2.321 389
3565	0.228 797	0.347 231	-0.149 083	2.329 104
4527	0.199 638	0.348 356	-0.148 964	2.338 534

Convergence orders for the lifting solution $\Psi_F = \psi_F + \Psi_\infty$ may now be obtained for the standard finite element method of the previous section, where the coefficient of the leading singularity is determined by (16.23a-16.23c) with $j = 1$, and for the Dual Singular Function Method of Theorem 11.

Theorem 12 *The following estimates hold:*

$$\|\Psi_F - P_h\Psi_F\|_{H^1(\Omega_h)} = O(h^{r_0}) \ ; \ \|\Psi_F - P_h\Psi_F\|_{L^2(\Omega_h)} = O(h^{2r_0}) \tag{16.34a}$$

and

$$\|\Psi_F - P_h^\infty\Psi_F\|_{H^1(\Omega_h)} = O(h) \ ; \ \|\Psi_F - P_h^\infty\Psi_F\|_{L^2(\Omega_h)} = O(h^2) \ , \tag{16.34b}$$

where $r_0 = \frac{\pi}{\omega_0} - \epsilon_0$, for any $\epsilon_0 > 0$.

Proof. Let $\psi^{(i)} = c_0^{(i)}\eta\, u_1 \circ \phi + w^{(i)}$ denote the variational solutions ($i = 0, 1$) of (16.6) and $\psi_h^{(i)} = P_h\psi^{(i)}$ the finite element approximation satisfying the Galerkin system (16.12) such that $c_0^{(i)}$ is determined by (16.23c) with $j = 1$. Then as $\Psi_F = \psi^{(0)} + \kappa\psi^{(1)}$, we obtain the estimate for the lifting solution ψ_F over the polygonal domain Ω_h:

$$\|\Psi_F - P_h\Psi_F\| \leq \|\psi^{(0)} - P_h\psi^{(0)}\| + |\kappa - \kappa_h|\ \|\psi^{(1)} - P_h\psi^{(1)}\| \ , \tag{16.35a}$$

for either the L^2 or H^1 norms, and $\kappa = -c_0^{(0)}/c_0^{(1)}$, $\kappa_h = -c_0^{1,(0)}/c_0^{1,(1)}$ are well–defined for sufficiently small h by Lemma 9 and Theorem 1. Noting that

$$|\kappa - \kappa_h| \leq c\left(|c_0^{(0)} - c_0^{1,(0)}| + |c_0^{(1)} - c_0^{1,(1)}|\right) , \tag{16.35b}$$

for some constant c, the estimate (16.34a) follows from (16.35a-16.35b), Theorems 6 and 8, and Lemma 9. The rates of convergence (16.34b) follow from Theorem 11.

16.6 NUMERICAL EXPERIMENTS

We present in this section a verification of method efficiencies based on the convergence rate of the circulation constant κ for the flow past a Kármán–Trefftz profile (cf. [31]). In particular, experiments are performed for a unit free–stream velocity (i.e., $\mathbf{q}_\infty = 1$) and angles of attack (relative to the zero–lift axis of the profile) of $0°$ and $10°$ (cf. Figure 1). Calculations follow for the standard finite element method and include the first perturbation term in the expansion for the coefficient of the leading singularity (i.e., equation (16.23a-16.23c) with $j = 1$ and $K(|\mathbf{x}|) = \frac{K_1}{|\mathbf{x}|}$). An auxiliary boundary of radius 5 is utilized in all trials as are cut–off function of radii $r_1 = 0.10$ and $r_0 = 0.05$.

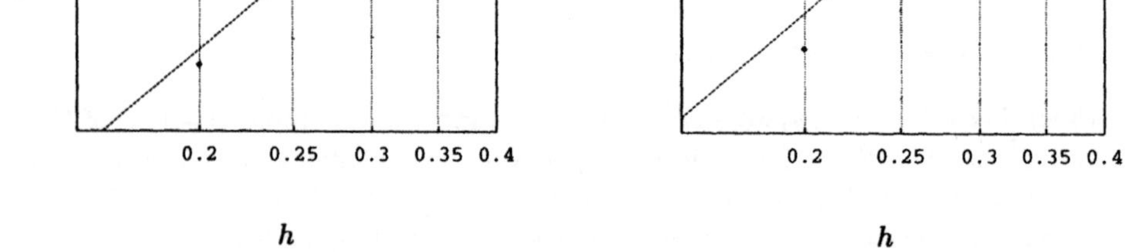

a. $\alpha = 0°$ **b.** $\alpha = 10°$

Figure 5. *Approximation of the circulation.*

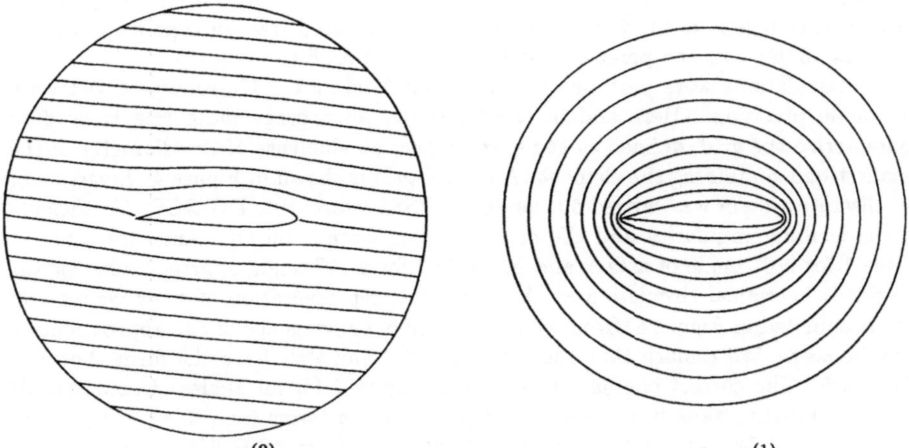

a. $\psi^{(0)}$ b. $\psi^{(1)}$

Figure 6. *Decoupled solutions,* $\alpha = 10^0$.

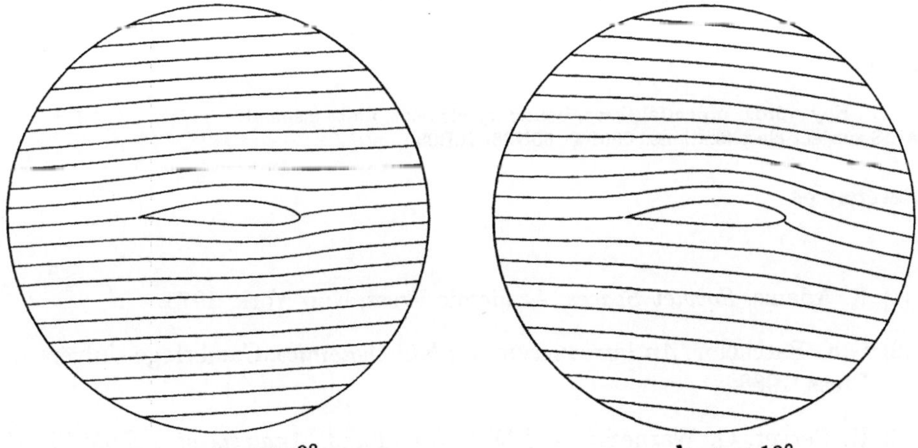

a. $\alpha = 0^\circ$ b. $\alpha = 10^\circ$

Figure 7. *The coupled solution.*

Integration between standard basis functions was exact, as was that between singular functions. Singular basis function/standard basis function interactions were likewise carried out analytically in the radial coordinate while Simpson's rule provided the means of integrating over ω. A difference of 10^{-7} between successive applications of the rule was considered satisfactory for convergence, a higher order not being sustainable, in general, due to the approximation of the boundary by linear elements for the courser mesh. A least-squares fit of a 9th–order polynomial was used to emulate the functions $\Theta_\pm$ for radius less than $2r_1$ for the transformation from the linear corner region K to $\overline{K}$ (see Figure 2). The enclosed trailing edge angle in this case was 18.091°, as compared to 18° for the actual profile.

All calculations were performed on the IBM 3090 at the University of Delaware in double precision. Here and in the following, an average mesh size is used to characterize the grid, defined as the average over all elements of the diameter of the smallest ball containing the element. A typical grid is shown in Figure 4. Inversion of the discrete system was carried out using the ESSL subroutine DGEICD. Coefficients of the leading singularity and the corresponding circulation constants are listed in Table 1 for a 0° angle of attack and in Table 2 for a 10° angle of attack, and for the specified theoretical circulation κ. A logarithmic representation of error versus grid size is depicted in Figure 5, from which the rate of convergence of the approximation may be seen. We remark that the solutions $\psi^{(0)}$ and $\psi^{(1)}$ have the form shown in Figures 6. The correct coupling is shown in Figure 7 for an angles of attack of 0° and 10°. Finally, we indicate that for no perturbation terms (i.e. $K(\mathbf{x}) \equiv 0$), c_1 was found to be -0.147 518, while the circulations were $\kappa_h = 0.002\ 378$ for $\alpha = 0°$ and $\kappa_h = 2.338\ 331$ when $\alpha = 10°$.

To conclude the paper, we remark that the present approach based on a combination of finite and boundary element method seems feasible for treating three-dimensional problems as well as for more general flow problems. In this connection, we refer to the recent papers [3],[34] for the transonic flow.

Notes

1. Key words: potential flows, boundary element, finite element.
AMS subject classifications: 65M60, 65N38, 76B05.

References

[1] R. Adams, *Sobolev Spaces*, Academic Press, New York, 1975.

[2] G.K. Batchelor, *An Introduction to Fluid Dynamics*, Cambridge University Press, 1988.

[3] H. Berger, G. Warnecke, and W.L. Wendland, *Analysis of a FEM/BEM coupling method for transonic flow computations*, Preprint 93-9, Mathematisches Institut A, Univ. Stuttgart, 1993.

[4] H. Blum, *A simple and accurate method for the determination of stress intensity factors and solutions for problems on domains with corners*, The Mathematics of Finite Elements and Applications IV (Ed. J.R. Whiteman), Academic Press, 1982.

[5] H. Blum, *Numerical treatment of corner and crack singularities*, FEM and BEM Techniques from Mathematical and Engineering Points of View, Saarbrücken, 1987.

[6] H. Blum and M. Dobrowolski, *A finite element method for elliptic equations in domains with corners*, manuscript, Univ. of Bonn, 1980.

[7] P.G. Ciarlet, *The Finite Element Method for Elliptic Problems*, North–Holland, Amsterdam, 1978.

[8] J.F. Ciavaldini, M. Pogu, and G. Tournemine, *Existence and regularity of stream functions for subsonic flows past profiles with a sharp trailing edge*, Arch. Rat. Mech. Anal., 93 (1986), pp. 1-14.

[9] M. Costabel and E. Stephen, *Curvature terms in the asymptotic expansions for solutions of boundary integral equations on curved polygons*, J. Integral Equations, 5 (1983), pp. 353–371.

[10] M. Dauge and M. Pogu, *Existence et régularité de la function potential pour des écoulements subcritiques s'établissant autour d'un corps à singularité conique*, Ann. Fac. des Sci. de'Toulouse, IX (1988), pp. 213–242.

[11] M. Dauge and J.L. Steux, *Problème de Dirichlet pour le Laplacian dans un polygone curviligne*, J. Diff. Equat., 70 (1987), pp. 93–113.

[12] M. Dauge, J. Lubuma, and J. Nicaise, *'Coefficients des singularitiés pour le problème de Dirichlet sur un polygone*, C.R. Acad. Sc. Paris 304 *Série I*, 16 (1987), pp. 483–486.

[13] M. Djaoua, *'A method of calculation of lifting flows around two-dimensional corner shaped bodies*, Math. Comput., 36 (1981), pp. 405–425.

[14] M. Dobrowolski, *Numerical approximation of elliptic interface and corner problems*, Habilitationsschrift, Univ. Bonn, 1981.

[15] M. Dobrowolski, *On finite element methods for nonlinear elliptic problems on domains with corners*, Singularities and Constructive Methods for their Treatment, Lecture Notes in Mathematics 1121, Springer–Verlag, 1985.

[16] M. Dobrowolski, *On quasilinear elliptic equations in domains with conical boundary points*, Report 8506, Univ. der Bundeswehr München, FB Informatik, 1985.

[17] M. Feistauer, *Mathematical Methods in Fluid Dynamics*, Pitman, New York, 1993.

[18] M. Feistauer and A. Zenísek, *Finite element solution of nonlinear elliptic problems*, Numer. Math., 50 (1987), pp. 451–475.

[19] M. Feistauer, J. Felcman, M. Rokyta, and Z. Vlásek, *Finite-element solution of flow problems with trailing conditions*, J. Comp. Appl. Math., 44 (1992), pp. 131–165.

[20] P. Grisvard, *Elliptic Problems in Nonsmooth Domains*, Pitman, London, 1985.

[21] S. Hilderbrandt and E. Wienholtz, *Constructive proofs of representation theorems in seperable Hilbert space*, Com. Pure and Applied Math., 17 (1964), pp. 369–373.

[22] G.C. Hsiao, *The coupling of boundary element and finite element methods*, Z. angew. Math. Mech., 70 (1990), pp. T493–T503.

[23] G.C. Hsiao, M.D. Marcozzi, and S. Zhang, *An efficient computational method for the flow past an airfoil*, Contemporary Math., 180 (1994), pp. 497–502.

[24] G.C. Hsiao and M.D. Marcozzi, *Variational mehtods for the potential flow past an airfoil*, Appl. Anal. 65, (1997), pp. 153–182.

[25] C. Johnson and J.C. Nedelec, *On the coupling of boundary integral and finite element methods*, Math. Comput., 35 (1980), pp. 1063–1079.

[26] V.A. Kondratiev, *Boundary value problems for elliptic equations in domains with conical or angular points*, Trudy Moskovkogo Mat. Obschetsva, 16 (1967), pp. 209–292 (and Transactions Moscow Mat. Soc., pp. 227–313).

[27] A. Kufner and A.M. Sändig, *Some Applications of Weighted Sobolev Spaces*, BSB Tueber, Leipzig, 1987.

[28] M.N. Le Roux, *Méthod d'éléments finis pour la résolution numérique de problèmes extérieurs en dimension 2*, R.A.I.R.O. Anal. Numér., 11 (1977), pp. 27–60.

[29] V.G. Maz'ya, *Boundary integral equations*, Encyclopedia of Mathematical Science, 27, pp. 127–222, Springer–Verlag, Berlin, 1991.

[30] V.A. Maz'ya and B.A. Plamenevskiĭ, *On the coefficients of the solutions of elliptic boundary value problems in domains with conical points*, Math. Nachr., 76 (1977), pp. 29–60 (and Transl., II, Am. Math. Soc., 123 (1984), pp. 57–88).

[31] R.Von Mises, *Theory of Flight*, Dover, 1959.

[32] M. Schiffer, *Analytic theory of subsonic and supersonic flows*, Encylopedia of Physics, 9, pp. 1-161, Springer–Verlag, New York, 1960.

[33] W.L. Wendland, *On asymptotic error estimates for the combined BEM and FEM*, Finite Element and Boundary Element Techniques from Mathematical and Engineering point of view, CISM Lecture Notes 301, Udine, Springer–Verlag (Wein–New York); E. Stein, W.L. Wendland, pp. 273–333, 1988.

[34] W.L. Wendland and C. Coclici,*On the convergence of a coupled finite element-boundary element method for the transonic full potential flow*, to appear.

[35] J. Wloka, *Partial Differential Equations*, Cambridge University Press, 1987.

[36] S.S. Zargaryan and V.G. Maz'ya, *On the asymptotes of the solutions of the integral equations of potential theory in a neighborhood of angular points of the contour*, Prikl. Mat. Mekh., 48 (1984), pp. 169–174 (and J. Appl. Math. Mech, 48 (1985), pp. 120–124.

17 BOUNDARY VALUE PROBLEMS OF NONLINEAR ELASTOSTATICS

K. Taira

Dedicated to Professor R.P. Gilbert on the occasion of his 65th birthday

17.1 INTRODUCTION

[1] This paper is devoted to an L^p approach to boundary value problems of nonlinear elastostatics in the case where solutions of the linearized problem correspond faithfully to those of the nonlinear problem, that is, in the case where there is no bifurcation. We show that if the linearized problem has unique solutions, then so does the nonlinear one, nearby. This is done by using the linear L^p theory and the inverse mapping theorem. Our boundary condition is a genuine mixed displacement-traction boundary condition; more precisely, it is a *linear combination* of displacement and traction boundary conditions, but is not equal to the pure traction boundary condition (Main Theorem). The crucial point is how to find a function space associated with the boundary condition in which the linearized problem has unique solutions. Our result can be applied to the St. Venant-Kirchhoff elastic material and the Hencky-Nadai elasto-plastic material (Theorem 1 and Theorem 2). Some previous results with pure displacement boundary condition are due to Ciarlet [4], Dinca [5], Marsden-Hughes [10] and Valent [14]. The results here extend and improve substantially those results in a *unified* theory. For detailed studies of boundary value problems of nonlinear elastostatics, the reader is referred to Taira [12].

H. Florian et al. (eds.), Generalized Analytic Functions, 183–195.

17.2 THE MAIN RESULTS

Let Ω be an open, connected subset of $\mathbb{R}^3$ with piecewise smooth boundary $\partial\Omega$. We think of the closure $\overline{\Omega}$ of Ω as representing the volume occupied by an undeformed body; so the set $\mathcal{B} = \overline{\Omega}$ is called the *reference configuration*. A *configuration* of $\mathcal{B}$ is a C^1 map $\phi : \mathcal{B} \to \mathbb{R}^3$ which is orientation-preserving and invertible. A configuration represents a deformed state of the body (see Figure 17.1 below). Points in $\mathcal{B}$ are denoted by $X = (X_1, X_2, X_3) \in \mathcal{B}$ and are called *material points*, while points in $\mathbb{R}^3$ are denoted by $x = (x_1, x_2, x_3) \in \mathbb{R}^3$ and are called *spatial points*. We write as

$$x = \phi(X); \quad x_i = \phi_i(X).$$

The 3×3 matrix of partial derivatives of ϕ is denoted by $\mathbf{F}(X) = D\phi(X)$ and is called the *deformation gradient*:

$$F_{ij} = \frac{\partial \phi_i}{\partial X_j}.$$

Note that $\det \mathbf{F} > 0$, since ϕ is orientation-preserving. The symmetric tensor

$$\mathbf{C} = {}^t\mathbf{F}\mathbf{F}; \quad C_{ij} = \sum_{k=1}^{3} F_{ki}F_{kj},$$

is called the Green *deformation tensor*.

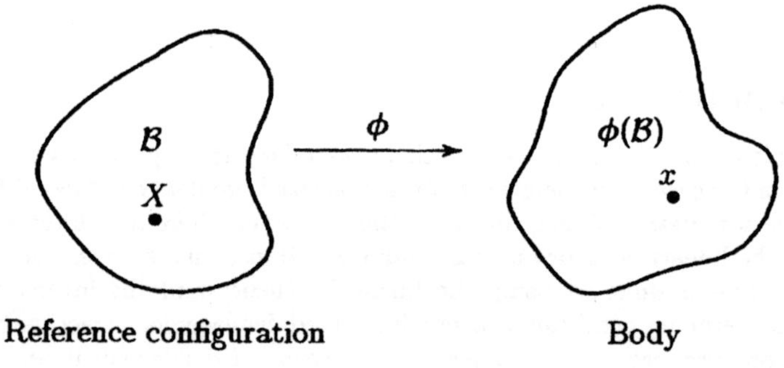

Figure 17.1

A body $\phi(\mathcal{B})$ is acted on by applied *body forces* $\mathbf{b}(x)$ in its interior and by applied *surface forces* $\boldsymbol{\tau}(x)$ on a portion of the boundary (see Figure 17.2 below). The pair $(\mathbf{b}, \boldsymbol{\tau})$ of forces is called the *load*, and is given or prescribed in advance.

In addition, the body generally experiences internal forces of stress across any given surface. Let $\mathbf{t}(x, \mathbf{n})$ be the force at position x across an oriented surface element with outward unit normal $\mathbf{n}$. The celebrated Cauchy theorem asserts (cf. [4, Chapter 2];

[10, Chapter 2]) that if the balance of momentum holds, then the stress vector $\mathbf{t}(x, \mathbf{n})$ depends linearly on $\mathbf{n}$, that is, there exists a two-tensor $\boldsymbol{\sigma}(x)$ such that

$$\mathbf{t}(x, \mathbf{n}) = \boldsymbol{\sigma}(x) \cdot \mathbf{n}; \quad t_i(x, \mathbf{n}) = \sum_{j=1}^{3} \sigma_{ij}(x) n_j.$$

The vector $\mathbf{t}(x, \mathbf{n})$ is called the *Cauchy stress vector* and the tensor $\boldsymbol{\sigma}(x)$ is called the *Cauchy stress tensor*. We remark that the balance of angular momentum leads to the symmetry of the tensor $\boldsymbol{\sigma}(x)$: $\sigma_{ij}(x) = \sigma_{ji}(x)$.

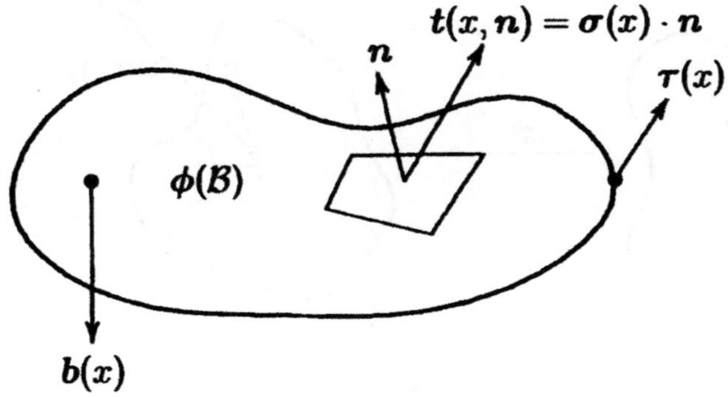

Figure 17.2

The vector $\mathbf{T}(X, \mathbf{N})$, defined by

$$\mathbf{T}(X, \mathbf{N}) = \mathbf{P}(X) \cdot \mathbf{N}, \tag{17.1}$$

$$\mathbf{P}(X) = \det(D\phi(X))\boldsymbol{\sigma}(\phi(X)) \cdot ({}^t D\phi(X))^{-1}, \tag{17.2}$$

is called the *first Piola-Kirchhoff stress vector*, where $\mathbf{N}$ is the outward unit normal at X. We remark that the vector $\mathbf{T}(X, \mathbf{N})$ is parallel to the Cauchy stress vector $\mathbf{t}(x, \mathbf{n})$, but measures the force per unit undeformed area with normal $\mathbf{N}$ (see Figure 17.3 below). The two-tensor $\mathbf{P}(X)$, which is the Piola transform of the Cauchy stress tensor $\boldsymbol{\sigma}(x)$, is called the *first Piola-Kirchhoff stress tensor*.

A material is said to be *elastic* if one can write the first Piola-Kirchhoff stress tensor $\mathbf{P}(X)$ as a function $\widehat{\mathbf{P}}(X, \mathbf{F})$ of points $X \in \mathcal{B}$ and 3×3 matrices $\mathbf{F}$ with $\det \mathbf{F} > 0$ such that

$$\mathbf{P}(X) = \widehat{\mathbf{P}}(X, D\phi(X)).$$

Specification of a function $\widehat{\mathbf{P}}(X, \mathbf{F})$ for an elastic material is called a *stress-strain law*, and $\widehat{\mathbf{P}}(X, \mathbf{F})$ is called a *constitutive function*.

An elastic material is said to be *hyperelastic* if there exists a function $W(X, \mathbf{F})$ of points $X \in \mathcal{B}$ and 3×3 matrices $\mathbf{F}$ with $\det \mathbf{F} > 0$ such that

$$\widehat{\mathbf{P}}(X, \mathbf{F}) = \frac{\partial W}{\partial \mathbf{F}}(X, \mathbf{F}); \quad \widehat{P}_{ij}(X, \mathbf{F}) = \frac{\partial W}{\partial F_{ij}}(X, \mathbf{F}).$$

The function $W(X, \mathbf{F})$ is called a *stored energy function*. The four-index tensor $\mathbf{A} = \partial\widehat{\mathbf{P}}/\partial\mathbf{F} = \partial^2 W/\partial\mathbf{F}\partial\mathbf{F}$, defined by

$$\mathrm{A}_{ij\ell m}(X, \mathbf{F}) = \frac{\partial \widehat{P}_{ij}}{\partial F_{\ell m}}(X, \mathbf{F}) = \frac{\partial^2 W}{\partial F_{ij} \partial F_{\ell m}}(X, \mathbf{F}),$$

is called the *first elasticity tensor*.

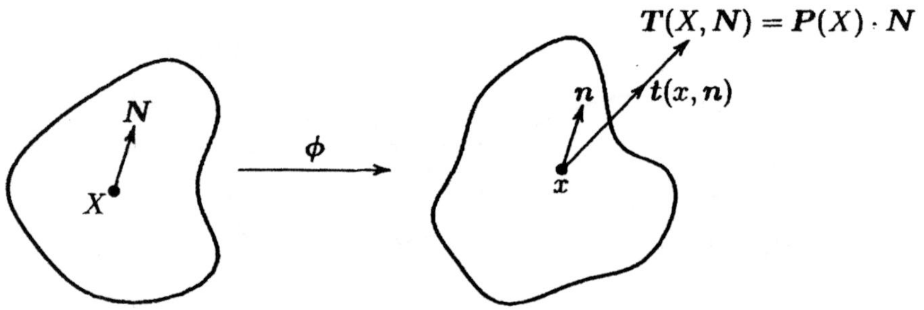

Figure 17.3

Some boundary conditions often encountered are the following:

(a) *Displacement*: $\phi(X) = \phi_d(X)$ is described for $X \in \partial\mathcal{B}$.

(b) *Traction*: $\widehat{\mathbf{P}}(X, D\phi(X)) \cdot \mathbf{N}(X) = \tau(X)$ is described for $X \in \partial\mathcal{B}$, where $\mathbf{N}(X)$ is the outward unit normal to $\partial\mathcal{B}$ at X.

We make the following two assumptions:

(H.1) The reference configuration is a *bounded* region $\mathcal{B} = \overline{\Omega} \subset \mathbb{R}^3$ with *smooth* boundary $\partial\Omega$.

(H.2) The material is *hyperelastic*; that is, we are given a smooth stored energy function $W(X, \mathbf{F})$ such that $\widehat{\mathbf{P}}(X, \mathbf{F}) = (\partial W/\partial\mathbf{F})(X, \mathbf{F})$.

We give two examples of stored energy functions for hyperelastic materials.

Example 1 (The Hencky-Nadai elasto-plastic material) The stored energy function has the form:

$$W(X, \mathbf{F}) = \frac{3}{4}\int_0^{\Gamma(\mathbf{F})} g(\xi)\, d\xi + \frac{K}{2}\left(\sum_{k=1}^{3} F_{kk} - 3\right)^2,$$

where $g \in C^\infty([0, \infty), \mathbb{R})$, K is the modulus of compression and

$$\Gamma(\mathbf{F}) = \frac{4}{3}\sum_{i,j=1}^{3}\left(\frac{1}{2}(F_{ij} + F_{ji}) - \frac{1}{3}\left(\sum_{k=1}^{3} F_{kk}F_{kk}\right)\delta_{ij}\right)^2.$$

The first Piola-Kirchhoff stress tensor is given by

$$\widehat{P}_{ij}(X,\mathbf{F}) = \left(K - \frac{2}{3}g(\Gamma(\mathbf{F}))\right)\left(\sum_{k=1}^{3} F_{kk} - 3\right)\delta_{ij} + g(\Gamma(\mathbf{F}))\,(F_{ij} + F_{ji} - 2\delta_{ij})\,.$$

Example 2 (The St. Venant-Kirchhoff isotropic material) The stored energy function has the form:

$$W(X,\mathbf{F}) = \frac{\lambda(X)}{8}\left(\sum_{k=1}^{3} C_{kk}(\mathbf{F}) - 3\right)^2 + \frac{\mu(X)}{4}\sum_{i,j=1}^{3}(C_{ij}(\mathbf{F}) - \delta_{ij})^2\,,$$

where $\lambda(X)$, $\mu(X)$ are smooth functions and

$$C_{ij}(\mathbf{F}) = \sum_{k=1}^{3} F_{ki}F_{kj}$$

is the deformation tensor.

The first Piola-Kirchhoff stress tensor is given by

$$\widehat{P}_{ij}(X,\mathbf{F}) = \left[\frac{\lambda(X)}{2}\left(\sum_{k=1}^{3} C_{kk}(\mathbf{F}) - 3\right) - \mu(X)\right]F_{ij} + \mu(X)\sum_{m=1}^{3} F_{im}C_{mj}(\mathbf{F}).$$

We remark that the first Piola-Kirchhoff stress tensors in Examples 1 and 2 are *not* linear functions of the deformation gradient $\mathbf{F}$, which leads to equations of nonlinear elastostatics.

Let $\mathbf{B} : \Omega \to \mathbb{R}^3$ be the density of a given body force per unit volume in the reference configuration and $\tau : \partial\Omega \to \mathbb{R}^3$ the density of a given surface force per unit area in the reference configuration (see Figure 17.4 below). This paper is devoted to an L^p approach to the following equilibrium equations for the unknown configuration ϕ:

$$\begin{cases} \mathrm{DIV}\,\widehat{\mathbf{P}}(X, D\phi(X)) + \mathbf{B}(X) = 0 & \text{in } \Omega, \\ \alpha(X)\widehat{\mathbf{P}}(X, D\phi(X)) \cdot \mathbf{N}(X) + (1-\alpha(X))\phi(X) = \tau(X) & \text{on } \partial\Omega, \end{cases} \qquad (*)$$

where α is a smooth function on $\partial\Omega$ such that

$$0 \le \alpha \le 1 \quad \text{on } \partial\Omega.$$

We remark that our boundary condition is a *linear combination* of displacement and traction boundary conditions.

We study the nonlinear problem $(*)$ in the framework of L^p spaces. If s is a non-negative integer and $1 < p < \infty$, we define the Sobolev space

$$H^{s,p}(\Omega) = \text{the space of (equivalence classes of) functions} \qquad (17.3)$$
$$u \in L^p(\Omega) \text{ whose derivatives } D^\alpha u,\ |\alpha| \le s, \text{ in the} \qquad (17.4)$$
$$\text{sense of distributions are in } L^p(\Omega). \qquad (17.5)$$

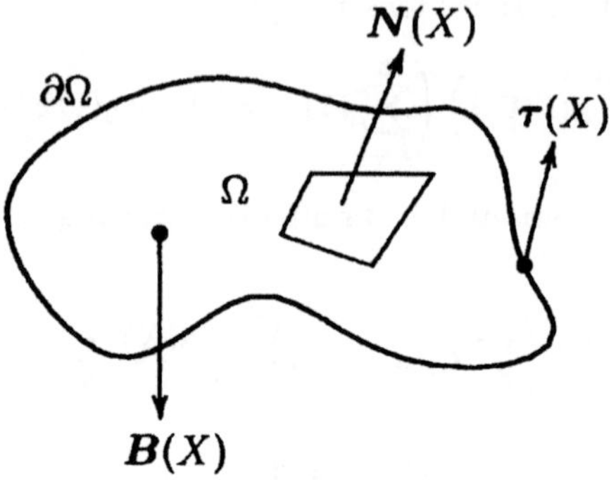

Figure 17.4

The space $H^{s,p}(\Omega)$ is a Banach space with the norm

$$\|u\|_{s,p} = \left(\sum_{|\alpha| \leq s} \int_\Omega |D^\alpha u(x)|^p \, dx \right)^{1/p} .$$

Further we define

$$B^{s-1/p,p}(\partial\Omega) = \text{the space of the boundary values } \varphi \text{ of functions} \tag{17.6}$$

$$u \in H^{s,p}(\Omega). \tag{17.7}$$

In the space $B^{s-1/p,p}(\partial\Omega)$, we introduce a norm

$$|\varphi|_{s-1/p,p} = \inf \|u\|_{s,p} ,$$

where the infimum is taken over all functions $u \in H^{s,p}(\Omega)$ which equal φ on the boundary. The space $B^{s-1/p,p}(\partial\Omega)$ is a Banach space with respect to this norm $|\cdot|_{s-1/p,p}$; more precisely, it is a Besov space (cf. [2], [13]).

Now we let

$$\mathbf{H}^{s,p}(\Omega, \mathbb{R}^3) = \text{ the space of all } H^{s,p} \text{ functions } \boldsymbol{\phi} : \Omega \to \mathbb{R}^3, \tag{17.8}$$

$$\mathbf{B}^{s-1/p,p}(\partial\Omega, \mathbb{R}^3) = \text{ the space of all } B^{s-1/p,p} \text{ functions } \boldsymbol{\varphi} : \partial\Omega \to \mathbb{R}^3. \tag{17.9}$$

We introduce a subspace of the Besov space $\mathbf{B}^{s-1-1/p,p}(\partial\Omega, \mathbb{R}^3)$ which is associated with the boundary condition

$$\alpha\widehat{\mathbf{P}}(D\boldsymbol{\phi}) \cdot \mathbf{N} + (1-\alpha)\boldsymbol{\phi} = \boldsymbol{\tau} \quad \text{on } \partial\Omega$$

as follows: We let

$$\mathbf{B}_{(\alpha)}^{s-1-1/p,p}(\partial\Omega, \mathbf{R}^3) = \Big\{\phi = \alpha\phi_1 + (1-\alpha)\phi_2 : \phi_1 \in \mathbf{B}^{s-1-1/p,p}(\partial\Omega, \mathbf{R}^3), \\ \phi_2 \in \mathbf{B}^{s-1/p,p}(\partial\Omega, \mathbf{R}^3)\Big\},$$

and define

$$\|\phi\|_{\alpha;s-1-1/p,p} = \inf\{\|\phi_1\|_{s-1-1/p,p} + \|\phi_2\|_{s-1/p,p} : \phi = \alpha\phi_1 + (1-\alpha)\phi_2\}.$$

Then it is easy to verify that the space $\mathbf{B}_{(\alpha)}^{s-1-1/p,p}(\partial\Omega, \mathbf{R}^3)$ is a Banach space with the norm $\|\cdot\|_{\alpha;s-1-1/p,p}$. We remark that

$$\mathbf{B}_{(\alpha)}^{s-1-1/p,p}(\partial\Omega, \mathbf{R}^3) = \mathbf{B}^{s-1/p,p}(\partial\Omega, \mathbf{R}^3) \quad \text{if } \alpha \equiv 0 \text{ on } \partial\Omega \text{ (displacement)}, \tag{17.10}$$

$$\mathbf{B}_{(\alpha)}^{s-1-1/p,p}(\partial\Omega, \mathbf{R}^3) = \mathbf{B}^{s-1-1/p,p}(\partial\Omega, \mathbf{R}^3) \quad \text{if } \alpha \equiv 1 \text{ on } \partial\Omega \text{ (traction)}. \tag{17.11}$$

If $s > 3/p + 1$, we let

$$\mathcal{X} = \text{the space } \mathbf{H}^{s,p}(\Omega, \mathbf{R}^3) \text{ of all } H^{s,p} \text{ vector functions } \phi,$$

and

$$\mathcal{C} = \text{the subspace of all } \textit{configurations } \phi \text{ in } \mathcal{X}.$$

We remark that the set $\mathcal{C}$ is open in the space $\mathcal{X}$. Indeed, this follows from an application of the inverse mapping theorem, since the Sobolev imbedding theorem tells us that the $H^{s,p}$ topology is stronger than the C^1 topology for $s > 3/p + 1$. We let

$$\mathcal{Y} = \text{the space of all pairs } (\mathbf{B}, \boldsymbol{\tau}) \in \mathbf{H}^{s-2,p}(\Omega, \mathbf{R}^3) \times \mathbf{B}_{(\alpha)}^{s-1-1/p,p}(\partial\Omega, \mathbf{R}^3).$$

We associate with problem (∗) a map between Banach spaces

$$F : \mathcal{X} \longrightarrow \mathcal{Y}$$

as follows:

$$F(\phi) = \left(-\mathrm{DIV}\,\widehat{\mathbf{P}}(D\phi), \alpha\widehat{\mathbf{P}}(D\phi)\cdot\mathbf{N} + (1-\alpha)\phi\Big|_{\partial\Omega}\right), \quad \phi \in \mathcal{C}.$$

It follows from an application of the ω-lemma (cf. [10, Chapter 3, Theorem 1.13]; [14, Chapter II, Section 4]) that the map F is of class C^1.

Now we can state our main existence and uniqueness result for problem (∗) of nonlinear elastostatics (cf. [10, Chapter 6, Theorem 4.2]; [14, Chapter IV, Theorem 4.5]):

Theorem 1 (Main Theorem) *We assume that conditions (H.1) and (H.2) are satisfied, and $s > 3/p + 1$ with $1 < p < \infty$. Furthermore assume that:*

(P) $\widehat{\mathbf{P}}(D \overset{\circ}{\rightarrow} \phi) = \mathbf{0}$ *when* $\overset{\circ}{\rightarrow} \phi = \mathbf{I}_\Omega$ *(identity map on* Ω*).*

(H) The elasticity tensor $\overset{\circ}{\rightarrow} \mathbf{A} = \partial^2 W / \partial \mathbf{F} \partial \mathbf{F}$ *evaluated at* $\overset{\circ}{\rightarrow} \phi = \mathbf{I}_\Omega$ *is smooth and* uniformly pointwise stable, *that is, there is a constant* $\eta > 0$ *such that*

$$\frac{1}{2}\mathbf{e} \cdot \overset{\circ}{\rightarrow} \mathbf{A} \cdot \mathbf{e} \geq \eta \|\mathbf{e}\|^2$$

for all symmetric two tensors $\mathbf{e}$.

(A) $\alpha \not\equiv 1$ *on* $\partial\Omega$.

Then there exist a neighborhood $\mathcal{U}$ *of the configuration* $\overset{\circ}{\rightarrow} \phi$ *in* $\mathcal{X}$ *and a neighborhood* $\mathcal{V}$ *of the point*

$$\left(-\mathrm{DIV}\, \widehat{\mathbf{P}}(D \overset{\circ}{\rightarrow} \phi), \alpha \widehat{\mathbf{P}}(D \overset{\circ}{\rightarrow} \phi) \cdot \mathbf{N} + (1 - \alpha) \overset{\circ}{\rightarrow} \phi|_{\partial\Omega} \right)$$

in $\mathcal{Y}$ *such that the map* $F : \mathcal{U} \to \mathcal{V}$ *is one-to-one and onto.*

Condition (P) implies that the undeformed state is *stress free*. We remark that the first Piola-Kirchhoff stress tensors in Examples 1 and 2 satisfy condition (P). Condition (A) implies that our boundary condition is not equal to the *pure traction* boundary condition. It is known (cf. [10, Chapter 7, Section 7.3]) that the pure traction problem may have non-unique solutions even for small loads and near a stress free state.

Rephrased, Main Theorem states that if the linearized problem is uniformly pointwise stable, then, for slight perturbations of the load or boundary conditions from their values at the undeformed state, the nonlinear problem (*) has a unique solution ϕ near $\overset{\circ}{\rightarrow} \phi = \mathbf{I}_\Omega$.

For the Hencky-Nadai elasto-plastic material, we have the following result (cf. [5, Théorème 2]):

Theorem 2 *We assume that conditions (H.1) and (H.2) are satisfied, and* $s > 3/p+1$ *with* $1 < p < \infty$. *Furthermore assume that:*

(A) $\alpha \not\equiv 1$ *on* $\partial\Omega$.

(G) $g(0) > 0$ *and* $K > 0$.

Then condition (H) is satisfied and so Main Theorem applies.

For the St. Venant-Kirchhoff isotropic material, we have the following result (cf. [4, Theorem 6.7-1]):

Theorem 3 *We assume that conditions (H.1) and (H.2) are satisfied, and* $s > 3/p+1$ *with* $1 < p < \infty$. *Furthermore assume that:*

(A) $\alpha \not\equiv 1$ *on* $\partial\Omega$.

(M) There exist constants $c_1 > 0$ *and* $c_2 > 0$ *such that*

$$\mu(X) \geq c_1 \quad \text{on } \Omega, \tag{17.12}$$

$$\lambda(X) + \frac{2}{3}\mu(X) \geq c_2 \quad \text{on } \Omega. \tag{17.13}$$

Then condition (H) is satisfied and so Main Theorem applies.

17.3 OUTLINE OF PROOF

(I) First we linearize problem (*) as follows:

$$\begin{cases} \mathrm{DIV}\,(\overset{\circ}{\to}\mathbf{A}\cdot\nabla\mathbf{V}) = -\mathrm{DIV}\,\widehat{\mathbf{P}}(\overset{\circ}{\to}\mathbf{F}) - \mathbf{B} & \text{in } \Omega, \\ \alpha(\overset{\circ}{\to}\mathbf{A}\cdot\nabla\mathbf{V}\cdot\mathbf{N}) + (1-\alpha)\mathbf{V} = \boldsymbol{\tau} - \alpha\widehat{\mathbf{P}}(\overset{\circ}{\to}\mathbf{F})\cdot\mathbf{N} - (1-\alpha)\overset{\circ}{\to}\boldsymbol{\phi} & \text{on } \partial\Omega, \end{cases} \quad (**)$$

where

$$\overset{\circ}{\to}\mathbf{A} = \frac{\partial\widehat{\mathbf{P}}}{\partial\mathbf{F}}(\overset{\circ}{\to}\mathbf{F}), \quad \overset{\circ}{\to}\mathbf{F} = D\overset{\circ}{\to}\boldsymbol{\phi},$$

is the first elasticity tensor evaluated at the given configuration $\overset{\circ}{\to}\boldsymbol{\phi}$.

Therefore we are reduced to the study of a problem of linear elastostatics for the unknown vector function $\mathbf{v}$:

$$\begin{cases} \mathbf{A}\mathbf{v} = \mathrm{div}\,(\mathbf{a}\cdot\nabla\mathbf{v}) = \mathbf{f} & \text{in } \Omega, \\ \mathbf{B}_\alpha\mathbf{v} = \alpha\,(\mathbf{a}\cdot\nabla\mathbf{v}\cdot\mathbf{n}) + (1-\alpha)\mathbf{v} = \varphi & \text{on } \partial\Omega, \end{cases} \quad (+)$$

where $\mathbf{a}$ is a smooth elasticity tensor and $\mathbf{n}$ is the outward unit normal to $\partial\Omega$.

(II) Next we study the linearized problem (+) in the framework of Sobolev spaces of L^p style, by using the L^p theory of pseudo-differential operators. Our fundamental existence and uniqueness theorem for problem (+) is stated as follows:

Theorem 4 *We assume that conditions (H.1) and (H.2) are satisfied. Furthermore assume that:*

(H) The elasticity tensor $\mathbf{a}$ *enjoys the property of* symmetry*:*

$$a_{ij\ell m} = a_{ji\ell m} = a_{ijm\ell} = a_{\ell mij},$$

and is uniformly pointwise stable, *that is, there is a constant* $\eta > 0$ *such that*

$$\frac{1}{2}\mathbf{e}\cdot\mathbf{a}\cdot\mathbf{e} \geq \eta\|\mathbf{e}\|^2$$

for all symmetric two tensors $\mathbf{e}$.

(A) $\alpha \not\equiv 1$ *on* $\partial\Omega$.

Then the operator

$$(\mathbf{A}, \mathbf{B}_\alpha) : \mathbf{H}^{s,p}(\Omega, \mathbf{R}^3) \longmapsto \mathbf{H}^{s-2,p}(\Omega, \mathbf{R}^3) \times \mathbf{B}_{(\alpha)}^{s-1-1/p,p}(\partial\Omega, \mathbf{R}^3)$$

is an algebraic and topological isomorphism, *for* $s > 1 + 1/p$.

(III) In order to prove Theorem 3, we show that problem (+) can be reduced to the study of a 3×3 matrix-valued pseudo-differential operator on the boundary. We explain more precisely the idea of our approach to problem (+).

(1-a) First we consider the displacement boundary value (Dirichlet) problem

$$\begin{cases} \mathbf{A}\mathbf{v} = \mathrm{div}\,(\mathbf{a}\cdot\nabla\mathbf{v}) = \mathbf{f} & \text{in } \Omega, \\ \mathbf{v} = \varphi & \text{on } \partial\Omega. \end{cases} \quad (D)$$

The existence and uniqueness theorem for problem (D) is well established in the framework of Sobolev spaces of L^p style. Thus one can introduce the *Poisson operator*

$$\mathcal{P} : \mathbf{B}^{s-1/p,p}(\partial\Omega, \mathbb{R}^3) \longrightarrow \mathbf{H}^{s,p}(\Omega, \mathbb{R}^3)$$

as follows: For any $\varphi \in \mathbf{B}^{s-1/p,p}(\partial\Omega, \mathbb{R}^3)$, the function $\mathcal{P}\varphi$ is the unique solution of the problem

$$\begin{cases} \mathbf{A}\mathbf{v} = \mathbf{0} & \text{in } \Omega, \\ \mathbf{v} = \varphi & \text{on } \partial\Omega. \end{cases}$$

(1-b) Next we consider the following mixed displacement-traction boundary value problem:

$$\begin{cases} \mathbf{A}\mathbf{v} = \operatorname{div}(\mathbf{a} \cdot \nabla\mathbf{v}) = \mathbf{f} & \text{in } \Omega, \\ (\mathbf{a} \cdot \nabla\mathbf{v} \cdot \mathbf{n}) + \mathbf{v} = \varphi & \text{on } \partial\Omega. \end{cases} \tag{M}$$

The existence and uniqueness theorem for problem (M) is also well established in the framework of Sobolev spaces of L^p style.

(1-c) Then, using problems (D) and (M), we show that problem (+) can be reduced to the study of a 3×3 matrix-valued pseudo-differential operator on the boundary, which is a generalization of the classical Fredholm integral equation.

Indeed, if we let

$$\mathbf{T}_\alpha\varphi = \mathbf{B}_\alpha\mathcal{P}\varphi,$$

then we have

$$\mathbf{T}_\alpha = \alpha\boldsymbol{\Pi} + (1-\alpha)\mathbf{I},$$

where

$$\boldsymbol{\Pi}\varphi = \mathbf{a} \cdot \nabla\mathcal{P}\varphi \cdot \mathbf{n}|_{\partial\Omega}.$$

It is known that the operator $\boldsymbol{\Pi}$ is a 3×3 matrix-valued, classical pseudo-differential operator of first order on $\partial\Omega$; hence the operator $\mathbf{T}_\alpha$ is a 3×3 matrix-valued, classical pseudo-differential operator of first order on $\partial\Omega$.

We can show that problem (+) can be reduced to the study of the system $\mathbf{T}_\alpha$ of pseudo-differential operators on the boundary $\partial\Omega$. We study of the pseudo-differential operator $\boldsymbol{\Pi}$ (and hence $\mathbf{T}_\alpha$) in question. In particular, we find that the operator $\boldsymbol{\Pi}$ is *strongly elliptic* on the boundary $\partial\Omega$.

(2) A regularity theorem for problem (+): More precisely, one can construct a *parametrix* $\mathbf{S}_\alpha$ for the operator $\mathbf{T}_\alpha$ in the Hörmander class $\mathbf{L}^0_{1,1/2}(\partial\Omega, \mathbb{R}^3)$ ([8]), and then apply a *Besov-space boundedness theorem* ([3]) to the parametrix $\mathbf{S}_\alpha$ to obtain the regularity theorem for problem (+).

(3) A uniqueness theorem for problem (+): We show that the operator

$$\mathcal{A}_\alpha = (\mathbf{A}, \mathbf{B}_\alpha) : \mathbf{H}^{s,p}(\Omega, \mathbb{R}^3) \longmapsto \mathbf{H}^{s-2,p}(\Omega, \mathbb{R}^3) \times \mathbf{B}^{s-1-1/p,p}_{(\alpha)}(\partial\Omega, \mathbb{R}^3)$$

is injective. In the proof we make good use of Korn's inequality ([6]) to show an inequality of Gårding type for problem (+).

(4) An existence theorem for problem (+): This is an essential step in the proof of Main Theorem.

By the uniqueness theorem, we know that the operator $\mathcal{A}_\alpha$ is injective. Hence, in order to prove the surjectivity, it suffices to show that the index of the operator $\mathcal{A}_\alpha$ is equal to *zero*.

(4-a) To do so, we replace the operator $\mathbf{A}$ by the operator $\mathbf{A} - \lambda\mathbf{I}$ with $\lambda \geq 0$, and consider instead of problem (+) the following boundary value problem:

$$\begin{cases} (\mathbf{A} - \lambda\mathbf{I})\mathbf{v} = \operatorname{div}(\mathbf{a} \cdot \nabla\mathbf{v}) - \lambda\mathbf{v} = \mathbf{f} & \text{in } \Omega, \\ \mathbf{B}_\alpha\mathbf{v} = \alpha(\mathbf{a} \cdot \nabla\mathbf{v} \cdot \mathbf{n}) + (1-\alpha)\mathbf{v} = \varphi & \text{on } \partial\Omega. \end{cases} \qquad ((+)_\lambda)$$

We remark that problem $(+)_\lambda$ coincides with problem (+) when $\lambda = 0$.

(4-b) To study problem $(+)_\lambda$, we shall make use of a method essentially due to Agmon [1]. This is a technique of treating a spectral parameter $\lambda\mathbf{I}$ as a second-order differential operator of an extra variable and relating the old problem to a new one with the additional variable, which we will explain more precisely.

We introduce an auxiliary variable y of the unit circle

$$S = \mathbf{R}/2\pi\mathbf{Z},$$

and replace the parameter $-\lambda\mathbf{I}$ by the second-order differential operator $\frac{\partial^2}{\partial y^2}\mathbf{I}$. That is, we replace the operator $\mathbf{A} - \lambda\mathbf{I}$ by the operator

$$\widetilde{\mathcal{A}} = \mathbf{A} + \frac{\partial^2}{\partial y^2}\mathbf{I},$$

and consider instead of problem $(+)_\lambda$ the following boundary value problem:

$$\begin{cases} \widetilde{\mathcal{A}}\tilde{\mathbf{v}} = \operatorname{div}(\mathbf{a} \cdot \nabla\tilde{\mathbf{v}}) + \frac{\partial^2\tilde{\mathbf{v}}}{\partial y^2} = \tilde{\mathbf{f}} & \text{in } \Omega \times S, \\ \mathbf{B}_\alpha\tilde{\mathbf{v}} = \alpha(\mathbf{a} \cdot \nabla\tilde{\mathbf{v}} \cdot \mathbf{n}) + (1-\alpha)\tilde{\mathbf{v}} = \tilde{\varphi} & \text{on } \partial\Omega \times S. \end{cases} \qquad (\widetilde{+})$$

Then the most fundamental relationship between problems $(+)_\lambda$ and problem $(\widetilde{+})$ is the following ([11]):

If the index of problem $(\widetilde{+})$ is finite, then there exists a finite subset K of $\mathbf{Z}$ such that problem $(+)_\lambda$ is uniquely solvable for all $\lambda' = \ell^2$ satisfying $\ell \in \mathbf{Z}\backslash K$. In particular, the index of $\mathcal{A}_\alpha$ is equal to zero.

(4-c) We show that one can construct a parametrix $\widetilde{\mathbf{S}}_\alpha$ for the operator $\widetilde{\mathbf{T}}_\alpha$ in the Hörmander class $\mathbf{L}^0_{1,1/2}(\partial\Omega \times S, \mathbf{R}^3)$. By virtue of the Besov-space boundedness theorem, we find that the index of problem $(\widetilde{+})$ is finite. In view of step (4-b), this proves that the index of $\mathcal{A}_\alpha$ is equal to zero.

(5) Proof of Main Theorem, Theorem 1 and Theorem 2: By Theorem 3, our Main Theorem follows from an application of the inverse mapping theorem. In the proof of Theorems 1 and 2, we calculate explicitly the first elasticity tensor $\overset{\circ}{\to}\mathbf{A}$, and verify that condition (G) or condition (M) implies condition (H).

17.4 OPEN PROBLEMS

In this final section we give two important open problems concerning the boundary value problems of nonlinear elastostatics (see Figure 17.5):

(I) The first problem is to generalize Main Theorem to the case where the domain Ω has *corner singularities*.

(II) The second problem is to study the case where the function $\alpha(x)$ is the *characteristic function* of a subset of the boundary $\partial\Omega$.

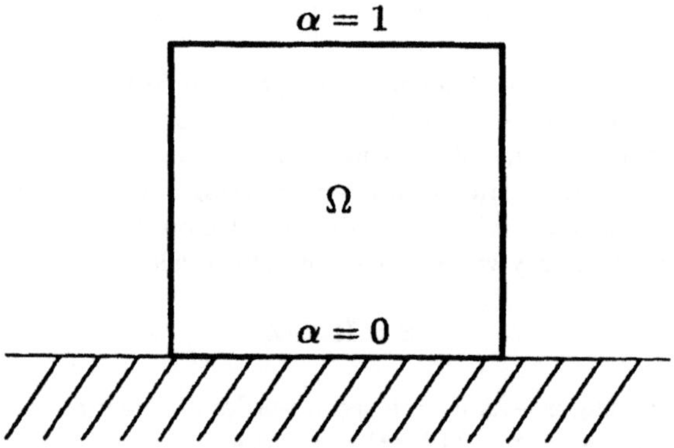

Figure 17.5

We remark that Ito [9] studied Problem (II) in the linear case (cf. [7]).

Notes

1. 1991 *Mathematics Subject Classification.* Primary 73C50, 73C35; Secondary 35J65

References

[1] S. Agmon. *Lectures on elliptic boundary value problems.* Van Nostrand, Princeton, 1965.

[2] J. Bergh and J. Löfström. *Interpolation spaces, an introduction.* Springer-Verlag, Berlin New York Heidelberg, 1976.

[3] G. Bourdaud. l^p-estimates for certain non-regular pseudo-differential operators. *Comm. in Partial Differential Equations*, 7:1023–1033, 1982.

[4] P. G. Ciarlet. *Mathematical elasticity, Vol. I.* Studies in mathematics and its applications No. 20. North-Holland, Amsterdam New York Oxford Tokyo, 1988.

[5] G. Dinca. Sur la monotonie d'après minty-browder de l'opérateur de la théorie de plasticité. *C. R. Acad. Sci. Paris*, 269:535–538, 1969.

[6] G. et J.-L. Lions Duvaut. *Les inéquations en mécanique et en physique*. Dunod, Paris, 1972.

[7] G.N. Gatica and G.C. Hsiao. *Boundary-field equation methods for a class of nonlinear problems*. Pitman Research Notes in Mathematics Series No. 331. Longman, Harlow, 1995.

[8] L. Hörmander. *The analysis of linear partial differential operators, Vol. III*. Springer-Verlag, New York Berlin Heidelberg Tokyo, 1983.

[9] H. Ito. On certain mixed-type boundary-value problems of elastostatics. *Tsukuba J. Math.*, 14:133–153, 1990.

[10] J.E. Marsden and T.J.R. Hughes. *Mathematical foundations of elasticity*. Prentice-Hall, Inc., Englewood Cliffs, New Jersey, 1983.

[11] K. Taira. *Diffusion processes and partial differential equations*. Academic Press, San Diego New York London Tokyo, 1988.

[12] K. Taira. On boundary value problems of nonlinear elastostatics. *Osaka J. Math.*, 33:353–383, 1996.

[13] H. Triebel. *Interpolation theory, function spaces, differential operators*. North-Holland, Amsterdam, 1978.

[14] Y. Valent. *Boundary value problems of finite elasticity*. Springer tracts in natural philosophy No. 31. Springer-Verlag, New York Berlin Heidelberg Tokyo, 1988.

18 ON SOME REGULARITY RESULTS FOR HEMIVARIATIONAL INEQUALITIES

Z. Naniewicz and P.D. Panagiotopoulos

Dedicated to Professor Robert P. Gilbert on the occasion of his 65th birthday

Abstract: The aim of the present paper is to investigate the regularity of the nonlinear term in hemivariational inequalities. This term expresses the virtual work of the nonmonotone multivalued stress-strain or reaction-displacement law which gives rise to the hemivariational inequality under consideration.

18.1 INTRODUCTION

The theory of hemivariational inequalities begun in the early eighties with the works of the second author of the present paper, and has as a main reason for its birth the need for the derivation of variational expressions for mechanical problems involving nonconvex, nonsmooth energy functions. For a complete information concerning the area of hemivariational inequalities we refer to [18, 13]. Until now a basic open problem was the problem of the regularity of the nonlinear term $j^0(u, v)$ appearing in every hemivariational inequality. This is a problem with concrete mechanical meaning as we shall see further. In order to explain better the arising question let us consider a simple hemivariational inequality studied in [16, 17]. Let V be a real Hilbert space with the property that

$$V \subset [L^2(\Omega)]^n \subset V^*,$$

where V^* denotes the dual space of V, Ω is an open bounded subset of R^n, and the injections are continuous and dense. We denote in this Section by $(\cdot,\cdot)$ the $[L^2(\Omega)]^n$ inner product and the duality pairing, by $\|\cdot\|$ the norm of V and by $|\cdot|_2$ the $[L^2(\Omega)]^n$-norm. We recall that the form $(\cdot,\cdot)$ extends uniquely from $V \times L^2[(\Omega)]^n$ to $V \times V^*$.

H. Florian et al. (eds.), Generalized Analytic Functions, 197–214.

Moreover let $L : V \to L^2(\Omega)$, $Lu = \hat{u}$, $\hat{u}(x) \in R$ be a linear continuous mapping. Further, assume that $l \in V^\star$, that

$$L : V \to L^2(\Omega) \qquad \text{is compact}$$

and that

$$\tilde{V} = \{v \in V : \hat{v} \in L^\infty(\Omega)\} \qquad \text{is dense in } V \text{ for the } V\text{-norm},$$

and has a Galerkin base. It is also assumed that $a(\cdot,\cdot)$: $V \times V \to R$ is a bilinear symmetric continuous form which is coercive, i.e. there exists $c > 0$ constant such that

$$a(v,v) \geq c||v||^2 \qquad \forall v \in V.$$

We denote by j: $R \to R$, a locally Lipschitz function defined as it follows: let $\beta \in L^\infty_{\rm loc}(R)$ and consider the functions $\bar{\beta}_\mu$ and $\bar{\bar{\beta}}_\mu$ defined by

$$\bar{\beta}_\mu(\xi) = \operatorname*{essinf}_{|\xi_1-\xi|\leq\mu} \beta(\xi_1) \text{ and } \bar{\bar{\beta}}_\mu(\xi) = \operatorname*{esssup}_{|\xi_1-\xi|\leq\mu} \beta(\xi_1).$$

They are decreasing and increasing functions of μ, respectively; therefore the limits for $\mu \to 0_+$ exist. We denote them by $\bar{\beta}(\xi)$ and $\bar{\bar{\beta}}(\xi)$ respectively; the multivalued function $\tilde{\beta}$ is defined by

$$\tilde{\beta}(\xi) = [\bar{\beta}(\xi), \bar{\bar{\beta}}(\xi)].$$

If $\beta(\xi_{\pm 0})$ exists for every $\xi \in R$ then a locally Lipschitz function j: $R \to R$ can be determined up to an additive constant such that

$$\tilde{\beta}(\xi) = \partial j(\xi).$$

Now we formulate the following coercive hemivariational inequality (problem P^C): Find $u \in V$ such that

$$a(u, v-u) + \int_\Omega j^0(\hat{u}, \hat{v} - \hat{u}) d\Omega \geq (l, v-u) \qquad \forall v \in V.$$

An element $u \in V$ is said to be a solution of P^C if there exists $\chi \in L^1(\Omega)$ with $L^\star\chi \in V^\star$ ($L^\star$ denotes the transpose operator of L) such that [13, 16]

$$a(u,v) + (L^\star\chi, v) = (l, v) \quad \forall v \in V$$

and

$$\chi(x) \in \partial j(u(x)) \quad \text{a.e. on } \Omega,$$

and where

$$(L^\star\chi, v) = \int_\Omega \chi L v d\Omega = \int_\Omega \chi \hat{v} d\Omega \quad \text{if } v \in \tilde{V}.$$

Therefore one can give the following definition. An element $u \in V$ is said to be a solution of P^C if there exists $\chi \in L^1(\Omega)$ such that (Problem P_1^C)

$$a(u,v) + \int_\Omega \chi \hat{v} d\Omega = (l,v) \quad \forall v \in \tilde{V}$$

and

$$\chi(x) \in \partial j(u(x)) \quad \text{a.e. on } \Omega,$$

hold. Obviously due to the density assumption the two above definitions are equivalent. Indeed one can show that the second problem has a solution [13, 16].

It arises naturally the following question: Which is the regularity of the product χv for $v \in V$ (and not for $v \in \tilde{V}$). Since v is a generalized displacement and χ is a generalized force for the corresponding mechanical problem, one expects that the product, i.e. the work, $v\chi$ must have an L^1-regularity at least and that the problem P^C and P_1^C are completely equivalent. This important result is proved in the present paper, under some mild growth assumptions following the methods of [15]. Further we deal with a more general hemivariational inequality related to a multidimensional superpotential law.

18.2 ABOUT THE PROBLEM

Let $V = H^1(\Omega; R^N)$, $N \geq 1$, be a vector valued Sobolev space compactly imbedded into $L^p(\Omega; R^N)$, $p > 1$, where Ω is a bounded domain in R^m, $m \geq 1$, with sufficiently smooth boundary $\partial\Omega$. We write $\|\cdot\|_V$ and $\|\cdot\|_{L^p(\Omega;R^N)}$ for the norms in V and $L^p(\Omega; R^N)$, respectively. For the the pairing over $V^* \times V$ the symbol $\langle\cdot,\cdot\rangle_V$ will be used, V^* being the dual of V.

Let $A : V \to V^*$ be a bounded operator in the sense that it maps bounded sets into bounded sets. Moreover, it will be assumed that A is pseudomonotone. This means that [3]:

(i) The effective domain of A coincides with the whole V;

(ii) Whenever $u_n \to u$ strongly in V and $Au_n \to B$ weakly in V^* for some $B \in V^*$, then $B = Au$;

(iii) Whenever $u_n \to u$ weakly in V and $\limsup\langle Au_n, u_n - u\rangle_V \leq 0$, then $Au_n \to Au$ weakly in V^* and $\langle Au_n, u_n\rangle_V \to \langle Au, u\rangle_V$.

Further, let $j : R^N \to R$ be a locally Lipschitz function from the Euclidean space R^N into R. Throughout the paper we assume j to satisfy the following unilateral growth restriction [12]:

$$j^0(\xi; \eta - \xi) \leq \alpha(r)(1 + |\xi|^\sigma) \quad \forall\ \xi, \eta \in R^N,\ |\eta| \leq r,\ r \geq 0, \tag{18.1}$$

where $1 \leq \sigma < p$ and $\alpha : R^+ \to R^+$ is assumed to be a nondecreasing function from R^+ into R^+. Here $j^0(\cdot,\cdot)$ stands for the generalized Clarke differential [4], i.e.

$$j^0(\xi;\eta) = \limsup_{\substack{h\to 0 \\ \lambda \to 0_+}} \frac{j(\xi + h + \lambda\eta) - j(\xi + h)}{\lambda}.$$

Let $g \in V^*$ be an arbitrary element of V^*. The aim of the present paper is to study the following hemivariational inequality problem:

Problem (P). Find $u \in V$ such that

$$\langle Au - g, v - u \rangle_V + \int_\Omega j^0(u; v - u) d\Omega \geq 0 \quad \forall\, v \in V, \tag{18.2}$$

where the integral above is assumed to take $+\infty$ as its values whenever $j^0(u; v-u) \notin L^1(\Omega)$.

To be precise, we are going to establish a stronger result, namely, to show the existence of a $\chi \in L^1(\Omega; R^N)$ with the properties that

$$\langle Au - g, v - u \rangle_V + \int_\Omega \chi \cdot (v - u)\, d\Omega = 0 \quad \forall\, v \in V \cap L^\infty(\Omega; R^N) \tag{18.3}$$

$$\chi \cdot u \in L^1(\Omega), \quad \chi \in \partial j(u) \ \text{ a.e. in } \Omega, \tag{18.4}$$

where the dot "$\cdot$" stands for the inner product in R^N, and then to derive the validity of hemivariational inequality (18.2) from (18.3) and (18.4).

The stated problem is not straightforward because the unilateral growth restriction (18.1) does not guarantee the finite integrability of $j^0(u; v - u)$ in Ω for each $v \in V$ even for $u \in V$ with $j(u) \in L^1(\Omega)$.

18.3 FINITE DIMENSIONAL APPROXIMATION

Let Λ be the family of all finite dimensional subspaces F of $V \cap L^\infty(\Omega; R^N)$, ordered by inclusion. Denote by $i_F : F \to V$ the inclusion mapping of F into V, and by $i_F^* : V^* \to F^*$ the dual projection mapping of V^* into F^*, F^* being the dual of F. The pairing over $F^* \times F$ will be denoted by $\langle \cdot, \cdot \rangle_F$.

For any $F \in \Lambda$ we formulate the finite dimensional hemivariational inequality problem.

Problem (P_F). Find $u_F \in F$ and $\chi_F \in L^1(\Omega; R^N)$ such that

$$\begin{gathered} \langle Au_F - g, v \rangle_V + \int_\Omega \chi_F \cdot v d\Omega = 0 \quad \forall\, v \in F \\ \chi_F \in \partial j(u_F) \ \text{ a.e. in } \Omega. \end{gathered} \tag{18.5}$$

Proposition 1 *Let us posit all the hypotheses stated before and suppose additionally that A is coercive, i.e.*

$$\langle Av, v\rangle_V \geq c(||v||_V)||v||_V \quad \forall\ v \in V,$$

for a certain coercivity function $c(\cdot)$ with $c(r) \to \infty$ as $r \to \infty$, and that j fulfills the estimation

$$j^0(\xi; -\xi) \leq k(1 + |\xi|) \quad \forall \xi \in R^N, \tag{18.6}$$

k being a constant. Then for any $F \in \Lambda$ the (P_F) problem possesses at least one solution. Moreover, $\{u_F\}_{F\in\Lambda}$ is bounded in V independently of F and the corresponding set $\{\chi_F\}_{F\in\Lambda}$ is weakly precompact in $L^1(\Omega; R^N)$.

The proof of Proposition 1 will follow from a sequence of lemmas below.

Recall first that j has been assumed to be locally Lipschitz. That is why for each $v_F \in F$ a constant C_F can be determined such that

$$|j^0(v_F, w)| \leq C_F|w| \leq C_F||w||_{L^\infty(\Omega)} \tag{18.7}$$

for any $w \in L^\infty(\Omega; R^N)$. It allows us to define for each $v_F \in F$ a mapping $\Gamma_F : F \to 2^{L^1(\Omega;R^N)}$ by the following formula

$$\Gamma_F(v_F) = \{\psi \in L^1(\Omega; R^N) : \int_\Omega \psi \cdot w d\Omega \leq \int_\Omega j^0(v_F, w) d\Omega \quad \forall\, w \in L^\infty(\Omega; R^N)\}. \tag{18.8}$$

Notice that if $\psi \in \Gamma_F(v_F)$, then $\psi \in \partial j(v_F)$ a.e. in Ω. It can be easily deduced that for each $v_F \in F$, $\Gamma_F(v_F)$ is a nonempty, convex and weakly compact subset of $L^1(\Omega; R^N)$ (for detailes see [13]).

For $F \in \Lambda$, let $i_F : F \to V$ denote the inclusion mapping of F into V, and let $i_F^\star : V^\star \to F^\star$ be the transpose of i_F. We denote by $\langle\cdot,\cdot\rangle_F$ the duality pairing over $F^\star \times F$. Further, let $\tau_F : L^1(\Omega; R^N) \to F^\star$ be an operator which assigns to any $\psi \in L^1(\Omega; R^N)$ the element $\tau_F\psi \in F^\star$ defined by

$$\langle \tau_F\psi, v\rangle_F := \int_\Omega \psi \cdot v d\Omega \quad \text{for any } v \in F. \tag{18.9}$$

Now let us consider a mapping $T_F : F \to 2^{F^\star}$ given by formula

$$T_F(v_F) := \tau_F\Gamma_F(v_F) \quad \text{for } v_F \in F. \tag{18.10}$$

The following properties of T_F can be pointed out.

Lemma 2 *T_F is an upper semicontinuous mapping from F into 2^{F^*} with nonempty, bounded, closed and convex values.*

Proof. Let us note that τ_F defined by (18.9) is linear and continuous from the weak topology on $L^1(\Omega; R^N)$ to the (unique) topology on F^*. Thus taking into account the already established properties of $\Gamma_F(v_F)$, $v_F \in F$, it is easily seen that $T_F(v_F)$ is a nonempty bounded, closed, convex subset of F^* for each $v_F \in F$. Thus it remains to show that T_F is upper semicontinuous from F into 2^{F^*}. For this purpose let us suppose that a sequence $\{v_{F_n}\} \subset F$ converges to v_F in F, and that the corresponding sequence $\{\tau_F \psi_n\} \subset F^*$ with $\psi_n \in \Gamma_F(v_{F_n})$, converges to some $\psi^* \in F^*$. We have to prove that there exists $\psi \in \Gamma_F(v_F)$ such that $\psi^* = \tau_F \psi$.

First we notice that the convergence of $\{v_{F_n}\}$ in F implies the existence of a finite upper bound for $\{||v_{F_n}||_{L^\infty(\Omega;R^N)}\}$, say $\hat{C}$, i.e.,

$$||v_{F_n}||_{L^\infty(\Omega;R^N)} \leq \hat{C} \quad \text{for } n = 1, 2, \dots . \tag{18.11}$$

The Lipschitz constant of j on the corresponding ball $\{\xi \in R^N; |\xi| \leq \hat{C}\}$ will be denoted by C. Taking into account that for any mesurable $\omega \subset \Omega$

$$\int_\Omega \psi_n \cdot w d\Omega \leq \int_\Omega j^0(v_{Fn}, w) d\Omega \quad \forall\, w \in L^\infty(\Omega; R^N), \tag{18.12}$$

we obtain the estimate

$$\int_\omega |\psi_n| d\Omega \leq \sqrt{N} \int_\omega C d\Omega, \quad \omega \subset \Omega, \tag{18.13}$$

which implies immediately that for any $\varepsilon > 0$ there exists a $\delta > 0$ with the property that

$$\int_\omega |\psi_n| d\Omega \leq \varepsilon \quad n = 1, 2, \dots , \tag{18.14}$$

provided mes $\omega < \delta$. Applying the Dunford-Pettis theorem we assert the weak precompactness of $\{\psi_n\}$ in $L^1(\Omega; R^N)$. Accordingly, a subsequence of $\{\psi_n\}$ can be extracted (again denoted by the same symbol) such that $\psi_n \to \psi$ weakly in $L^1(\Omega; R^N)$ for some $\psi \in L^1(\Omega; R^N)$. This fact implies that $\tau_F \psi_n \to \tau_F \psi = \psi^*$ in F^*. We claim that $\psi \in \Gamma_F(v_F)$. Indeed, taking into account the upper semicontinuity of the function

$$F \ni v_{F_n} \to \int_\Omega j^0(v_{F_n}, w) d\Omega, \quad \forall\, w \in L^\infty(\Omega; R^N), \tag{18.15}$$

by means of (12)we are led directly to the relation

$$\int_\Omega \psi \cdot w d\Omega \leq \int_\Omega j^0(v_F, w) d\Omega \quad \forall\, w \in L^\infty(\Omega; R^N), \tag{18.16}$$

implying immediately the desired $\psi \in \Gamma_F(v_F)$. The proof is complete. □

Lemma 3 *For every $F \in \Lambda$ the problem (P_F) has at least one solution.*

Proof. Define $A_F = i_F^\star A i_F$ and $f_F = i_F^\star f$. Notice that in order to obtain the solvability of (P_F) it suffices to show that f_F belongs to the range of the multivalued mapping $A_F + T_F$. We shall show that, in fact, Range $(A_F + T_F) = F^\star$. Indeed, due to Lemma 2, for each $v \in F, A_F(v) + T_F(v)$ is a nonempty bounded closed convex subset of $F^\star$. Further, the pseudomonotonicity of A implies the continuity of A_F from F into $F^\star$. Therefore the mapping $A_F + T_F$ is upper semicontinuous from F into $2^{F^\star}$. Moreover, the coercivity of A and (18.6) yield

$$\begin{aligned}
\langle A_F v, v\rangle_F + \int_\Omega \psi \cdot v d\Omega &\geq \langle Av, v\rangle_V - \int_\Omega j^0(v, -v) d\Omega \\
&\geq c(\|v\|_V)\|v\|_V - \int_\Omega k|v| d\Omega \\
&\geq c(\|v\|_V)\|v\|_V - k(\operatorname{mes}\Omega)^{\frac{p-1}{p}}\|v\|_{L^p(\Omega)} \\
&\geq c(\|v\|_V)\|v\|_V - k\gamma(\operatorname{mes}\Omega)^{\frac{p-1}{p}}\|v\|_V, \qquad (18.17) \\
& v \in F,\ \psi \in \Gamma_F(v),
\end{aligned}$$

where γ is a positive constant resulting from the imbedding $V \subset L^p(\Omega)$ being compact. This means that $A_F + T_F$ is also coercive and consequently, by the well known result (cf. [3]), Range $(A_F + T_F) = F^\star$. Thus for $g \in V^\star$ there exist $u_F \in F$ and $\chi_F \in \Gamma_F(u_F)$ such that $g_F = A_F(u_F) + \tau_F \chi_F$, from which (5) is easily deduced. The proof is complete. □

Lemma 4 *There exist a positive constant M not depending on $F \in \Lambda$ such that*

$$\|u_F\|_V \leq M, \quad F \in \Lambda. \qquad (18.18)$$

Proof. To show that the family $\{u_F\}$ is uniformly bounded we make use of the coercivity of A and (18.6). The estimations follow

$$
\begin{aligned}
\|g\|_{V^*}\|u_F\|_V &\geq \langle g, u_F\rangle_V = \langle g_F, u_F\rangle_F = \langle A_F u_F, u_F\rangle_F + \int_\Omega \chi_F \cdot u_F d\Omega \\
&\geq \langle A u_F, u_F\rangle_V - \int_\Omega j^0(u_F, -u_F) d\Omega \geq c(\|u_F\|_V)\|u_F\|_V \\
&\quad - \int_\Omega k|u_F| d\Omega \\
&\geq c(\|u_F\|_V)\|u_F\|_V - k(\operatorname{mes}\Omega)^{\frac{p-1}{p}}\|u_F\|_{L^p(\Omega)} \\
&\geq c(\|u_F\|_V)\|u_F\|_V - k\gamma(\operatorname{mes}\Omega)^{\frac{p-1}{p}}\|u_F\|_V,
\end{aligned}
$$

which thanks to the behaviour of $c = c(\cdot)$ at infinity lead easily to the assertion. □

The next lemma concerns the compactness property of the set $\{\chi_F : F \in \Lambda\}$ in $L^1(\Omega; R^N)$.

Lemma 5 *Let for some $F \in \Lambda$ a pair $(u_F, \chi_F) \in F \times L^1(\Omega; R^N)$ be a solution of (P_F). Then the set $\{\chi_F \in L^1(\Omega; R^N) : (u_F, \chi_F)$ is a solution of (P_F) for some $u_F \in F$, $F \in \Lambda\}$ is weakly precompact in $L^1(\Omega; R^N)$.*

Proof. According to the Dunford-Pettis theorem it suffices to show that for each $\varepsilon > 0$ a $\delta_\varepsilon > 0$ can be determined such that for any $\omega \subset \Omega$ with $\operatorname{mes}\omega < \delta_\varepsilon$,

$$
\int_\omega |\chi_F| d\Omega < \varepsilon, \quad F \in \Lambda. \tag{18.19}
$$

Fix $r > 0$ and let $\eta \in R^N$ be such that $|\eta| \leq r$. Then we have

$$
\chi_F \cdot (\eta - u_F) \leq j^0(u_F, \eta - u_F)
$$

from which, by virtue of (18.6) it results that

$$
\chi_F \cdot \eta \leq \chi_F \cdot u_F + \alpha(r)(1 + |u_F|^\sigma)
$$

a.e. in Ω. Let us set

$$
\eta = \frac{r}{\sqrt{N}}(\operatorname{sgn}\chi_{F_1}, \ldots, \operatorname{sgn}\chi_{F_N}),
$$

where $\chi_{F_i}, i = 1, 2, \ldots, N$, are the components of χ_F and where

$$
\operatorname{sgn} y = \begin{cases} 1 & \text{if } y > 0 \\ 0 & \text{if } y = 0 \\ -1 & \text{if } y < 0 \end{cases} .
$$

It is not difficult to verify that $|\eta| \leq r$ for almost all $x \in \Omega$ and that

$$\chi_F \cdot \eta \geq \frac{r}{\sqrt{N}} |\chi_F|.$$

Therefore by virtue of (18.1) we are led to the estimate

$$\frac{r}{\sqrt{N}} |\chi_F)| \leq \chi_F \cdot u_F + \alpha(r)(1 + |u_F|^\sigma).$$

Integrating this inequality over $\omega \subset \Omega$ yields

$$\begin{aligned} \int_\omega |\chi_F| d\Omega &\leq \frac{\sqrt{N}}{r} \int_\omega \chi_F \cdot u_F d\Omega + \frac{\sqrt{N}}{r} \alpha(r) \operatorname{mes} \omega \\ &+ \frac{\sqrt{N}}{r} \alpha(r) (\operatorname{mes} \omega)^{\frac{p-\sigma}{p}} \|u_F\|^\sigma_{L^p(\Omega)}. \end{aligned} \tag{18.20}$$

Thus, from (18.18) we obtain

$$\begin{aligned} \int_\omega |\chi_F| d\Omega &\leq \frac{\sqrt{N}}{r} \int_\omega \chi_F \cdot u_F d\Omega \\ &+ \frac{\sqrt{N}}{r} \alpha(r) \operatorname{mes} \omega + \frac{\sqrt{N}}{r} \alpha(r) (\operatorname{mes} \omega)^{\frac{p-\sigma}{p}} \gamma^\sigma \|u_F\|^\sigma_V \\ &\leq \frac{\sqrt{N}}{r} \int_\omega \chi_F \cdot u_F d\Omega \\ &+ \frac{\sqrt{N}}{r} \alpha(r) \operatorname{mes} \omega + \frac{\sqrt{N}}{r} \alpha(r) (\operatorname{mes} \omega)^{\frac{p-\sigma}{p}} \gamma^\sigma M^\sigma. \end{aligned} \tag{18.21}$$

Now we show that

$$\int_\omega \chi_F \cdot u_F d\Omega \leq C \tag{18.22}$$

for some positive constant C not depending on $\omega \subset \Omega$ and $F \in \Lambda$. Indeed, from (18.6) one can easily deduce that

$$\chi_F \cdot u_F + k(1 + |u_F|) \geq 0 \text{ a.e. in } \Omega.$$

Thus it follows

$$\int_\omega [\chi_F \cdot u_F + k(1 + |u_F|)] d\Omega \leq \int_\Omega [\chi_F \cdot u_F + k(1 + |u_F|)] d\Omega,$$

and consequently

$$\int_\omega \chi_F \cdot u_F d\Omega \le \int_\Omega \chi_F \cdot u_F d\Omega + k \operatorname{mes}\Omega + k\gamma(\operatorname{mes}\Omega)^{\frac{p-1}{p}} ||u_F||_V \le \int_\Omega \chi_F \cdot u_F d\Omega + k \operatorname{mes}\Omega + k\gamma(\operatorname{mes}\Omega)^{\frac{p-1}{p}} M. \tag{18.23}$$

But A maps bounded sets into bounded sets. Therefore, by means of (18.5) and (18.18) we conclude that

$$\int_\Omega \chi_F \cdot u_F d\Omega = -\langle Au_F - g, u_F\rangle_V \le ||Au_F - g||_{V^*} ||u_F||_V \le \hat{C}, \quad \hat{C} = \text{const.}$$

From the last two estimates (18.22) is easily obtained. Further, from (18.20) and (18.22), for $r > 0$ one gets

$$\int_\omega |\chi_F| d\Omega \le \frac{\sqrt{N}}{r} C + \frac{\sqrt{N}}{r}\alpha(r) \operatorname{mes}\omega + \frac{\sqrt{N}}{r}\alpha(r)(\operatorname{mes}\omega)^{\frac{p-\sigma}{p}} \gamma^\sigma M^\sigma. \tag{18.24}$$

This estimation is crucial for (18.19) to be obtained. Namely, let $\varepsilon > 0$. Fix $r > 0$ with

$$\frac{\sqrt{N}}{r} C < \frac{\varepsilon}{2} \tag{18.25}$$

and determine $\delta_\varepsilon > 0$ small enough such that

$$\frac{\sqrt{N}}{r}\alpha(r) \operatorname{mes}\omega + \frac{\sqrt{N}}{r}\alpha(r)(\operatorname{mes}\omega)^{\frac{p-\sigma}{p}} \gamma^\sigma M^\sigma \le \frac{\varepsilon}{2} \tag{18.26}$$

whenever $\operatorname{mes}\omega < \delta_\varepsilon$. Then from (18.24), (18.25) and (18.26) it follows

$$\int_\omega |\chi_F| d\Omega \le \varepsilon \quad F \in \Lambda,$$

for any $\omega \subset \Omega$ with $\operatorname{mes}\omega < \delta_\varepsilon$. Accordingly, the weak precompactness of $\{\chi_F : F \in \Lambda\}$ in $L^1(\Omega; R^N)$ is proved. □

Combining the results of Lemmas 2 – 5 establishes the proof of Proposition 1.

18.4 INFINITE DIMENSIONAL CASE

Now we are in a position to formulate our main result.

Theorem 1 *Let us posit all the hypotheses stated before. Then there exists at least one $u \in V$ and $\chi \in L^1(\Omega; R^N)$ with the following properties*

$$\langle Au - g, v - u\rangle_V + \int_\Omega \chi \cdot (v - u) d\Omega = 0 \quad \forall\, v \in V \cap L^\infty(\Omega; R^N) \tag{18.27}$$

$$\chi \in \partial j(u), \quad \chi \cdot u \in L^1(\Omega). \tag{18.28}$$

Proof. The proof is divided into a sequence of steps.
Step 1. For $F \in \Lambda$ let

$$W_F = \bigcup_{\substack{F' \in \Lambda \\ F' \supset F}} \{(u_{F'}, \chi_{F'}) \in V \times L^1(\Omega; R^N) : (u_{F'}, \chi_{F'}) \text{ satisfies (18.5)}\}.$$

We use the symbol weakcl (W_F) to denote the closure of W_F in the weak topology of $V \times L^1(\Omega; R^N)$. Moreover, let

$$Z = \bigcup_{F \in \Lambda} \{\chi_F \in L^1(\Omega; R^N) : (u_F, \chi_F) \text{ satisfies (18.5) for some } u_F \in F\}.$$

Denoting by weakcl (Z) the closure of Z in the weak topology of $L^1(\Omega; R^N)$ we get

$$\text{weakcl}\,(W_F) \subset B_V(O, M) \times \text{weakcl}\,(Z) \quad \forall\, F \in \Lambda.$$

Since $B_V(O, M)$ is weakly compact in V and, by Lemma 5, weakcl (Z) is weakly compact in $L^1(\Omega; R^N)$, the family $\{\text{weakcl}(W_F) : F \in \Lambda\}$ is contained in the weakly compact set $B_V(O, M) \times \text{weakcl}(Z)$ in $V \times L^1(\Omega; R^N)$. Now let us notice that for any $F_1, \dots, F_k \in \Lambda$, $k = 1, 2, \dots$, we have the inclusion $W_{F_1} \cap \dots \cap W_{F_k} \supset W_F$, with $F = F_1 + \dots + F_k$, from which it follows by Lemma 4 that the family $\{\text{weakcl}(W_F) : F \in \Lambda\}$ has the finite intersection property. Thus the intersection

$$\bigcap_{F \in \Lambda} \text{weakcl}\,(W_F)$$

is not empty. Let (u, χ) be an element of this intersection.

Step 2. Let us fix $v \in V \cap L^\infty(\Omega; R^N)$ arbitrarily. We choose $F \in \Lambda$ such that $v \in F$. Thus there exists a sequence $\{(u_{F_n}, \chi_{F_n})\} \subset W_F$ (for simplicity of notations it will be denoted by (u_n, χ_n)) with the properties that

$$\begin{gathered} \langle Au_n - g, w - u_n\rangle + \int_\Omega \chi_n \cdot (w - u_n) d\Omega = 0 \quad \forall\, w \in F_n \\ \chi_n \in \partial j(u_n) \end{gathered} \tag{18.29}$$

and

$$\begin{aligned} u_n &\to u \quad \text{weakly in } V \\ \chi_n &\to \chi \quad \text{weakly in } L^1(\Omega; R^N). \end{aligned} \tag{18.30}$$

Taking into account (18.29) and the fact that $v \in F \subset F_n$ we have

$$\langle Au_n - g, v\rangle + \int_\Omega \chi_n \cdot v d\Omega = 0, \quad n = 1, 2 \ldots . \tag{18.31}$$

Now we are in a position to pass to the limit in (18.31) as $n \to \infty$. The boundedness of $\{Au_n\}$ allows us to conclude that for some $B \in V^\star$, $Au_n \to B$ weakly in $V^\star$ (by passing to a subsequence, if necessary). Since $v \in V \cap L^\infty(\Omega; R^N)$ has been chosen arbitrarily, the equality

$$\langle B - g, v\rangle + \int_\Omega \chi \cdot v d\Omega = 0 \tag{18.32}$$

is valid for any $v \in V \cap L^\infty(\Omega; R^N)$.

Step 2. We prove that the first claim in (18.24), i.e. $\chi \in \partial j(u)$, holds. Since V is compactly imbedded into $L^p(\Omega; R^N)$, from (18.30) we obtain (by passing to a subsequence, if necessary)

$$u_n \to u \quad \text{strongly in } L^p(\Omega; R^N). \tag{18.33}$$

This implies that for a subsequence of $\{u_n\}$ (again denoted by the same symbol) one gets

$$u_n \to u \quad \text{a.e. in } \Omega.$$

Thus Egoroff's theorem can be applied from which it follows that for any $\varepsilon > 0$ a subset $\omega \subset \Omega$ with $\text{mes}\,\omega < \varepsilon$ can be determined such that

$$u_n \to u \quad \text{uniformly in } \Omega \setminus \omega$$

with $u \in L^\infty(\Omega \setminus \omega; R^N)$. Let $v \in L^\infty(\Omega \setminus \omega; R^N)$ be an arbitrary function. From the estimate

$$\int_{\Omega\setminus\omega} \chi_n \cdot v d\Omega \le \int_{\Omega\setminus\omega} j^0(u_n, v) d\Omega$$

combined with the weak convergence in $L^1(\Omega; R^N)$ of χ_n to χ, (18.33) and with the upper semicontinuity of

$$L^\infty(\Omega \setminus \omega; R^N) \ni w \longmapsto \int_{\Omega\setminus\omega} j^0(w, v) d\Omega$$

we obtain

$$\int_{\Omega\setminus\omega} \chi \cdot v d\Omega \leq \int_{\Omega\setminus\omega} j^0(u,v) d\Omega \quad \forall\, v \in L^\infty(\Omega\setminus\omega; R^N).$$

But the last inequality amounts to

$$\chi \in \partial j(u) \quad \text{a.e. in } \Omega\setminus\omega.$$

Since $\operatorname{mes}\omega < \varepsilon$ and ε was chosen arbitrarily,

$$\chi \in \partial j(u) \quad \text{a.e. in } \Omega, \tag{18.34}$$

as claimed.

Step 3. Now we show that $\chi \cdot u \in L^1(\Omega)$. For this purpose we shall need the followig truncation result for vector-valued Sobolev spaces.

Theorem 2 *(Naniewicz, [15]). For each $v \in H^1(\Omega; R^N)$ there exists a sequence of functions $\{\varepsilon_n\} \subset L^\infty(\Omega)$ with $0 \leq \varepsilon_n \leq 1$ such that*

$$\begin{aligned} &\{(1-\varepsilon_n)v\} \subset H^1(\Omega; R^N) \cap L^\infty(\Omega; R^N) \\ &(1-\varepsilon_n)v \to v \quad \textit{strongly in } H^1(\Omega; R^N). \end{aligned} \tag{18.35}$$

According to the aforementioned theorem for $u \in V$ determined in Step 1 of the proof of Theorem 1 one can find a sequence $\{\varepsilon_k\} \in L^\infty(\Omega)$ with $0 \leq \varepsilon_k \leq 1$ such that $\hat{u}_k := (1-\varepsilon_k)u \in V \cap L^\infty(\Omega; R^N)$ and $\hat{u}_k \to u$ in V as $k \to \infty$. By the compact argument without loss of generality we can assume that $\hat{u}_k \to u$ a.e. in Ω. Since we already know that $\chi \in \partial j(u)$, one can apply (18.6) to obtain

$$\chi \cdot (-u) \leq j^0(u; -u) \leq k(1+|u|).$$

Hence

$$\chi \cdot u \geq -k(1+|u|) \tag{18.36}$$

and consequently,

$$\chi \cdot \hat{u}_k = (1-\varepsilon_k)\chi \cdot u \geq -k(1+|u|). \tag{18.37}$$

This implies that the sequence $\{\chi \cdot \hat{u}_k\}$ is bounded from below by a function which is integrable in Ω. On the other hand, due to (18.32) we get

$$C \geq \langle -B+g, \hat{u}_k \rangle = \int_\Omega \chi \cdot \hat{u}_k d\Omega$$

for a positive constant C. Thus by Fatou's lemma $\chi \cdot u \in L^1(\Omega)$, as required.

Step 4. In this step we are aimed at establishing the estimate

$$\liminf \int_\Omega \chi_n \cdot u_n d\Omega \geq \int_\Omega \chi \cdot u d\Omega, \tag{18.38}$$

where a sequence (u_n, χ_n) is assumed to satisfy (18.29) and (18.30). We can suppose here again by the compactness argument that $u_n \to u$ a.e. in Ω. Fix $v \in L^\infty(\Omega; R^N)$ arbitrarily. Since $\chi_n \in \partial j(u_n)$, one gets by (18.1)

$$\chi_n \cdot (v - u_n) \leq j^0(u_n; v - u_n) \leq \alpha(\|v\|_{L^\infty(\Omega;R^N)})(1 + |u_n|^\sigma).$$

Thus by Fatou's lemma and by the semicontinuity of $V \ni w \mapsto \int_\Omega j^0(u; w - u)d\Omega$, we obtain

$$\liminf \int_\Omega \chi_n \cdot u_n d\Omega \geq \int_\Omega \chi \cdot v d\Omega - \int_\Omega j^0(u; v - u)d\Omega \quad \forall\, v \in V \cap L^\infty(\Omega; R^N). \tag{18.39}$$

On substituting $v = \hat{u}_k$ (with $\hat{u}_k$ being described by the truncation argument of Theorem 2) into the right hand side of (18.39) one gets

$$\liminf_n \int_\Omega \chi_n \cdot u_n d\Omega \geq \liminf_k \int_\Omega \chi \cdot \hat{u}_k d\Omega - \limsup_k \int_\Omega j^0(u; \hat{u}_k - u)d\Omega, \tag{18.40}$$

which is valid for each number k. Further, it is easy to check that

$$j^0(u; \hat{u}_k - u) = \varepsilon_k j^0(u; -u) \leq \varepsilon_k k(1 + |u|) \leq k(1 + |u|). \tag{18.41}$$

Therefore due to the fact that $\hat{u}_k \to u$ a.e. in Ω we are allowed to apply Fatou's lemma to deduce

$$\limsup_k \int_\Omega j^0(u; \hat{u}_k - u)d\Omega \leq 0,$$

whereas from (18.37) we get

$$\liminf_k \int_\Omega \chi \cdot \hat{u}_k d\Omega \geq \int_\Omega \chi \cdot u d\Omega.$$

Finally, combining the last two inequalities with (18.40) yields (18.38), as desired.

Step 5. We proceed to establish (18.27). Our first claim is that

$$\langle B - g, u\rangle + \int_\Omega \chi \cdot u d\Omega = 0. \tag{18.42}$$

Indeed, (18.32) yields

$$\langle B - g, \hat{u}_k \rangle + \int_\Omega \chi \cdot \hat{u}_k d\Omega = 0.$$

Recall that $\chi \cdot u \in L^1(\Omega)$ and we have the following estimates

$$-k(1 + |u|) \leq \chi \cdot \hat{u}_k = (1 - \varepsilon_k)\chi \cdot u \leq |\chi \cdot u|.$$

Thus we are allowed to apply the dominated convergence to deduce

$$\int_\Omega \chi \cdot \hat{u}_k d\Omega \to \int_\Omega \chi \cdot u d\Omega,$$

which means that (18.42) has to hold by the limit procedure.

Step 6. Finally the pseudo-monotonicity of A will permit the derivation of (18.27). To this aid we first check that (18.29) (when substituting $w = 0$) and (18.42) yield

$$\langle Au_n, u_n - u \rangle_V = \langle g, u_n \rangle_V - \int_\Omega \chi_n \cdot u_n d\Omega - \langle Au_n, u \rangle_V + \langle B - g, u_n \rangle_V + \int_\Omega \chi \cdot u_n d\Omega.$$

Hence by virtue of (18.38)

$$\limsup \langle Au_n, u_n - u \rangle_V \leq - \liminf \int_\Omega \chi_n \cdot u_n d\Omega + \int_\Omega \chi \cdot u d\Omega \leq 0.$$

Thus we are allowed to benefit the pseudo-monotonicity of A which are provided with the properties $\langle Au_n, u_n \rangle_V \to \langle Au, u \rangle_V$ and $Au_n \to B = Au$ weakly in V^* as $n \to \infty$. Thus from (18.29) and (18.38) we get (18.27) immediately. □

Now we establish the last result.

Theorem 3 *Let us posit all the hypotheses of Theorem 1 Then there exists at least one $u \in V$ such as to satisfy the hemivariational inequality*

$$\langle Au - g, v - u \rangle_V + \int_\Omega j^0(u; v - u) d\Omega \geq 0 \quad \forall\, v \in V, \tag{18.43}$$

where the integral above is assumed to take $+\infty$ as its values whenever $j^0(u; v - u) \notin L^1(\Omega)$.

Proof. Let us first choose $v \in V \cap L^\infty(\Omega; R^N)$ arbitrarily. From (1) we have

$$\chi \cdot (v - u) \leq j^0(u; v - u) \leq \alpha(\|v\|_{L^\infty(\Omega; R^N)})(1 + |u|^\sigma)$$

with $\chi \cdot (v-u) \in L^1(\Omega)$ and $\alpha(||v||_{L^\infty(\Omega;R^N)})(1+|u|^\sigma) \in L^1(\Omega)$. Hence $j^0(u; v-u)$ is finite integrable and (18.43) follows immediately from (18.27).

Now let us consider the case $j^0(u; v-u) \in L^1(\Omega)$ with $v \notin V \cap L^\infty(\Omega; R^N)$. According to Theorem 2 one can find a sequence $\hat{v}_k = (1-\varepsilon_k)v$ such that $\{\hat{v}_k\} \subset V \cap L^\infty(\Omega; R^N)$ and $\hat{v}_k \to v$ strongly in V. As we already know

$$\langle Au - g, \hat{v}_k - u \rangle_V + \int_\Omega j^0(u; \hat{v}_k - u)d\Omega \geq 0, \tag{18.44}$$

so in order to establish (44) it remains to show that

$$\limsup_k \int_\Omega j^0(u; \hat{v}_k - u)d\Omega \leq \int_\Omega j^0(u; v-u)d\Omega. \tag{18.45}$$

For this purpose let us notice that $\hat{v}_k - u = (1-\varepsilon_k)(v-u) + \varepsilon_k(-u)$ and the convexity of $j^0(u; \cdot)$ yields

$$\begin{aligned} j^0(u; \hat{v}_k - u) &\leq (1-\varepsilon_k)j^0(u; v-u) + \varepsilon_k j^0(u; -u) \\ &\leq |j^0(u; v-u)| + k(1+|u|). \end{aligned}$$

Finally, (18.45) is obtained by Fatou's lemma . The proof of Theorem 3 is complete. □

References

[1] H. Antes, P.D. Panagiotopoulos, *The Boundary Integral Approach to Static and Dynamic Contact Problems. Equality and Inequality Methods*, Birkhäuser Verlag, Basel, Boston 1992.

[2] H. Brézis, F.E. Browder, Some Properties of Higher Order Sobolev Spaces, *J. Math. Pures et Appl.* 61 (1982), 245-259.

[3] F.E. Browder, P. Hess, Nonlinear Mappings of Monotone Type in Banach Spaces. *J. Funct. Anal.*, 11 (1972) 251-294

[4] F.H. Clarke, *Optimization and Nonsmooth Analysis*, Wiley, New York 1983.

[5] I. Ekeland, R. Temam, *Convex Analysis and Variational Problems*, North Holland, Amsterdam and American Elsevier, New York 1976.

[6] L.I. Hedberg, Two Approximation Problems in Function Spaces, *Ark. Mat.* 16 (1978), 51-81.

[7] D. Kinderlehrer, G. Stampacchia, *An Introduction to Variational Inequalities and their Applications*, Academic Press, New York, San Francisco 1980.

[8] D. Motreanu, Z. Naniewicz, Discontinuous Semilinear Problems in Vector - Valued Function Spaces, *Differential and Integral Equations* 9 (1996), 581-598.

[9] D. Motreanu, P.D. Panagiotopoulos, *Hysteresis: The Eigenvalue Problem for Hemivariational Inequalities. Models of Hysteresis* (ed. by A. Visintin) Pitman Research Notes in Mathematics, Longman, Harlow (1993).

[10] Z. Naniewicz, On the Pseudo-Monotonicity of Generalized Gradients of Nonconvex Functions, *Applicable Analysis* 47 (1992), 151-172.

[11] Z. Naniewicz, Hemivariational Inequality Approach to Constrained Problems for Star-Shaped Admissible Sets, *J. Opt. Theory Appl.* 83 (1994), 97-112.

[12] Z. Naniewicz, Hemivariational Inequalities with Functionals which are not Locally Lipschitz, *Nonlin. Anal.* 25 (1995), 1307-1320.

[13] Z. Naniewicz, P.D. Panagiotopoulos, *Mathematical Theory of Hemivariational Inequalities and Applications*, Marcel Dekker, Inc. New York 1995.

[14] Z. Naniewicz, On Variational Aspects of Some Nonconvex Nonsmooth Global Optimization Problem, *Journal of Global Optimization* 6 (1995), 383-400.

[15] Z. Naniewicz, Hemivariational inequalities as necessary conditions for optimality for a class of nonsmooth nonconvex functionals, *Nonlinear World* 4 (1997), 117-133.

[16] P.D. Panagiotopoulos, *Inequality Problems in Mechanics and Applications. Convex and Nonconvex Energy Functions*, Birkhäuser Verlag, Basel, Boston 1985. (Russian Translation MIR Publ. Moscow 1989).

[17] P.D. Panagiotopoulos, Coercive and Semicoercive Hemivariational Inequalities, *Nonlin. Anal.* 16 (1991), 209-231.

[18] P.D. Panagiotopoulos, *Hemivariational Inequalities. Applications in Mechanics and Engineering*, Springer Verlag, Berlin 1993.

[19] P.D. Panagiotopoulos, *Hemivariational Inequalities and their Applications*, In: Topics in Nonsmooth Mechanics (ed. by J.J.Moreau, P.D.Panagiotopoulos and G. Strang) Birkhäuser Verlag, Boston 1988, 75-142

[20] P.D. Panagiotopoulos, *Nonconvex Superpotentials and Hemivariational Inequalities. Quasidifferentiability in Mechanics*, In: Nonsmooth Mechanics and Applications (ed. by J.J. Moreau and P.D. Panagiotopoulos) CISM Lect. Notes, Vol. 302, 83-176, Springer Verlag, Wien, N.York 1988.

[21] P.D. Panagiotopoulos, J. Haslinger, *Optimal Control of Systems governed by Hemivariational Inequalities* In: Math. Models for Phase Change Problems, (ed. by J.F.Rodrigues), Vol. 88, Birkhäuser Verlag, Basel Boston ISNM, 1989.

[22] J. Rauch, Discontinuous Semilinear Differential Equations and Multiple Valued Maps, *Proc. Amer. Math. Soc.* 64 (1977), 277-282.

[23] J.R.L. Webb, Boundary Value Problems for Strongly Nonlinear Elliptic Equations, *The Journal of the London Math. Soc.* 21 (1980) 123-132

19 UNDERWATER ACOUSTICS*

R. P. Gilbert and Z. J. Lin

Abstract: In this paper we discuss various schemes for computing approximate acoustic Green's functions in an ocean environment. Our ambition is to use these to solve certain inverse problems which are natural in some sense. In particular, the undetermined coefficient problem and the unknown inclusion problem. We find that knowledge of the acoustic far-field can be helpful for this purpose. The time-harmonic, acoustic far-field is computed using a normal modal expansions and **transmutation** to obtain an explicit characteristic equation. The modal solutions are computed exactly using a computer algebra system, and the far field is then found using an integral representation. The expressions are quite complicated; indeed, these problems belong to a class that would probably not have been solved prior to the availability of a computer algebra system. The unknown inclusion problem is investigated by a variant on a method due to Scotti and Wirgin [12], [13].

19.1 THE ELASTIC SEABED

In this section we consider the inverse coefficient problem associated with a shallow ocean with an elastic seabed. In a shallow ocean the acoustic waves interact with the seabed, and hence, the acoustic far-field must contain information about the elastic seabed parameters. As these are related to the compressional and shear wave speeds in the seabed it suggests they may be determined from measurements of the acoustic

*This research was supported in part by National Science Foundation through grant BES-9402539.

H. Florian et al. (eds.), Generalized Analytic Functions, 215–228.

pressure far-field. Our procedure will be based on finding a representation of the acoustic pressure in terms of the seabed constants, that is to find a representation of the direct problem in terms of these parameters.

In the finite-depth, elastic seabed, we assume that it is tightly packed, and hence, homogeneous. Moreover, as the ocean is shallow, and of uniform depth, we assume it is also homogeneous. For our method to work a representation for the acoustic pressure is necessary. Let c_0 be the sound speed in the water column, with c_l and c_t being the compressional wave speed and shear wave speeds respectively in the seabed. These can be related to the Lamé coefficients, and the sediment and water densities, namely λ, μ, and ρ_s and ρ_w. We assume that b is the sediment thickness and ω is the frequency of the source. The corresponding wave numbers are related by

$$k_0 = \frac{\omega}{c_0}, \quad k_l = \frac{\omega}{c_l}, \quad k_t = \frac{\omega}{c_t} \quad ;$$

whereas, the related sound speeds are given by

$$c_l^2 = \frac{\lambda + 2\mu}{\rho_s}, \quad \text{and } c_t^2 = \frac{\mu}{\rho_s}.$$

As we wish to compute the Green function, and the coefficients are independent of the range, we may ignore the azimuthal angles of the *source* and *observation* points, θ and ϕ, which appear only in the form $G(|r \exp\theta - \rho \exp\phi|, z, \zeta)$. Hence, the acoustic pressure P equation is given by

$$(\partial_r + \frac{1}{r})\partial_r P + \partial_z^2 P - k_0^2 P + \frac{\delta(r)\,\delta(z - z0)}{2\,\pi\, r} = 0; \tag{19.1}$$

whereas, the displacement is related to the acoustic pressure by

$$\omega^2 \rho_w u_{zo} + \partial_z P = 0, \quad 0 < z < h. \tag{19.2}$$

Following the ideas used in Gilbert-Lin [8], [9] we simplify our problem if we introduce, for the homogeneous sea-bed, the displacement potentials, ϕ and ψ, which may easily be seen to satisfy the equations

$$(\partial_r + \frac{1}{r})\partial_r \phi + \partial_z^2 \phi - k_l^2 \phi = 0, \tag{19.3}$$

$$\partial_r(\partial_r + \frac{1}{r})\psi + \partial_z^2 \psi - k_t^2 \psi = 0, \quad h < z < \infty. \tag{19.4}$$

On the oceanic surface we assume a pressure release condition

$$P\big|_{z=0} = 0;$$

the transitional conditions on the ocean-sediment interface are taken to be

1. $u_{zo} = u_z$, continuity of material,
2. $P = \tau_{zz}$, continuity of pressure,

3. $\tau_{zr} = 0$, shear free at interface.

For our finite-depth sediment model, we assume that the sediment lies on a rigid rock foundation, for which we use the boundary conditions

$$\partial_z \tau_{zz}\big|_{z=b} = \partial_z u_z\big|_{z=b} = 0. \tag{19.5}$$

By using the computer algebra system, Macsyma, to solve the system of ocean-seabed equations for the Hankel transform of the pressure, $\hat{P}$, we obtain

$$\begin{aligned}\hat{P}(k^2) &= \frac{1}{2\pi}\left(\frac{L(k^2; h-z_0; U)}{L(k^2, h; U)}\frac{\sinh(\sqrt{k^2-k_0^2}z)}{\sqrt{k^2-k_0^2}}\right.\\ &\left. -H(z-z_0)\frac{\sinh(\sqrt{k^2-k_0^2}(z-z_0))}{\sqrt{k^2-k_0^2}}\right).\end{aligned} \tag{19.6}$$

where, $H(\cdot)$ is the Heaviside function, and where $L(k^2, z; U)$ is given by

$$\begin{aligned}L(k^2; z; U) &= \left((k_t^2 - 2k^2)^2 \cosh\left(\sqrt{k^2-k_l^2}(h-b)\right)\sinh\left(\sqrt{k^2-k_t^2}(h-b)\right)\right.\\ &- 4k^2\sqrt{k^2-k_l^2}\sqrt{k^2-k_t^2}\sinh\left(\sqrt{k^2-k_l^2}(h-b)\right)\\ &\left.\times\cosh\left(\sqrt{k^2-k_t^2}(h-b)\right)\right)\cosh\left(\sqrt{k^2-k_0^2}z\right)\\ &+\frac{k_t^4\rho_s}{\rho_w}\sqrt{k^2-k_l^2}\sinh\left(\sqrt{k^2-k_l^2}(h-b)\right)\\ &\times\sinh\left(\sqrt{k^2-k_t^2}(h-b)\right)\frac{\sinh\left(\sqrt{k^2-k_0^2}z\right)}{\sqrt{k^2-k_0^2}},\end{aligned}$$

b finite.

By using the Mittag Leffer decomposition, a residue calculation, and the integral representation, the case of finite seabed depth we get the following modal representation for the pressure [8], [9]

$$P(r,z) = \frac{i}{4}\sum_{L(k^2;h;U)=0}\frac{L(k^2; h-z_0; U)}{\partial_{k^2}L(k^2, h; b; U)}\frac{\sinh(\sqrt{k^2-k_0^2}z)}{\sqrt{k^2-k_0^2}}H_0^{(1)}(\sqrt{k^2}r). \tag{19.7}$$

This representation permits us to calculate the far-field F as

$$F(r,z) := \frac{i}{4}\sqrt{\frac{2}{\pi}}\sum_{n=0}^{N}\frac{L(k_n^2; h-z_0; U)}{\partial_{k^2}L(k_n^2, h; U)}\frac{\sinh(\sqrt{k_n^2-k_0^2}z)}{\sqrt{k_n^2-k_0^2}}\frac{e^{i(k_n r - \pi/4)}}{\sqrt{k_n}} \tag{19.8}$$

The Undetermined Coefficient Problem:

1. Let the measured acoustic far-field in the ocean at the range r, be given as $F\text{meas}(r, z_j)$, $j = 1, 2, ..., M$.

2. Introduce the functional

$$G(\vec{U}) \;:=\; \sum_{j=1}^{M}\left|\text{Fmeas}(R,z_j) - \frac{i}{4}\sqrt{\frac{2}{\pi}}\sum_{n=0}^{M}\frac{L(k_n^2;h-z_0;U)}{\partial_{k^2}L(k_n^2,h;U)}\right.$$
$$\left.\times\frac{\sinh(\sqrt{k_n^2-k_0^2}z_j)}{\sqrt{k_n^2-k_0^2}}\,\frac{e^{i(k_nR-\pi/4)}}{\sqrt{k_n}}\right|^2 \tag{19.9}$$

3. Minimizing $G(\vec{U})$ over the parameter space, using say, Newton's method, that is we compute

$$U - \text{Hessian}^{-1}(G)\text{grad}\,G(U) \to U$$

where grad $G(U)$ can be easy computed from the modal representation for the pressure with the help of Macsyma.

Let the measurements be $V(r_j)$, $j = 1, 2 \ldots M$. For this pupose we define

$$G(\vec{U}) := \sum_{j=1}^{M}\left|V(r_j) + \frac{i}{4\omega^2\rho_s}\sqrt{\frac{2}{\pi}}\sum_{n=0}^{N}\frac{L(k_n^2;h-z_0;b;U)}{\partial_{k^2}L(k_n^2,h;b;U)}\,\frac{e^{i(k_nr_j-\pi/4)}}{\sqrt{k_nr_j}}\right|^2, \tag{19.10}$$

and similarly, we may use Newton's method to find a minimizer for this function. Numerical experiments concerning this method have appeared in the papers [10] [11]

19.2 THE PORO-ELASTIC SEABED

The methodology used for the elastic seabed is easily generalized to to the case of poro-elastic seabeds. In the past section we considered only the case of a homogeneous ocean. Using **transmutation** we may also treat the case of variable index of refraction in the ocean. As in the previous ection it is expedient to use cylindrical coordinates. As we are interested only in computing the Green function we repress the azimuthal variable. The Biot-Stoll model [1], [2], [18], [3] takes the form

$$(\lambda + 2\mu)\partial_r e - \mu\partial_{rz}u_z + \mu'\partial_r u_z + \partial_z(\mu\partial_z u_r)$$
$$+Q\partial_r\epsilon + p_{11}u_r + p_{12}U_r = 0, \tag{19.11}$$
$$Q\partial_r e + R\partial_r\epsilon + p_{12}u_r + p_{22}U_r = 0, \tag{19.12}$$
$$\mu\,\Delta\,u_z + \mu\partial_z e + 2\mu'\partial_z u_z + \partial_z(\lambda e) + \partial_z(Q\epsilon) + p_{11}u_z + p_{12}U_z = 0, \tag{19.13}$$
$$\partial_z(Qe) + \partial_z(R\epsilon) + p_{12}u_z + p_{22}U_z = 0, \tag{19.14}$$

where

$$p_{11} := \omega^2\rho_{11} - i\omega b_1, \quad p_{12} := \omega^2\rho_{12} + i\omega b_1, \quad p_{22} := \omega^2\rho_{22} - i\omega b_1 \tag{19.15}$$

and

$$\Delta := \partial_{rr} + \frac{1}{r}\partial_r + \partial_{zz}$$

is the azimuthally independent Laplacian. In the ocean these equations simplify as there is only a fluid displacement and the Lamé coefficient $\mu = 0$. The above system may be manipulated into the form [3]

$$\Delta((\lambda + 2\mu)e + Q\epsilon) + p_{11}e + p_{12}\epsilon = 0, \tag{19.16}$$

$$\Delta(Qe + R\epsilon) + p_{12}e + p_{22}\epsilon + = 0, \tag{19.17}$$

$$\mu \Delta u_z + \mu \partial_z e + \partial_z(\lambda e) + \partial_z(Q\epsilon) + p_{11}u_z + p_{12}U_z = 0, \tag{19.18}$$

$$\partial_z(Qe) + \partial_z(R\epsilon) + p_{12}u_z + p_{22}U_z = 0. \tag{19.19}$$

As the modal pressure equation in the ocean has a depth dependent index of refraction a Gelfand-Levitan type transmutation [8] is found useful to reduce this to a constant coefficient equation, namely

$$\begin{aligned} \hat{P}(z) &= \hat{P}'(0)\left(\frac{\sinh(x_k z)}{x_k} + \int_0^z G(z,s)\frac{\sinh(x_k s)}{x_k}ds\right) \\ &\quad - \frac{H(z-z_0)}{2\pi}\left(\frac{\sinh(x_k(z-z_0))}{x_k}\right. \\ &\quad \left. + \int_0^{z-z_0} K(z-z_0,s)\frac{\sinh(x_k s)}{x_k}ds\right) \end{aligned} \tag{19.20}$$

where

$$x_k := \sqrt{k^2 - k_0^2},$$

$$G(z,s) := G\left(z, s, k_0^2(1 - n^2(\cdot))\right),$$

$$K(z,s) := G\left(z, s, k_0^2(1 - n^2(z_0 + \cdot))\right).$$

In particular, $G(z,s) = K(z,s) = 0$ when $n(z) \equiv 1$, which corresponds to the homogeneous ocean.

We now solve the system of the equations using MACSYMA to get

$$\hat{P}'(0) = \frac{s_1(k^2)T_4(k^2) + s_2(k^2)T_2(k^2)}{s_1(k^2)T_3(k^2) + s_2(k^2)T_1(k^2)} =: \frac{B(k^2)}{A(k^2)}, \tag{19.21}$$

where

$$s_2 = \det\begin{pmatrix} w_{1,1,1}\,x_\tau & w_{1,1,2}\,x_\eta & w_{0,1,3} \\ w_{2,3,1}\,x_\tau^2 + w_{0,3,1} & w_{2,3,2}\,x_\eta^2 + w_{0,3,2} & w_{1,3,3}\,x_\gamma \\ w_{3,4,1}\,x_\tau^3 + w_{1,4,1}\,x_\tau & w_{3,4,2}\,x_\eta^3 + w_{1,4,2}\,x_\eta & w_{2,4,3}\,x_\gamma^2 + w_{0,4,3} \end{pmatrix}$$

$$s_3 = \det\begin{pmatrix} w_{1,1,1}\,x_\tau & w_{1,1,2}\,x_\eta & w_{0,1,3} \\ w_{1,2,1}\,x_\tau & w_{1,2,2}\,x_\eta & w_{0,2,3} \\ w_{3,4,1}\,x_\tau^3 + w_{1,4,1}\,x_\tau & w_{3,4,2}\,x_\eta^3 + w_{1,4,2}\,x_\eta & w_{2,4,3}\,x_\gamma^2 + w_{0,4,3} \end{pmatrix}$$

in which $x_\tau = \sqrt{k^2 - \tau}$, $x_\eta = \sqrt{k^2 - \eta}$, $x_\gamma = \sqrt{k^2 - \gamma}$ and the constants $\tau, \eta, \gamma, \{w_{ikj}\}$ are uniquely expressed in terms of the physical coefficients. It can be shown that except for an accountable set of ω values, the principal part of $\hat{P}(h - z)$ is an infinite sum of partial fractions

$$\hat{P}(h-z) \sim \sum_{n=1}^{\infty} \frac{c_n(z)}{k^2 - k_n^2},$$

where

$$\begin{aligned} c_n(z) &= Res_{k^2=k_n^2} \hat{P}(h-z) \\ &= \frac{B(k_n^2)}{A'(k_n^2)} \left(\frac{\sinh(\sqrt{k_n^2 - k_0^2} z)}{\sqrt{k_n^2 - k_0^2}} + \int_0^z G(z,s) \frac{\sinh(\sqrt{k_n^2 - k_0^2} s)}{\sqrt{k_n^2 - k_0^2}} \right) \end{aligned}$$

and hence the contribution to the far field from the discrete spectrum is

$$\begin{aligned} \hat{P}(r, h-z) \sim & \frac{\pi i}{2} \sum_{A(k^2)=0} \frac{B(k^2)}{A'(k^2)} \left(\frac{\sinh\left(\sqrt{k^2 - k_0^2} z\right)}{\sqrt{k^2 - k_0^2}} \right. \\ & \left. + \int_0^z G(z,s) \frac{\sinh\left(\sqrt{k^2 - k_0^2} s\right)}{\sqrt{k^2 - k_0^2}} \right) H_0^{(1)}(kr) \end{aligned}$$

Basically one may proceed now as in the elastic case by posing the inverse problem as an extremal problem involving the far-field.

19.3 THE UNKNOWN INCLUSION PROBLEM FOR A COMPLETELY REFLECTING SEABED

In this section we report on a recent work by the author with Scotti, Wirgin, and Xu [12], [13]. The work is based on an idea used earlier by Scotti and Wirgin [15], [16], [17] to reconstruct three-dimensional objects in $\mathbb{R}^3$. We wish to determine the shape of an unknown inclusion from measured field data on a cylinder of radius r_b enclosing the object. By using the *normal* waves introduced earlier by Gilbert-Xu [7], we are able to formulate this inverse problem in a convenient form. These waves have the form

$$u^i(r, \theta, z) = \sum_{n=1}^{N} \phi_n(z_0)\, \phi_n(z)\, e^{ik a_n \vec{\alpha} \cdot \vec{r}}, \tag{19.22}$$

where

$$
\begin{aligned}
N &= \left[\frac{1}{2}+\frac{kh}{\pi}\right], \text{ and } [\cdot] \text{ denotes the integer part of the term,}\\
\phi_n(z) &= \sqrt{\frac{2}{h}}\,\sin\left(\frac{(2n-1)\,\pi z}{2h}\right), \quad n=1,2,\dots.\\
a_n &= \sqrt{1-\frac{(n-\frac{1}{2})^2\,\pi^2}{k^2h^2}},\\
z_0 & \quad \text{is a fixed number}, \quad 0<z_0<h,\\
k &= \text{wave number},\\
\vec{\alpha} &= (\alpha_1,\alpha_2), \quad |\vec{\alpha}|=1, \quad \vec{\alpha}=(\cos\theta_0 \ \sin\theta_0),\\
\vec{r} &= (x_1,x_2), \quad |\vec{r}|=r, \quad \vec{r}=r(\cos\theta\,,\,\sin\theta).
\end{aligned}
$$

The fundamental idea behind this method is that regardless of the shape of the scatterer, outside any cylinder D_d enclosing the body the total wave field can be represented in $\mathbf{R}^3_h\backslash D_d$,

$$D_d=\{(r,\theta,z)\,|\,r=d\geq \max_{\theta,z}\{\rho(\theta,z)\}\,;0\leq\theta<2\pi\,;0<z<h\},$$

by

$$
\begin{aligned}
u(r,\theta,z) &= u^i(r,\theta,z)+\sum_{n=1}^{\infty}\sum_{m=0}^{\infty}\phi_n(z)\,H_m^{(1)}(ka_nr)\\
&\quad\times\left[A_{nm}\cos(m\theta)+B_{nm}\sin(m\theta)\right],
\end{aligned}
\tag{19.23}
$$

where the *plane* incident wave may be represented as [12]

$$u^i(r,\theta,z)=\sum_{n=1}^{N}\sum_{m=0}^{\infty}\phi_n(z_0)\,\phi_n(z)\,\epsilon_m i^m\,\cos(m(\theta-\theta_0))\,J_m(ka_nr)\,. \tag{19.24}$$

where $J_m, H_m^{(1)}$ are the mth order Bessel and Hankel functions of the 1st kind respectively

$$\epsilon_m=1 \quad \text{for} \quad m=0, \quad \epsilon_m=2 \quad \text{for} \quad m>0\,.$$

Inverse problem: Our idea can be used for three-dimensional bodies which are represented by a shape function $\rho(\theta,z)$ as

$$D=\{(r,\theta,z)\,|\,r=\rho(\theta,z), \quad 0\leq\theta<2\pi, \quad 0\leq z\leq h\}$$

where $\rho(\theta,z)\geq 0$.

Suppose we are able to measure the acoustic field at the points (r_b,θ^q,z^p) where the indices range over $q\in\{1,2,\dots,Q\}$ and $p\in\{1,2,\dots,P\}$. Then the inverse problem consists of reconstructing the shape of the scatterer from this measured data. Denoting the measured field at these points by $u^*(r_b, \theta^q, z^p)$ where $r_b > a$ is some constant, we wish to determine the **shape function** $\eta^{qp} = \eta(\theta^q, z^p)$ for $q \in \{1, \dots, Q\}$ and $p \in \{1, \dots, P\}$. To do this we match the measured data to the mathematical representation of the acoustic pressure by setting the

$$\begin{aligned} u^*(r_b, \theta^q, z^p) \quad &- \quad [u^i(r_b, \theta^q, z^p) \\ &+ \sum_{n=1}^{N} \sum_{m=0}^{M} c_{nm}(\rho(\theta^q, z^p)) \cos m(\theta^q - \theta_0)\, \phi_n(z^p)\, H_m^{(1)}(ka_n r_b)] \simeq 0., \end{aligned}$$

Note that $\rho(\theta^q, z^p)$ can not be identically equal to the radius at any point we pass a cylinder through, as we have necessarily truncated the m-series for computational purposes. Now for each (p, q), (19.24) yields at least one solution for the *radius* of the unknown cylinder passing through this point. **Remark.** Ω is an open set whose topological boundary is a subset of D. D may contain a number of line segments which are not boundary points. On account of (19.24), the coefficients $c_{nm}(\rho(\theta^q, z^p))$ are given by

$$c_{nm}(\rho(\theta^q, z^p)) = -\epsilon_m\, i^m\, \phi_n(z_0)\, \frac{J_m(ka_n \rho(\theta^q, z^p))}{H_m^{(1)}(ka_n \rho(\theta^q, z^p))}. \tag{19.25}$$

Since $J_n(\xi)$ is bounded as $\xi \to 0$, and $H_m^{(1)}(\xi) \to \infty$ as $\xi \to 0$ we have

$$c_{nm}(\rho) \to 0 \quad \text{as} \quad \rho \to 0 \quad \text{for all} \quad n, m. \tag{19.26}$$

Remark. The above equations imply that

(1) If $u(r, \theta^q, z^p) \approx u^i(r, \theta^q, z^p)$, then it follows that $c_{nm} = 0$ and $\rho(\theta^q, z^p) = 0$.

(2) This method is reasonable only if from $\rho(\theta^q, z^p) = 0$ it follows that $u(r, \theta^q, z^p) \approx u^i(r, \theta^q, z^p)$. An example of such a region where this happens is a cylinder–like region where $\rho(\theta^q, z^p) \geq a > 0$ for some a.

We formulate the inverse problem as determining the discretized shape function $\rho^{qp} = \rho(\theta^q, z^p)$ on the index set $q \in \{1, \dots, Q\}$ and $p \in \{1, \dots, P\}$ such that

$$u^*(r_b, \theta^q, z^p) - [u^i(r_b, \theta^q, z^p) + \sum_{n=1}^{N} \sum_{m=0}^{M} c_{nm}(\rho^{qp}) \cdot \cos m(\theta^q - \theta_0) \cdot$$

$$\phi_n(z^p)\, H_m^{(1)}(ka_n r_b)] \simeq 0. \tag{19.27}$$

Here $u^*(r_b, \theta^q, z^q)$ are given data and the sum represents the total field expression for a cylinder with radius ρ^{qp}.

19.4 THE UNKNOWN INCLUSION PROBLEM FOR AN ELASTIC SEABED

For the shallow ocean with an elastic seabed, in addition to the Helmholtz equation holding in the ocean

$$(\Delta + k^2)\, u = 0 \quad \text{in} \quad \mathbb{R}^3_h \backslash \overline{\Omega},$$

we have to add the acoustic equations for the elastic seabed. These are

$$\begin{aligned} \partial_r \sigma_{rr} + \partial_z \sigma_{rz} + \frac{1}{r}(\sigma_{rr} - \sigma_{\theta\theta}) + \rho\omega\, u_r &= 0 \\ \partial_r \sigma_{rz} + \partial_z \sigma_{zz} + \frac{1}{r}\sigma_{rz} + \rho\omega^2\, u_z &= 0 . \end{aligned} \tag{19.28}$$

Replacing the stresses σ_{rr}, σ_{zz}, and shear σ_{rz} by Hooke's law for an isotropic material.

$$\begin{aligned} \sigma_{rr} &= \lambda e + 2\mu\, \partial_r u_r \\ \sigma_{\theta\theta} &= \lambda e + \frac{2\mu}{r}\, u_r \\ \sigma_{rz} &= \mu(\partial_z u_r + \partial_r u_z) , \end{aligned} \tag{19.29}$$

we obtain the Navier displacement equations.

Actually these equations hold for the entire ocean–seabed system if we set $\mu = 0$ in the ocean. The constitutive equation for the ocean becomes, once we recognize $\sigma_{rr} = \sigma_{\theta\theta} = \sigma_{zz} := p$,

$$p = \lambda e ,$$

The boundary, transition conditions are given by

$$p(r, 0) = 0 ,$$

$$\begin{aligned} u_r(r, h^-) &= u_r(r, h^+) , \\ \sigma_{zz}(r, h^+) &= p(r, h^-) , \\ \sigma_{rz}(h^+) &= 0 , \\ u_z(r, b^-) &= \sigma_{rz}(r, b^-) = 0 . \end{aligned} \tag{19.30}$$

The last two conditions represent the seabed lying over a rigid, smooth, rock formation. The method we outline below carries over for different conditions at $z = b^-$, such as the condition for complete reflection at a very rigid surface

$$u_z(r, b^-) = \partial_z \sigma_{zz}(r, b^-) = 0 .$$

If we seek seabed solutions in the separated form

$$u_r = \alpha(z)\, H_1(cr) , \quad u_z = \beta(z)\, H_0(cr) , \tag{19.31}$$

then we obtain a system of ordinary differential equations,

$$\begin{aligned} (\mu\alpha')' \; &+ \; [\rho\omega^2 - (\lambda + 2\mu)\, c^2]\, \alpha - c\mu'\beta - c(\lambda + \mu)\, \beta' = 0 , \\ ((\lambda + 2\mu)\beta')' \; &+ \; [\rho\omega^2 - \mu c^2]\, \beta + c\lambda'\alpha + c(\lambda + \mu)\, \alpha' = 0 . \end{aligned} \tag{19.32}$$

19.4.1 A Suitable Inner Product

We construct a suitable inner-product under which the set of eigen-functions of the system (19.32) are orthogonal [14].

Let $\{\xi_n\}$ be the set of eigenvalues and $\{(\phi_n, \alpha_n, \beta_n)\}$ be the corresponding eigen-functions normalized by $\phi_n'(0) = 1$. Then from the Navier equations we have

$$\phi_n'' + (k_0^2 n^2(z) - \xi_n)\phi_n = 0, \quad \text{for } 0 < z < h,$$

$$(\mu\alpha_n')' + [\rho\omega^2 - (\lambda + 2\mu)\xi_n]\alpha_n - \mu'\beta_n -$$
$$(\lambda + \mu)\beta_n' = 0,\ h < z < b$$
$$((\lambda + 2\mu)\beta_n')' + (\rho\omega^2 - \mu\xi_n^2)\beta_n + \xi_n\lambda'\alpha_n +$$
$$\xi_n(\lambda + \mu)\alpha_n' = 0.$$
$$\phi_n(0) = 0, \text{ pressure release}$$
$$\phi_n'(h) + \omega^2\rho_w\beta_n(h) = 0, \text{ first transition cond.}$$
$$\phi_n(h) = \xi_n\lambda(h)\alpha_n(h) + (\lambda(h) + 2\mu(h))\beta_n'(h)$$
$$\alpha_n'(h) - \beta_n(h) = 0, \text{ third transition cond.}$$
$$\alpha_n'(b) = \beta_n(b) = 0 \text{ bottom condition.}$$

For each equation above, we have a conjugate form and also we can change the index from n to m. Let us use the shorter notation $\int_b^a f(z)dz = \int_b^a f$. Multiplying the first equation by $\bar{\phi}_m$, integration by parts over $[0, h]$ and using the pressure release condition and the first transition condition, we obtain

$$-\omega^2\rho_w\beta_n(h)\bar{\phi}_m(h) - \int_0^h \phi_n'\bar{\phi}_m' - \xi_n\int_0^h \phi_n\bar{\phi}_m(z) +$$
$$k_0^2\int_0^h n^2\phi_n\bar{\phi}_m = 0.$$

By multiplying the second equation by $\bar{\alpha}_m$ and integrating by parts over $[h, b]$ and using the third transition condition and the bottom condition, we have

$$-\int_h^b \mu\alpha_n'\bar{\alpha}_m' + \int_h^b \rho\omega^2\alpha_n\bar{\alpha}_m - \xi_n\int_h^b (\lambda + 2\mu)\alpha_n\bar{\alpha}_m$$
$$+\int_h^b (\mu\beta_n\bar{\alpha}_m' - \lambda\beta_n'\bar{\alpha}_m) = 0.$$

By multiplying the third equation by $\bar{\beta}_m$ and performing the same procedure as the above, but making use of the first transition condition, we obtain

$$-\phi_n(h)\bar{\beta}_m(h) - \int_h^b (\lambda + 2\mu)\beta_n'\bar{\beta}_m' + \int_h^b (\rho\omega^2 - \mu\xi_n)\beta_n\bar{\beta}_m$$
$$+\xi_n\int_h^b (\mu\alpha_n'\bar{\beta}_m - \lambda\alpha_n\bar{\beta}_m') = 0.$$

It is convenient to use the following notation for integrals that are repeated in the calculations

$$R_{nm} = \int_0^h (-\phi'_n\bar{\phi}'_m + k_0^2 \int_0^h n^2\phi_n\bar{\phi}_m,$$

$$S_{nm} = \int_0^h \phi_n\bar{\phi}_m,$$

$$T_{nm} := \int_h^b (-\mu\alpha'_n\bar{\alpha}'_m + \rho\omega^2\alpha_n\bar{\alpha}_m),$$

$$U_{nm} := \int_h^b (\lambda + 2\mu)\alpha_n\bar{\alpha}_m,$$

$$V_{nm} := \int_h^b (\mu\beta_n\bar{\alpha}'_m - \lambda\beta'_n\bar{\alpha}_m),$$

$$W_{nm} := \int_h^b (-(\lambda + 2\mu)\beta'_n\bar{\beta}'_m + \rho\omega^2\beta_n\bar{\beta}_m),$$

$$X_{nm} := \int_h^b \mu\beta_n\bar{\beta}_m,$$

$$Y_{nm} = \beta_n(h)\bar{\phi}_m(h).$$

The above integral identities can be rephrased as

$$-\omega^2\rho_w Y_{nm} + R_{nm} - \xi_n S_{nm} = 0,$$

$$T_{nm} - \xi_n U_{nm} - V_{nm} = 0,$$

$$W_{nm} - \xi_n X_{nm} + \xi_n \bar{V}_{mn} - \bar{Y}_{mn} = 0.$$

By interchanging the indices n and m, conjugating each term and using the symmetry property of some of these symbols we obtain

$$-\omega^2\rho_w \bar{Y}_{mn} + R_{nm} - \bar{\xi}_m S_{nm} = 0,$$

$$T_{nm} - \bar{\xi}_m U_{nm} - \bar{V}_{mn} = 0,$$

$$W_{nm} - \bar{\xi}_m X_{nm} + \bar{\xi}_m V_{nm} - Y_{nm} = 0.$$

It is useful to introduce the vector functions $\Phi_n := \langle\phi_n, \alpha_n, \beta_n\rangle$, $\Phi_m := \langle\phi_m, \alpha_m, \beta_m\rangle$ we can combine the above expressions, using a computer algebra, to obtain [14]

$$(\xi_n - \bar{\xi}_m)(\Phi_n, \Phi_m) = 0,$$

where

$$(\Phi_n, \Phi_m) := \int_0^h \phi_n\bar{\phi}_m + \omega^2\rho_0 \int_h^b \mu\beta_n\bar{\beta}_m + \omega\rho_0 \int_h^b (\mu\alpha'_n\bar{\alpha}'_m - \rho_s\omega^2\alpha_n\bar{\alpha}_m). \tag{19.33}$$

Therefore if $(\Phi_n, \Phi_n) \neq 0$, then ξ_n is real. But even though it is uncertain whether this condition is true for an arbitrary frequency, we are sure it is true for small frequencies, while for $\omega = 0$ it is obvious. The answer is that $(\Phi_n, \Phi_n) \neq 0$ for all n and for all but a countable number of frequencies. The components of the eigenvector functions Φ_n, Φ_m are defined in the ocean $[0, h]$; whereas, α_n, β_n are defined in the seabed $[h, b]$. In order to find ϕ_n one must also find α_n, β_n. We have shown that if $n \neq m$ $(\Phi_n, \Phi_m) = 0$; except for a countable number of frequencies ω we have shown $(\Phi_m, \Phi_m) = \|\Phi_m\| \neq 0$.

19.4.2 The Inversion Procedure

Recall that in terms of the inner product (19.33) we have shown [14] that we can express the Green function due to a source at z_0, $(0 < z_0 < h)$ for the ocean–seabed system as

$$G(r, z, z_0) = \frac{i}{4} \sum_{j=0}^{\infty} \frac{1}{(\Phi_n, \Phi_n)} \phi_n(z) \phi_n(z_0) H_0^{(1)}(\sqrt{c_n} r).$$

We can now proceed to develop the procedure for the unknown target problem for the case of the elastic seabed [12].

If we choose our incoming wave as before

$$u^i(r, \theta, z; \vec{\alpha}) = \sum_{n=0}^{N} \frac{\phi_n(z) \phi_n(z_0}{(\Phi_n, \Phi_n)}) e^{ika_n \vec{\alpha} \cdot \vec{r}} \tag{19.34}$$

$$= \sum_{n=1}^{N} \sum_{m=0}^{\infty} \frac{\phi_n(z) \phi_n(z_0)}{(\Phi_n, \Phi_n)} \epsilon_m i^m \cos(m[\theta - \theta_0]) J_m(ka_n r).$$

then we may still write the solution in the water column as

$$u(r, \theta, z) = u^i(r, \theta, z) + \sum_{n=1}^{\infty} \sum_{m=0}^{\infty} \phi_n(z) H_m^{(1)}(ka_n r)[R_{nm} \cos(m\theta) + S_{nm} \sin(m\theta)].$$

Then we clearly obtain the same formula for the coefficients R_{nm}, S_{nm} or for that matter

$$c_{nm}(a) = -\frac{\epsilon_m i^m \phi_n(z_0)}{(\Phi_n, \Phi_n)} \frac{J_m(ka_n \theta)}{H_m^{(1)}(ka_n a)},$$

except that here the ϕ_n are the first term of the triplet $\Phi_n := \langle \phi_n, \alpha_n, \beta_n \rangle$. The method works formally for the elastic seabed. We believe an inner product may be found for the poro–elastic case too, but we have not been successful in obtaining it.

References

[1] Biot, M. A.: *Theory of propagation of elastic waves in a fluid-saturated porous solid. I. Lower frequency range*, and *II. Higher frequency range*, J. Acoust Soc. Am. 28(2)(1956),168-178,179-191.

[2] Biot, M. A.: *Mechanics of deformation and acoustic propagation in porous media*, Jour. Applied Physics 33 (1962), 1482-1498.

[3] Buchanan, J. and Gilbert, R. : *Transmission loss in the far field for a* BIOT *sediment model*, Jour. Computational Acoustics, 1996.

[4] Gilbert, R.P., Xu, Y.: *An inverse problem for harmonic acoustics in stratified oceans*, J. Math. Anal. Appl. 17(1) (1993), 121-137.

[5] Gilbert, R.P. and Xu, Y.: *Dense sets and the projection theorem for acoustic harmonic waves in homogeneous finite depth oceans*, Math. Methods Appl. Scis. **12**(1989)69-76.

[6] Gilbert, R. P.; Xu, Y.: *The propagation problem and far-field patterns in a stratified finite-depth ocean*, Math. Methods in the Appl. Sciences, 12,(1990), 199-208.

[7] Gilbert, R.P. and Xu, Y.: *Generalized Herglotz functions and inverse scattering problems in finite depth oceans* in *Inverse Problems*, SIAM, (1992).

[8] Gilbert, R.P. and Zhongyan Lin: *Acoustic waves in shallow inhomogeneous ocean with a layer of sediment*, Acustica

[9] Gilbert, R.P.,Zhongyan Lin, and Buchanan, J.L.:*Acoustic waves in shallow inhomogeneous ocean with an interactive seabed*, Proceedings of the Conference on Porous Media, St-Etienne, France.

[10] Buchanan, J.L., Gilbert, R.P. and Zhongyan Lin: Direct and Inverse Problems in Ocean Acoustics in Proceedings of the Conference on Nonlinear Analysis, Athens, 1996.

[11] Buchanan, J.L., and Gilbert, R. P.: *Determination of the Coefficients of an Elastic Seabed* (preprint)

[12] Gilbert, R.P., Xu, s. Y., Scotti, T., and Wirgin, A. *The Unidentified Object Problem in a Shallow Ocean*, submitted to JASA

[13] Gilbert, R.P., Xu, S. Y., Scotti, T., and Wirgin, A.: *Identification of a 3D object in a shallow sea from scattered sound* submitted to *Comptes Rendues*

[14] Gilbert, R.P. and Zhongyan Lin: *Fundamental Singularity in a Shallow Ocean* (to appear) *Applicable Analysis*

[15] Scotti, T., and Wirgin, A.: *Location and shape reconstruction of a soft body by means of canonical solutions and measured scattered sound fields, C.R.Acad.Sci.Paris II b*, 330, p. 641.

[16] Scotti, T., and Wirgin, A.: *Shape reconstruction using diffracted waves and canonical solutions, Inverse Probs.*, 11, p.1097.

[17] Scotti, T., and Wirgin, A.: *Shape reconstruction of a penetrable homogeneous 3D scattering body via the ICBA*, in *Proc. Conf. Problemes Inverses*

de Propagation et Diffraction d'Ondes, P.C. Sabatier, ed., Springer, Berlin, 1997 (to appear).

[18] Stoll, R.D.: *Acoustic waves in saturated sediments*, in Physics of Sound in Marine Sediments, L. Hampton (Ed.), Plenum, New York (1974).

20 ASYMPTOTIC METHODS IN UNDERWATER ACOUSTICS

K. Hackl

Dedicated to Robert P. Gilbert on the occasion of his 65th birthday.

Abstract: Starting from the equations of three–dimensional poroelasticity a rescaling technique and asymptotic expansions are applied in order to derive completely two–dimensional models for thin layers. It is shown that in the asymptotic limit the problem splits into two independent ones described by a fourth order and a second order partial differential equation respectively. Some implications of the results and their relevance to ocean acoustics are discussed.

20.1 INTRODUCTION

Gilbert and Xu in [8] and [9] and later Gilbert and Lin in [6] and [7] study wave propagation in a finite–depth ocean. I.e. they are looking for solutions of the Helmholtz–equation

$$\Delta\phi + k(x_3)^2\phi = 0 \tag{20.1}$$

in a domain of the form $\mathbb{R}^2 \times [0, b]$, $b > 0$. We refer to $\{x_3 = 0\}$ as "bottom" and to $\{x_3 = b\}$ as "surface" of the ocean. $k(x_3)$ denotes the depth–dependent index of refraction and ϕ is a displacement–potential, i.e. the displacement $\mathbf{u}_o$ of fluid particles in the ocean is given by

$$\mathbf{u}_o = \nabla\phi. \tag{20.2}$$

H. Florian et al. (eds.), Generalized Analytic Functions, 229–240.

The pressure in the ocean is then given by

$$p_o = \lambda(x_3)k(x_3)^2\phi, \tag{20.3}$$

where $\lambda(x_3)$ is the bulk–modulus of the fluid. At the surface the pressure is zero which leads to the boundary condition

$$\phi = 0 \qquad \text{at} \quad x_3 = b. \tag{20.4}$$

In the papers by Gilbert and Xu, [8], [9], the bottom was initially assumed to be hard, i.e. perfectly reflecting. Thus the vertical displacement is set to be zero. This leads to the boundary condition

$$\frac{\partial \phi}{\partial x_3} = 0 \qquad \text{at} \quad x_3 = 0. \tag{20.5}$$

While there is no dispute about condition (20.4), condition (20.5) was increasingly considered to be unrealistic on physical grounds. For this reason the author, [10], and Gilbert and Lin, [6], attempted to model the ocean bottom as elastic medium. Later Gilbert and Lin, [7], extended this approach to poroelastic material.

This procedure, however, leads to serious mathematical problems. Essentially the operators associated with those equations are non-selfadjoint, which makes it very hard to demonstrate the completeness of the corresponding system of normal modes. Normal mode expansions, however, are one of the essential tools for solving the direct and inverse scattering problem in the finite–depth case.

The objective of this paper is to establish a better conditioned problem which is still more realistic than equation (20.5) by modelling the bottom as infinitely thin layer using the asymptotic method developed by Ciarlet and his coworkers, [3]. We will do this for the case of poroelastic material. Equation (20.5) will then be replaced by a condition of the form

$$\nabla\phi = \mathbf{u} + \mathbf{w}, \quad p = h \qquad \text{at} \quad x_3 = 0, \tag{20.6}$$

where $\mathbf{u}(x_1, x_2)$, $\mathbf{w}(x_1, x_2)$ and $h(x_1, x_2)$ satisfy certain partial differential equations in x_1 and x_2 to be derived in the following.

20.2 THE BOUNDARY VALUE PROBLEM OF THREE-DIMENSIONAL POROELASTICITY

20.2.1 Geometry of the bottom–layer

Let the middle surface of the poroelastic layer considered be given by the bounded open set $\omega \subset \mathbb{R}$ with boundary γ. The layer of thickness 2ϵ is then given by

$$\Omega^\epsilon := \omega \times (-\epsilon, +\epsilon) \tag{20.7}$$

with boundaries

$$\Gamma^\epsilon_\pm := \omega \times \{\pm\epsilon\}, \qquad \Gamma^\epsilon_0 := \gamma \times (-\epsilon, +\epsilon). \tag{20.8}$$

$\Gamma^\epsilon_\pm$ are called the top and bottom face of the layer, Γ^ϵ_0 the edge of the layer.

To Ω^ϵ we assign a Cartesian $\{x_1, x_2, x_3\}$–coordinate system, ω situated in the x_1, x_2–plane. As usual latin indices will range over $\{1, 2, 3\}$ while Greek indices will range over $\{1, 2\}$.

20.2.2 Constitutive relations

We consider a linear poro-elastic medium as introduced by Biot, [1], [2], i.e. an elastic porous frame filled by a Newtonian fluid.

Let $\mathbf{u}$ denote the overall displacement of the macroscopic body and $\mathbf{u}_f$ the (absolute) displacement of fluid. Then the vector

$$\mathbf{w} = \beta(\mathbf{u}_f - \mathbf{u}) \tag{20.9}$$

denotes the amount and direction of fluid flow relative to the frame. β is the porosity of the material, i.e. the volume fraction occupied by fluid. The overall strain is given by

$$\mathbf{e} = \frac{1}{2}(\nabla\mathbf{u} + \nabla\mathbf{u}^T) \tag{20.10}$$

and the variation of fluid content by

$$\zeta = -\nabla \cdot \mathbf{w}. \tag{20.11}$$

The constitutive relations are then

$$\mathbf{e} = \chi : \boldsymbol{\sigma} + \frac{1}{3H}\, p\,\mathbf{I}, \tag{20.12}$$

$$\zeta = \frac{1}{3H}\,\mathrm{tr}\boldsymbol{\sigma} + \frac{1}{R}\,p. \tag{20.13}$$

Here H and R are the Biot-parameters and χ is the elastic compliance-tensor given by

$$\chi : \boldsymbol{\sigma} = \frac{1+\nu}{E}\,\boldsymbol{\sigma} - \frac{\nu}{E}\,\mathrm{tr}\boldsymbol{\sigma}\,\mathbf{I}, \tag{20.14}$$

where E is Young's modulus and ν Poisson's ratio. $\mathbf{I}$ is the identity-tensor and $\mathrm{tr}\mathbf{T}$ denotes the trace of a tensor $\mathbf{T}$.

20.2.3 Balance equations

We will now assume all quantities to be periodic in time with angular–frequency ω. Then the balance of momentum in the macroscopical body becomes

$$\nabla \cdot \boldsymbol{\sigma} + \mathbf{f} = \rho\ddot{\mathbf{u}} + \rho_f\ddot{\mathbf{w}} = -\omega^2(\rho\mathbf{u} + \rho_f\mathbf{w}). \tag{20.15}$$

Here ρ and ρ_f denote overall and fluid density respectively. $\mathbf{f}$ is a body–force. Balance of momentum in the fluid yields

$$-\nabla p + \mathbf{f}_f = \rho_f\ddot{\mathbf{u}} + \mu\ddot{\mathbf{w}} + \kappa\dot{\mathbf{w}} = -\omega^2\rho_f\mathbf{u} - (\omega^2\mu - i\omega\kappa)\mathbf{w}. \tag{20.16}$$

κ/ρ_f is the permeability coefficient which occurs in Darcy's law of fluid conduction while μ measures the effect of tortuosity, that is of microscopic forces acting between fluid and frame. $\mathbf{f}_f$ is the part of body–force acting on the fluid only.

For convenience we will assume the edge of the layer being clamped and impenetrable for fluid. This leads to the boundary conditions

$$\mathbf{u} = 0, \quad w_\alpha = 0 \quad \text{on} \quad \Gamma_0^\epsilon. \tag{20.17}$$

On the top and bottom face we will assume a traction $\mathbf{g}$ acting on the macroscopic body and a pressure h acting on the fluid only. This leads to the boundary conditions

$$\sigma_{i3} = \pm g_i, \quad p = \pm h \quad \text{on} \quad \Gamma_+^\epsilon \cup \Gamma_-^\epsilon. \tag{20.18}$$

20.2.4 Variational formulation

Equations (20.12),(20.13),(20.15),(20.16),(20.17) and (20.18) can now be restated in a week form via the following Hellinger-Reissner variational principle:

$$\int_{\Omega^\epsilon} \boldsymbol{\tau} : \boldsymbol{\chi} : \boldsymbol{\sigma}\, dV + \frac{1}{3H} \int_{\Omega^\epsilon} \operatorname{tr} \boldsymbol{\tau}\, p\, dV - \int_{\Omega^\epsilon} \boldsymbol{\tau} : \mathbf{e}(\mathbf{u})\, dV = 0 \quad \forall \boldsymbol{\tau} \in \Sigma^\epsilon, \tag{20.19}$$

$$\frac{1}{3H} \int_{\Omega^\epsilon} q \operatorname{tr} \boldsymbol{\sigma}\, dV + \frac{1}{R} \int_{\Omega^\epsilon} q\, p\, dV - \int_{\Omega^\epsilon} q\, \zeta(\mathbf{w})\, dV = 0 \quad \forall q \in \Pi^\epsilon, \tag{20.20}$$

$$\begin{aligned} \int_{\Omega^\epsilon} \mathbf{e}(\mathbf{v}) : \boldsymbol{\sigma}\, dV &= \omega^2 \int_{\Omega^\epsilon} \mathbf{v} \cdot (\rho\, \mathbf{u} + \rho_f \mathbf{w})\, dV \\ &\quad + \int_{\Omega^\epsilon} \mathbf{v} \cdot \mathbf{f}\, dV + \int_{\Gamma_+^\epsilon \cup \Gamma_-^\epsilon} \mathbf{v} \cdot \mathbf{g}\, dS \qquad \forall \mathbf{v} \in V^\epsilon, \end{aligned} \tag{20.21}$$

$$\begin{aligned} \int_{\Omega^\epsilon} \zeta(\mathbf{r}) p\, dV &= \int_{\Omega^\epsilon} \mathbf{r} \cdot (\rho_f \omega^2 \mathbf{u} + (\mu\omega^2 - i\omega\kappa)\mathbf{w})\, dV \\ &\quad + \int_{\Omega^\epsilon} \mathbf{r} \cdot \mathbf{f}_f\, dV - \int_{\Gamma_+^\epsilon \cup \Gamma_-^\epsilon} r_3\, h\, dS \qquad \forall \mathbf{r} \in W^\epsilon. \end{aligned} \tag{20.22}$$

The various function spaces are defined as

$$\begin{aligned} V^\epsilon &:= \{\mathbf{v} \in (H^1(\Omega^\epsilon))^3;\ \mathbf{v} = 0 \quad \text{on} \quad \Gamma_0^\epsilon\}, \\ W^\epsilon &:= \{\mathbf{w} \in (H^1(\Omega^\epsilon))^3;\ w_\alpha = 0 \quad \text{on} \quad \Gamma_0^\epsilon\}, \\ \Sigma^\epsilon &:= (L^2(\Omega^\epsilon))^9_s, \\ \Pi^\epsilon &:= L^2(\Omega^\epsilon). \end{aligned}$$

20.3 RESCALING AND ASYMPTOTIC EXPANSION

Following the procedure developed by Ciarlet and Destuynder, [3], we introduce a scaling transformation of the layer to unit thickness:

$$F^\epsilon : \Omega \to \Omega^\epsilon,$$

$$(x_1, x_2, x_3) \mapsto (x_1, x_2, \epsilon x_3),$$

where now the rescaled body and boundaries are given by

$$\Omega := \omega \times (-1, +1), \tag{20.23}$$

$$\Gamma_\pm := \omega \times \{\pm 1\}, \qquad \Gamma_0 := \gamma \times (-1, +1). \tag{20.24}$$

According to [3] we will assume that the displacements $\mathbf{u}$ and $\mathbf{w}$ transform as

$$v_\alpha^\epsilon = v_\alpha \circ F^\epsilon, \quad v_3^\epsilon = \epsilon v_3 \circ F^\epsilon \tag{20.25}$$

and the stresses as

$$\begin{aligned} \sigma_{\alpha\beta}^\epsilon &= \sigma_{\alpha\beta} \circ F^\epsilon, && (20.26)\\ \sigma_{\alpha 3}^\epsilon = \sigma_{3\alpha}^\epsilon &= \epsilon^{-1}\, \sigma_{\alpha 3} \circ F^\epsilon, && (20.27)\\ \sigma_{33}^\epsilon &= \epsilon^{-2}\, \sigma_{33} \circ F^\epsilon, && (20.28)\\ p^\epsilon &= \epsilon^{-2}\, p \circ F^\epsilon. && (20.29) \end{aligned}$$

For the external loads we assume

$$f_i^\epsilon = f_i \circ F^\epsilon, \quad g_i^\epsilon = g_i \circ F^\epsilon, \quad h^\epsilon = h \circ F^\epsilon. \tag{20.30}$$

Finally, in order to obtain a non–trivial rescaled model we also have to assume certain relations for the constitutive coefficients, compare Gilbert and Hackl, [5], for this approach:

$$H^\epsilon = \epsilon^{-2} H \circ F^\epsilon, \qquad R^\epsilon = \epsilon^{-4} R \circ F^\epsilon, \tag{20.31}$$

$$\rho^\epsilon = \epsilon^{-2}\rho, \qquad \rho_f^\epsilon = \epsilon^{-2}\rho_f, \qquad \mu^\epsilon = \epsilon^{-2}\mu, \tag{20.32}$$

$$\kappa^\epsilon \qquad \epsilon^{-2}\kappa. \tag{20.33}$$

Next we are going to expand all variables, which have become ϵ–dependent now, with respect to powers of the thickness of the layer:

$$\begin{aligned} \boldsymbol{\sigma}^\epsilon &= \epsilon^{-2}\boldsymbol{\sigma}^{-2} + \epsilon^{-1}\boldsymbol{\sigma}^{-1} + \boldsymbol{\sigma}^0 + \epsilon\boldsymbol{\sigma}^1 + \cdots,\\ \mathbf{u}^\epsilon &= \epsilon^{-2}\mathbf{u}^{-2} + \epsilon^{-1}\mathbf{u}^{-1} + \mathbf{u}^0 + \epsilon\mathbf{u}^1 + \cdots,\\ p^\epsilon &= \epsilon^{-2}p^{-2} + \epsilon^{-1}p^{-1} + p^0 + \epsilon p^1 + \cdots,\\ \mathbf{w}^\epsilon &= \epsilon^{-2}\mathbf{w}^{-2} + \epsilon^{-1}\mathbf{w}^{-1} + \mathbf{w}^0 + \epsilon\mathbf{w}^1 + \cdots. \end{aligned} \tag{20.34}$$

We substitute now (20.25)–(20.34) into (20.19)–(20.22). Since we are interested in the limit as $\epsilon \to 0$ we will retain only the lowest order terms in ϵ. By doing all this we obtain the following set of variational equations:

$$\int_\Omega \tau_{\alpha\beta}\, \chi_{\alpha\beta\gamma\delta}\, \sigma_{\gamma\delta}^{-2}\, dV + \frac{1}{3H^\epsilon}\int_\Omega \tau_{\gamma\gamma}\, p^{-2}\, dV - \int_\Omega \tau_{\alpha\beta}\, u_{\beta,\alpha}^{-2}\, dV = 0$$

$$\forall \tau_{\alpha\beta} \in L^2(\Omega), \tag{20.35}$$

$$\int_\Omega \tau_{\alpha 3}(u_{3,\alpha}^{-2} + u_{\alpha,3}^{-2})\, dV = 0 \qquad \forall \tau_{\alpha 3} \in L^2(\Omega), \tag{20.36}$$

$$\int_\Omega \tau_{33}\, u_{3,3}^{-2}\, dV = 0 \qquad \forall \tau_{33} \in L^2(\Omega), \tag{20.37}$$

$$\frac{1}{3H^\epsilon}\int_\Omega q\,\sigma_{\gamma\gamma}^{-2}\, dV + \frac{1}{R^\epsilon}\int_\Omega q\, p^{-2}\, dV + \int_\Omega q\, w_{3,3}^{-2}\, dV = 0 \qquad \forall q \in \Pi, \tag{20.38}$$

$$\int_\Omega (v_{\beta,\alpha}\,\sigma_{\alpha\beta}^{-2} + v_{\beta,3}\,\sigma_{3\beta}^{-2})\, dV = 0 \qquad \forall v_\beta \in \mathcal{V}, \tag{20.39}$$

$$\int_\Omega (v_{3,\alpha}\,\sigma_{\alpha 3}^{-2} + v_{3,3}\,\sigma_{33}^{-2})\, dV = \omega^2 \int_\Omega v_3(\rho^\epsilon\, u_3^{-2} + \rho_f^\epsilon\, w_3^{-2})\, dV + \int_{\Gamma_+\cup\Gamma_-} v_3\, g_3^\epsilon\, dS$$

$$\forall v_3 \in \mathcal{V}, \tag{20.40}$$

$$-\int_\Omega r_{3,3}\, p^{-2}\, dV = \int_{\Omega^\epsilon} r_3 \cdot (\rho_f^\epsilon\, \omega^2 u_3^{-2} + (\mu^\epsilon \omega^2 - i\omega\kappa^\epsilon) w_3^{-2})\, dV - \int_{\Gamma_+\cup\Gamma_-} r_3\, h^\epsilon\, dS$$

$$\forall r_3 \in H^1(\Omega), \tag{20.41}$$

where $\mathcal{V} \;:=\; \{v \in H^1(\Omega);\; v = 0 \;\text{ on }\; \Gamma_0^\epsilon\}$.

Equations (20.35)–(20.41) constitute the system we are going to discuss in the following. The striking feature which distinguishes them from the original system (20.19)–(20.22) is that they are to a large extent decoupled. We will show how these equations can be successively solved and lead to a completely two-dimensional problem stated on the set ω.

20.4 PRELIMINARY CONSIDERATIONS

20.4.1 *Kirchhoff-Love-displacements*

For brevity we will omit the index "−2" in the following.

From equation (20.37) it follows immediately that u_3 is independent of x_3:

$$u_3 = u_3(x_1, x_2). \tag{20.42}$$

Now (20.36) states that $u_{\alpha,3} = -u_{3,\alpha}$, i.e.

$$u_\alpha = \overset{\circ}{u}_\alpha (x_1, x_2) - x_3 u_{3,\alpha}(x_1, x_2) \tag{20.43}$$

with some functions $\overset{\circ}{u}_\alpha \in H_0^1(\omega)$.

(20.42) and (20.43) constitute a so called Kirchhoff-Love-displacement. Mechanically this means that cross–sections of the layer remain straight lines under deformation.

Note that (20.43) (together with the boundary conditions (20.17)) requires that $u_3 \in H_0^2(\omega)$.

20.4.2 *Split into two independent subproblems*

We are going to split all quantities into an even and an odd part with respect to x_3, e.g. $\mathbf{u} = \mathbf{u}^e + \mathbf{u}^o$. Applying this split to the test functions as well, it is obvious that (20.35)–(20.41) separate into two independent set of equations, one for the variables $\{\sigma^e_{\alpha\beta}, \sigma^o_{\alpha 3}, \sigma^e_{33}, p^e, \overset{\circ}{u}_\alpha, w^o_3\}$ and one for the variables $\{\sigma^o_{\alpha\beta}, \sigma^e_{\alpha 3}, \sigma^o_{33}, p^o, u_3, w^e_3\}$. For reasons which will become apparent later we will call those the "in-plane problem" and the "bending problem" respectively.

Before we will treat those problems in the following sections we have to introduce some further notions which are well–known in solid mechanics.

Let the section forces $n_{\alpha\beta}, n_p$ and the moments $m_{\alpha\beta}, m_p$ be defined by

$$n_{\alpha\beta} = \int_{-\epsilon}^{+\epsilon} \sigma_{\alpha\beta}\, dx_3, \qquad n_p = \int_{-\epsilon}^{+\epsilon} p\, dx_3, \tag{20.44}$$

$$m_{\alpha\beta} = \int_{-\epsilon}^{+\epsilon} x_3\, \sigma_{\alpha\beta}\, dx_3, \qquad m_p = \int_{-\epsilon}^{+\epsilon} x_3\, p\, dx_3. \tag{20.45}$$

It is clear that the integrands can be replaced by their even parts in (20.44) and by their odd parts in (20.45).

Now from (20.35) there follow the constitutive relations

$$\sigma^e_{\alpha\beta} = \frac{E}{1-\nu^2}\left[\frac{1}{2}(1-\nu)\,(\overset{\circ}{u}_{\alpha,\beta} + \overset{\circ}{u}_{\beta,\alpha}) + \nu\, \overset{\circ}{u}_{\gamma,\gamma}\, \delta_{\alpha\beta}\right] - \frac{1}{3H^\epsilon}\,\frac{E}{1-\nu}\, p^e\, \delta_{\alpha\beta} \tag{20.46}$$

and

$$\overset{o}{\sigma}_{\alpha\beta} = -x_3 \frac{E}{1-\nu^2}\left[(1-\nu)\, u_{3,\alpha\beta} + \nu\, \Delta u_3\, \delta_{\alpha\beta}\right] - \frac{1}{3H^\epsilon}\frac{E}{1-\nu} p^o\, \delta_{\alpha\beta}. \tag{20.47}$$

This yields for the section forces and moments the relations

$$n_{\alpha\beta} = \frac{2E}{1-\nu^2}\left[\frac{1}{2}(1-\nu)\,(\overset{o}{u}_{\alpha,\beta} + \overset{o}{u}_{\beta,\alpha}) + \nu\ \overset{o}{u}_{\gamma,\gamma}\ \delta_{\alpha\beta}\right] - \frac{1}{3H^\epsilon}\frac{E}{1-\nu} n_p\, \delta_{\alpha\beta}, \tag{20.48}$$

$$m_{\alpha\beta} = -\frac{2}{3}\frac{E}{1-\nu^2}\left[(1-\nu)\, u_{3,\alpha\beta} + \nu\, \Delta u_3\, \delta_{\alpha\beta}\right] - \frac{1}{3H^\epsilon}\frac{E}{1-\nu} m_p\, \delta_{\alpha\beta}. \tag{20.49}$$

From (20.38) we obtain

$$\frac{1}{3H^\epsilon}\sigma_{\gamma\gamma} + \frac{1}{R^\epsilon} p + w_{3,3} = 0, \tag{20.50}$$

and (20.41) yields

$$p_{,3} = \omega^2 \rho_f^\epsilon\, u_3 + (\omega^2 \mu^\epsilon - i\omega\kappa^\epsilon) w_3 \tag{20.51}$$

together with the boundary conditions

$$p^e(1) = h^{e\epsilon}, \qquad p^e(-1) = h^{e\epsilon}, \tag{20.52}$$

$$p^o(1) = h^{o\epsilon}, \qquad p^o(-1) = -h^{o\epsilon}, \tag{20.53}$$

where $h^{e\epsilon} = \frac{1}{2}(h^\epsilon_{|\Gamma_+} - h^\epsilon_{|\Gamma_-})$ and $h^{o\epsilon} = \frac{1}{2}(h^\epsilon_{|\Gamma_+} + h^\epsilon_{|\Gamma_-})$.

Elimination of $\sigma_{\gamma\gamma}$ from equation (20.50) via relations (20.46) and (20.47) gives together with (20.51) the system of first order ordinary differential equations

$$w_{3,3} = -A^{\epsilon 2}\, p + C^\epsilon\, (\Delta u_3\, x_3 - \overset{o}{u}_{\gamma,\gamma}), \tag{20.54}$$

$$p_{,3} = -B^{\epsilon 2}\, w_3 + D^\epsilon\, u_3, \tag{20.55}$$

where we have introduced the abbreviations

$$A^{\epsilon 2} = \frac{1}{R^\epsilon} - \frac{2}{9}\frac{E}{1-\nu}\frac{1}{H^{\epsilon 2}}, \qquad B^{\epsilon 2} = i\omega\kappa^\epsilon - \omega^2\mu^\epsilon, \tag{20.56}$$

$$C^\epsilon = \frac{1}{3}\frac{E}{1-\nu}\frac{1}{H^\epsilon}, \qquad D^\epsilon = \omega^2 \rho_f^\epsilon. \tag{20.57}$$

Now for B^ϵ we can choose any of the two complex roots of the defining equation. As for A^ϵ: Set $\zeta = 0$ in equation (20.13). This corresponds to an experiment where any change of fluid content in the body is inhibited. Now elimination of $\operatorname{tr}\boldsymbol{\sigma}$ between (20.12) and (20.13) yields

$$\operatorname{tr}\mathbf{e} = \left(\frac{1}{H} - \frac{3(1-2\nu)}{E}\frac{H}{R}\right)p. \tag{20.58}$$

The coefficient in front of p in equation (20.58) has to be positive for physical reasons. Hence we get the estimate

$$\frac{1}{R^\epsilon} - \frac{2}{9}\frac{E}{1-\nu}\frac{1}{H^{\epsilon 2}} > \epsilon^4\left(\frac{1}{R} - \frac{E}{3(1-2\nu)}\frac{1}{H^2}\right) > 0. \tag{20.59}$$

This allows us to define A^ϵ as the positive, real root of its defining equation.

The system (20.54) and (20.55) together with the boundary conditions (20.52) respectively (20.53) has the unique solutions

$$p^e = \left(h^{e\epsilon} + \frac{C^\epsilon}{A^{\epsilon 2}}\,\overset{\circ}{u}_{\gamma,\gamma}\right)\frac{\cosh A^\epsilon B^\epsilon x_3}{\cosh A^\epsilon B^\epsilon} - \frac{C^\epsilon}{A^{\epsilon 2}}\,\overset{\circ}{u}_{\gamma,\gamma}, \tag{20.60}$$

$$w_3^o = -\left(h^{e\epsilon} + \frac{C^\epsilon}{A^{\epsilon 2}}\,\overset{\circ}{u}_{\gamma,\gamma}\right)\frac{A^\epsilon}{B^\epsilon}\,\frac{\sinh A^\epsilon B^\epsilon x_3}{\cosh A^\epsilon B^\epsilon}, \tag{20.61}$$

respectively

$$p^o = \left(h^{o\epsilon} - \frac{C^\epsilon}{A^{\epsilon 2}}\,\Delta u_3\right)\frac{\sinh A^\epsilon B^\epsilon x_3}{\sinh A^\epsilon B^\epsilon} + \frac{C^\epsilon}{A^{\epsilon 2}}\,\Delta u_3\, x_3, \tag{20.62}$$

$$\begin{aligned} w_3^e &= -\left(h^{o\epsilon} - \frac{C^\epsilon}{A^{\epsilon 2}}\,\Delta u_3\right)\frac{A^\epsilon}{B^\epsilon}\,\frac{\cosh A^\epsilon B^\epsilon x_3}{\sinh A^\epsilon B^\epsilon} \\ &\quad - \frac{C^\epsilon}{A^{\epsilon 2} B^{\epsilon 2}}\,\Delta u_3 + \frac{D^\epsilon}{B^{\epsilon 2}}\,u_3. \end{aligned} \tag{20.63}$$

This gives for the section force

$$n_p = 2\left(h^{e\epsilon} + \frac{C^\epsilon}{A^{\epsilon 2}}\,\overset{\circ}{u}_{\gamma,\gamma}\right)\frac{\tanh A^\epsilon B^\epsilon}{A^\epsilon B^\epsilon} - 2\,\frac{C^\epsilon}{A^{\epsilon 2}}\,\overset{\circ}{u}_{\gamma,\gamma}, \tag{20.64}$$

and for the moment

$$m_p = 2\left(h^{o\epsilon} - \frac{C^\epsilon}{A^{\epsilon 2}}\,\Delta u_3\right)K^\epsilon + \frac{2}{3}\,\frac{C^\epsilon}{A^{\epsilon 2}}\,\Delta u_3 \tag{20.65}$$

with the abbreviation

$$K^\epsilon = \frac{1}{A^\epsilon B^\epsilon \tanh A^\epsilon B^\epsilon} - \frac{1}{A^{\epsilon 2} B^{\epsilon 2}}. \tag{20.66}$$

It is noteworthy that as in the case of the thermoelastic plate, see Gilbert, Hackl, [5], this problem involves the solution of ordinary differential equations in x_3.

20.5 THE IN-PLANE PROBLEM

Choose v_β in equation (20.39) to be of the form $v_\beta = v(x_1, x_2)$ with $v \in H_0^1(\omega)$. This yields

$$\int_\omega v_{,\alpha}\, n_{\alpha\beta}\, dS = 0. \tag{20.67}$$

Now substitution of (20.48) and (20.64) gives

$$\int_\omega v_{,\alpha} \left[\frac{E}{1+\nu}\, \overset{\circ}{u}_{\beta,\alpha} + \left(\frac{E}{1-\nu^2} + 2\, \frac{C^{\epsilon 2}}{A^{\epsilon 2}} (1 - \frac{\tanh A^\epsilon B^\epsilon}{A^\epsilon B^\epsilon}) \right) \overset{\circ}{u}_{\alpha,\beta} \right] dS$$
$$= 2\, C^\epsilon\, \frac{\tanh A^\epsilon B^\epsilon}{A^\epsilon B^\epsilon} \int_\omega v_{,\beta} h^{e\epsilon}\, dS \qquad \forall v \in H_0^1(\omega). \tag{20.68}$$

Equation (20.68) is equivalent to the well known problem of plane (two-dimensional) elasticity and has a unique solution $\overset{\circ}{u}_\beta \in H_0^1(\omega) \cap H^2(\omega)$ for $h^{e\epsilon} \in L^2(\omega)$ provided γ is smooth enough, see for example the book by Ciarlet, [4]. Once $\overset{\circ}{u}_\alpha$ is found all other quantities can be calculated by back–substitution.

20.6 THE BENDING PROBLEM

Choose v_β in equation (20.39) to be of the form $v_\beta = x_3\, v_{,\beta}(x_1, x_2)$ with $v \in H_0^2(\omega)$. This yields

$$\int_\omega v_{,\alpha\beta}\, m_{\alpha\beta}\, dS = - \int_\Omega v_{,\beta}\, \sigma_{3\beta}\, dV. \tag{20.69}$$

Now choose $v_3 = v$ in equation (20.40) in order to obtain

$$\int_\Omega v_{,\alpha}\, \sigma_{\alpha 3}\, dV = 2\omega^2 \int_\omega v \left[(\rho^\epsilon + \rho_f^\epsilon \frac{D^\epsilon}{B^{\epsilon 2}}) u_3 - \frac{\rho_f^\epsilon}{B^{\epsilon 2}}\, h^{o\epsilon} \right] dS$$
$$+ \int_\omega v\, (g_{3+}^\epsilon + g_{3-}^\epsilon)\, dS \qquad \forall v \in H_0^2(\omega), \tag{20.70}$$

where we substituted relation (20.63). Here $g_{3\pm}^\epsilon$ denotes the value of g_3^ϵ on Γ_+ or Γ_- respectively.

Now finally substitution of (20.49), (20.65) and (20.69) into (20.70) yields

$$\int_\omega \Delta v \left[\left(\frac{2}{3}\, \frac{E}{1-\nu^2} + \frac{C^{\epsilon 2}}{A^{\epsilon 2}} (\frac{2}{3} - 2K^\epsilon) \right) \Delta u_3 + 2C^\epsilon\, K^\epsilon\, h^{o\epsilon} \right] dS$$

$$= \quad 2\,\omega^2 \int_\omega v \left[(\rho^\epsilon + \rho_f^\epsilon \frac{D^\epsilon}{B^{\epsilon 2}}) u_3 - \frac{\rho_f^\epsilon}{B^{\epsilon 2}}\, h^{o\epsilon} \right] dS + \int_\omega v\, (g_{3+}^\epsilon + g_{3-}^\epsilon)\, dS$$
$$\forall v \in H_0^2(\omega), \tag{20.71}$$

Equation (20.71) is equivalent to the well-known problem of bending of an elastic plate. In [3] it is shown that a variational equation of the form (20.71) has a unique solution $u_3 \in H_0^2(\omega) \cap H^3(\omega)$ for $g_{3\pm}^{\epsilon}, h^{o\epsilon} \in L^2(\omega)$ provided γ is smooth enough (cp. also Kondrat'ev, [11], and Nečas, [12]). Once u_3 is found all other quantities can be calculated by back-substitution.

With equation (20.71) we have succeeded in reducing the bending problem to a completely two-dimensional description.

20.7 THE DIFFERENTIAL EQUATIONS

We translate now the variational equations (20.68) and (20.71) into their strong forms. Simultaneously we go back to the original, unscaled quantities. This yields

$$\frac{E}{1+\nu}\,\overset{\circ}{u}_{\beta,\alpha\alpha} + \left(\frac{E}{1-\nu^2} + 2\frac{C^2}{A^2}\left(1 - \frac{\tanh \epsilon AB}{\epsilon AB}\right)\right)\overset{\circ}{u}_{\alpha,\alpha\beta} = 2\,C\,\frac{\tanh \epsilon AB}{\epsilon AB}\,h^{\epsilon}_{,\beta} \tag{20.72}$$

for the in-plane problem and

$$\left(\frac{2E\epsilon^3}{3(1-\nu^2)} + \frac{C^2\epsilon^3}{A^2}\left(\frac{2}{3} - 2K^{\epsilon}\right)\right)\Delta\Delta u_3$$
$$= 2\omega^2\epsilon\left(\rho + \rho_f\frac{D}{B^2}\right)u_3 - 2\frac{D}{B^2}h^o - 2\epsilon^2 CK^{\epsilon}\Delta h^o + g_{3+} + g_{3-} \tag{20.73}$$

for the bending problem, where the various coefficients are now given by

$$A^2 = \frac{1}{R} - \frac{2}{9}\frac{E}{1-\nu}\frac{1}{H^2}, \quad B^2 = i\omega\kappa - \omega^2\mu, \quad C = \frac{1}{3}\frac{E}{1-\nu}\frac{1}{H}, \quad D = \omega^2\rho_f, \tag{20.74}$$

and

$$K^{\epsilon} = \frac{1}{\epsilon\,A\,B\tanh\epsilon\,A\,B} - \frac{1}{\epsilon^2\,A^2\,B^2}. \tag{20.75}$$

Now equation (20.72) can be recognized as Navier's equation of two-dimensional elasticity and equation (20.73) is the well-known equation for a vibrating Kirchhoff-plate. Membrane-stiffness, bending-stiffness, density, in-plane and normal loading are modified according to effects due to the porosity of the material. Note especially that not only the surface pressure h but also its derivatives enter the formulation. Also it is surprising that equation (20.72) is free of inertia-terms.

20.8 CONCLUSION

With equations (20.72)-(20.75) we have succeeded in deriving appropriate analytical models for thin poroelastic media. Both equations (20.72) and (20.73) may be suitable to describe the ocean floor under certain physical situations. The coupled ocean-bottom problem will be treated in a subsequent paper.

References

[1] Biot, M.A.; Willis, D.G.: The elastic coefficients of the theory of consolidation. *Journal of Applied Mechanics*, **24**, 594–601, 1957.

[2] Biot, M.A.: Mechanics of deformation and acoustic propagation in porous media. *Journal of Applied Physics*, **33**, 1482–1498, 1962.

[3] Ciarlet, P.G.; Destuynder, P.: A justification of the two-dimensional linear plate model, *J. de Méchanique* **18**(2), 315-344, 1979.

[4] Ciarlet, P.G.: Mathematical Elasticity, Volume I: Three-Dimensional Elasticity, North-Holland, Amsterdam, 1988.

[5] Gilbert, R.P.; Hackl, K.: The two-dimensional, linear thermoelastic plate. In: R.P. Gilbert and K. Hackl Ed.s: Asymptotic theories for plates and shells. Pitman Research Notes in Mathematics Series 319. Longman, UK, 67-80, 1995.

[6] Gilbert, R.P.; Lin, Z.Y.: Acoustic wave in the shallow ocean with a layer of elastic sediment, *ACOUSTICA*, to appear.

[7] Gilbert, R.P.; Lin, Z.Y.: Acoustic field in a shallow stratified ocean with a poro-elastic seabed, preprint, (1996).

[8] Gilbert, R. P.; Xu, Y.: Dense Sets and the Projection Theorem for Acoustic Waves in a Homogeneous Finite Depth Ocean, *Math. Methods in the Appl. Sciences*, **12**, 69-76, 1990.

[9] Gilbert, R. P.; Xu, Y.: The Propagation Problem and Far-field Patterns in a Stratified Finite-depth Ocean, *Math. Methods in the Appl. Sciences*, **12**, 199-208, 1990.

[10] Hackl, K.: A Normal Mode Expansion for a Unified Ocean-Bottom-System, in: D. Lee, R. Vichnevetsky, A.R. Robinson Ed.s: Computational Acoustics, Vol. 2: Acoustic Propagation, proceedings of the 3rd IMACS symposium on computational acoustics, Elsevier, Amsterdam, 259-268, 1993.

[11] Kondrat'ev, V.A.: Boundary value problems for elliptic equations in domains with conical or angular points, *Trudy Moskov. Math. Obšč* **16**, 209-292, 1967.

[12] Nečas, J.: Les méthodes directes en théorie des équations elliptiques, Masson, Paris, 1967.

21 COMPARISON OF THE PARABOLIC APPROXIMATION WITH RESIDUE CALCULUS IN OCEAN ACOUSTICS

James L. Buchanan †
and Robert P. Gilbert ‡

Abstract: The usual implementation of the numerical technique for approximating far field acoustic pressure in the ocean known as parabolic approximation will be unstable for some poroelastic sediments. This instability can be remedied if a certain conjecture is true. Moreover the resulting solutions are in good agreement with analytical solutions obtained by an application of the residue calculus.

21.1 INTRODUCTION

A numerical technique called *parabolic approximation* for determining the acoustic pressure in an ocean has been very successful when the underlying seabed is modelled as an elastic slab. See for instance Collins [5],[6]. An attempt by the authors several years ago to apply parabolic approximation when the underlying sediment is poroelastic, rather than elastic, was not uniformly successful because the technique

†Supported in part by a U.S. Naval Academy Recognition Grant.
‡This research was supported in part by the National Science Foundation through grant BES-9402539.

H. Florian et al. (eds.), Generalized Analytic Functions, 241–253.

was found to be unstable for coarse-grained sediments. In [10] Wetton and Brooke indicate instability may arise in parabolic approximation when the eigenvalues for the problem have imaginary parts lying below the real axis. In this article we discuss the nature of this instability and show that, providing a certain conjecture is true, the instability can be remedied. We then show that the resulting approximations are reasonably accurate by comparing the solution to an exact modal solution that we shall also derive.

21.2 THE BIOT EQUATIONS FOR A POROELASTIC SEDIMENT

The Biot model for a poroelastic seabed [1], [2] treats the seabed as an elastic frame with interstitial pore fluid. Two displacement vectors $\mathbf{u} := (u_{rs}, u_{zs})$ and $\mathbf{U} := (U_{rs}, U_{zs})$ track the motion of the frame and the fluid respectively. The *dilatations* for the frame and pore fluid are

$$e_s := \nabla \cdot \mathbf{u} = \partial_r u_{rs} + \frac{u_{rs}}{r} + \partial_z u_{zs}, \; \epsilon_s := \nabla \cdot \mathbf{U} = \partial_r U_{rs} + \frac{U_{rs}}{r} + \partial_z U_{zs}.$$

In [3] Buchanan and Gilbert show that assuming time harmonic oscillations $e_s(r,z,t) = e_s(r,z)e^{i\omega t}, \ldots$ the Biot model leads to the following system of differential equations for the unknowns e_s, ϵ_s, u_{zs}, and U_{zs}

$$\begin{aligned} \Delta((\lambda_s + 2\mu_s)e_s + Q\epsilon_s) + 2\mu_s' \,\Delta\, u_{zs} - 2\partial_z(\mu_s' e_s) + 2\mu_s'' \partial_z u_{zs} & \\ + p_{11} e_s + p_{12}\epsilon_s + p_{11}' u_{zs} + p_{12}' U_{zs} &= 0 \\ \Delta(Qe_s + R\epsilon_s) + p_{12} e_s + p_{22}\epsilon_s + p_{12}' u_{zs} + p_{22}' U_{zs} &= 0 \\ \mu_s \,\Delta\, u_{zs} + \mu_s \partial_z e_s + 2\mu_s' \partial_z u_{zs} + \partial_z(\lambda_s e_s) + \partial_z(Q\epsilon_s) + p_{11} u_{zs} + p_{12} U_{zs} &= 0 \\ \partial_z(Qe_s) + \partial_z(R\epsilon_s) + p_{12} u_{zs} + p_{22} U_{zs} &= 0. \end{aligned}$$

Here λ_s and μ_s are the Lamé coefficients of the porous skeleton. Biot gave the following physical interpretation to the parameters R and Q: R measures the pressure on the fluid required to force a certain volume of fluid into the sediment at constant volume; Q measures the coupling of changes in the volume of the solid and fluid. Also

$$p_{11} := \omega^2 \rho_{11} - i\omega b, \quad p_{12} := \omega^2 \rho_{12} + i\omega b, \quad p_{22} := \omega^2 \rho_{22} - i\omega b$$

where ρ_{11} and ρ_{22} are density parameters, ρ_{12} is a density coupling parameter, and b is a dissipation parameter.

We make the change of dependent variables

$$\tau := (\lambda + 2\mu)e_s + Q\epsilon_s, \quad \sigma := Qe_s + R\epsilon_s$$

the inverse transformation for which is

$$e_s = a_{11}\tau - a_{12}\sigma, \quad \epsilon_s = -a_{12}\tau + a_{22}\sigma$$

where

$$a_{11} := R/d, \quad a_{12} := Q/d, \quad a_{22} := (\lambda_s + 2\mu_s)/d, \quad d := R(\lambda_s + 2\mu_s) - Q^2.$$

The fourth equation can be solved to obtain

$$U_{zs} = -\frac{1}{p_{22}}\left(p_{12}u_{zs} + \partial_z\sigma\right).$$

Substituting this into the first three equations, changing to the variables τ and σ and using the third equation to eliminate the term $2\mu' \Delta u_z$ in the first equation yields a 3×3 system

$$\Delta \begin{pmatrix} \tau \\ \sigma \\ u_{zs} \end{pmatrix} + A\partial_z \begin{pmatrix} \tau \\ \sigma \\ u_{zs} \end{pmatrix} + B \begin{pmatrix} \tau \\ \sigma \\ u_{zs} \end{pmatrix} = 0.$$

When the seabed coefficients depend on depth the elements in the matrices A and B are complicated. See [3] for their formulas. The equations are considerably simpler in the constant coefficient case, however, and as we shall see can be solved analytically.

21.3 THE GEO-ACOUSTICAL MODEL

Take the surface of the ocean to be $z = 0$ and the bottom to be at $z = h$. Let $e_o(r,z)$, and $u_{zo}(r,z)$ denote dilatation and vertical displacement in the ocean. For a time-harmonic point source at depth $z = z_0$ the equations in the ocean are

$$\begin{aligned} \Delta(\lambda_o e_o) + \rho_o\omega^2 e_o + \rho_o'\omega^2 u_{zo} &= -\frac{1}{2\pi r}\delta(z - z_0)\delta(r) \\ \partial_z(\lambda_o e_o) + \rho_o\omega^2 u_{zo} &= 0. \end{aligned}$$

The following boundary and interface conditions are imposed
Ocean surface: Requiring pressure to be zero at the surface gives

$$e_o(r,0) = 0$$

Ocean-sediment interface: Continuity of vertical displacement, normal stress and tangential stress give

$$\begin{aligned} u_{zo}(r,h-) &= u_{zs}(r,h+) = U_{zs}(r,h+) \\ \lambda_o(h-)e_o(r,h-) &= (\lambda_s(h+) + Q(h+))e_s(r,h+) + 2\mu_s(h+)\partial_z u_{zs}(r,h+) \\ &\quad +(Q(h+) + R(h+))\epsilon_s(r,h+) \\ \partial_z u_{rs}(r,h+) &+ \partial_r u_{zs}(r,h+) = 0 \end{aligned}$$

Asymptotic condition at infinity:

$$e_s, \epsilon_s, u_{rs}, u_{zs}, U_{rs}, U_{zs} \to 0 \text{ as } z \to \infty.$$

21.4 MODAL SOLUTION IN THE CONSTANT COEFFICIENT CASE

In the case where all of the coefficients in the ocean and seabed are constant the equations in the seabed simplify and an analytic solution is possible. We seek solutions of the form

$$\begin{aligned}
e_s(r,z) &= e_s(z)H_0^{(1)}(k_0\sqrt{\kappa}r),\ \epsilon_s(r,z) = \epsilon_s(z)H_0^{(1)}(k_0\sqrt{\kappa}r)\\
u_{zs}(r,z) &= u_{zs}(z)H_0^{(1)}(k_0\sqrt{\kappa}r),\ U_{zs}(r,z) = U_{zs}(z)H_0^{(1)}(k_0\sqrt{\kappa}r)\\
u_{rs}(r,z) &= u_{rs}(z)H_1^{(1)}(k_0\sqrt{\kappa}r),\ U_{rs}(r,z) = U_{rs}(z)H_1^{(1)}(k_0\sqrt{\kappa}r)
\end{aligned}$$

where $H_j^{(1)}, j = 0, 1$ are outgoing Hankel functions and $\text{Im}(\kappa) \geq 0$ is required for solutions to approach zero as $r \to \infty$. Here $k_0 = \varpi/c_o$ and c_o is the sound speed in the ocean. Substituting these representations into the sediment equations gives

$$\begin{aligned}
\tau''(z) + a_\tau^2\tau(z) + B_{12}\sigma(z) &= 0,\\
\sigma''(z) + B_{21}\tau(z) + a_\sigma^2\sigma(z) &= 0\\
u_{zs}''(z) + A_{31}\tau'(z) + A_{32}\sigma'(z) + a_u^2 u_{zs}(z) &= 0
\end{aligned} \tag{21.1}$$

where

$$\begin{aligned}
a_\tau^2 &: = B_{11} - k_0^2\kappa = p_{11}a_{11} - p_{12}a_{12} - k_0^2\kappa,\\
a_\sigma^2 &: = B_{22} - k_0^2\kappa = p_{22}a_{22} - p_{12}a_{12} - k_0^2\kappa\\
a_u^2 &: = B_{33} - k_0^2\kappa = \frac{p_{11}}{\mu_s} - \frac{p_{12}^2}{p_{22}\mu_s} - k_0^2\kappa.
\end{aligned}$$

This gives as solutions that satisfy the asymptotic condition

$$\begin{aligned}
\tau(z) &= C_2e^{im_+(z-h)} + C_3e^{im_-(z-h)}\\
\sigma(z) &= -\frac{{a_\tau}^2e^{im_+(z-h)}C_2}{B_{12}} + \frac{e^{im_+(z-h)}{m_+}^2C_2}{B_{12}} - \frac{{a_\tau}^2e^{im_-(z-h)}C_3}{B_{12}}\\
&\quad + \frac{e^{im_-(z-h)}{m_-}^2C_3}{B_{12}}
\end{aligned} \tag{21.2}$$

where

$$m_\pm := \sqrt{\frac{a_\tau^2 + a_\sigma^2 \pm \sqrt{(a_\tau^2 - a_\sigma^2)^2 + 4B_{12}B_{21}}}{2}}$$

and the branch cut for the square root function is chosen so that $\text{Im}(m_\pm) \geq 0$. Using (21.2) the solution for u_{zs} now can be found from the third equation in (21.1).

In the ocean acoustic pressure is $p_o(r,z) = \lambda_o e_o(r,z)$ whence the differential equations in the ocean can be written

$$\Delta p_o + k_0^2 p_o = -\frac{1}{2\pi r}\delta(r)\delta(z - z_0), \quad \partial_z p_o + \lambda_o k_0^2 u_{zo} = 0.$$

We seek a Green's function representation

$$p_o(r,z,z_0) = \frac{k_0^2}{8\pi}\oint_{C_0} H_0^{(1)}(k_0\sqrt{\kappa}r)G_2(z,z_0,\kappa)d\kappa.$$

The contour C_0 must enclose all singularities of G_2 and exclude those of $H_0^{(1)}(k_0\sqrt{\kappa r})$. We choose it to be the slit cut enclosing the positive real axis, oriented counter clockwise. The depth Green's function G_2 is given by

$$G_2(z,z_0,\kappa) = C_1 \sin a_o z - \frac{U(z-z_0)}{a_o}\sin a_o(z-z_0),\ 0 \le z \le h$$

where U is the unit step function. Substituting this and the solutions to the sediment equations into the ocean sediment interface conditions gives a system of equations

$$\mathcal{M}\begin{pmatrix} \frac{C_1}{\lambda_o} \\ C_2 \\ C_3 \\ C_4 \end{pmatrix} = \begin{pmatrix} -\frac{1}{\lambda_o k_0^2}\cos a_o(h-z_0) \\ -\frac{1}{\lambda_o k_0^2}\cos a_o(h-z_0) \\ \frac{1}{a_o}\sin a_o(h-z_0) \\ 0 \end{pmatrix}$$

which in turn gives

$$C_1 = \lambda_0 \frac{\Delta_1}{\Delta_0}$$

where $\Delta_0 = \det(\mathcal{M})$ and $\Delta_1 = \det(\mathcal{M}_1)$ with

$$\mathcal{M}_1 = \begin{pmatrix} \frac{-1}{\lambda_o k_0^2}\cos(a_o(\kappa)(h-z_0)) & M_{12} & M_{13} & M_{14} \\ \frac{-1}{\lambda_o k_0^2}\cos(a_o(\kappa)(h-z_0)) & M_{22} & M_{23} & M_{24} \\ \frac{1}{a_o(\kappa)}\sin(a_o(\kappa)(h-z_0)) & M_{32} & M_{33} & M_{34} \\ 0 & M_{42} & M_{43} & M_{44} \end{pmatrix}.$$

The singularities of C_1 in the κ-plane turn out to be the zeroes$\{\kappa_n\}$ of $\Delta_0(\kappa)$ (the eigenvalues) and the branch cuts of the coefficients m_+, m_- and a_τ. See [4] for more detail. Applying the residue theorem gives

$$\begin{aligned} p_o(r,z,z_0) &= \sum_n \frac{k_0^2\lambda_0 i}{4}\frac{\Delta_1(\kappa_n)}{\frac{d\Delta_0}{d\kappa}(\kappa_n)}\sin(a_o(\kappa_n)z)H_0^{(1)}(k_0\sqrt{\kappa_n}r) \\ &\quad -\frac{k_0^2}{8\pi}\oint_{C_{m_+}+C_{m_-}+C_{a_u}} G_2(z,z_0,\kappa)H_0^{(1)}(k_0\sqrt{\kappa r})d\kappa. \end{aligned}$$

In the far field only those modes corresponding to eigenvalues near the real axis between $\kappa = 0$ and $\kappa = 1$ make a significant contribution in the summation part of the representation and the integrals along the branch make negligible contribution. The eigenvalues are computed numerically by finding the minima of $|\Delta_0(\kappa)|$. The locations of the eigenvalues of importance to the far field are determined by the location of the m_+ branch cut which runs left from the point

$$m_{0+} = \frac{B_{11} + B_{22} + \sqrt{(B_{11}-B_{22})^2 + 4B_{12}B_{21}}}{2k_0^2}.$$

Figure 21.1 shows the eigenvalue maps for two sediments, a sediment consisting of fine sand at a source frequency of 106.5 Hz and one consisting of coarse sand and fine

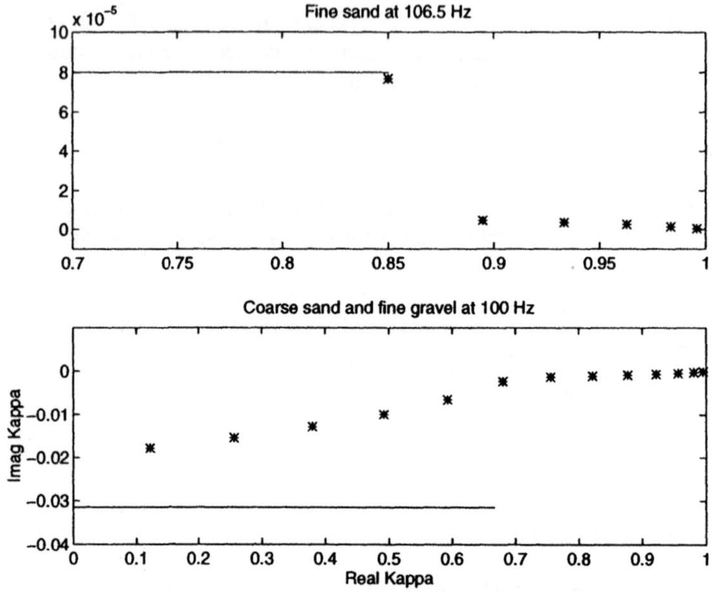

Figure 21.1 Eigenvalue maps for two sediments. Eigenvalues emerge from the m_+ branch cut (horizontal line) at certain frequencies.

gravel at a source frequency of 100 Hz. The receiver depth was $z_0 = 25$ m and the ocean depth was $h = 100$ m in both cases. The values of the Biot parameters were calculated from estimates given in Holland and Brunson [8]. See [3] for more detail on the calculation of the Biot parameters. In both these sediments and indeed for all sediments surveyed thus far, eigenvalues of consequence in the far field emerge from the m_+ branch cut and migrate toward $\kappa = 1$ as frequency increases. For fine-grained sediments the m_+ branch cut seems to lie on the positive side of the real κ-axis while it is on the negative side for coarse-grained sediments. Eigenvalues are observed to stay on the side of the real κ-axis on which the m_+ branch cut lies.

21.5 NUMERICAL SOLUTION OF THE BIOT EQUATIONS BY PARABOLIC APPROXIMATION

Parabolic approximation, a numerical technique for finding acoustic pressure in the far field, is applicable when the ocean and seabed coefficients are depth dependent and the depth of the ocean is range dependent. We illustrate it by considering the constant coefficient equations, however. In the constant coefficient case the differential equations are

Ocean:

$$\lambda_o \, \Delta \, e_o + \rho_o \omega^2 e_o = 0, \quad \lambda_o \partial_z e_o + \rho_o \omega^2 u_{zo} = 0$$

Sediment:

$$(\lambda_s + 2\mu_s) \Delta e_s + Q \Delta \epsilon + p_{11} e_s + p_{12} \epsilon = 0$$
$$Q \Delta e_s + R \Delta \epsilon + p_{12} e_s + p_{22} \epsilon = 0$$
$$\mu_s \Delta u_{zs} + \mu_s \partial_z e_s + \lambda_s \partial_z e_s + Q \partial_z \epsilon + p_{11} u_{zs} + p_{12} U_{zs} = 0$$
$$Q \partial_z e_s + R \partial_z \epsilon + p_{12} u_{zs} + p_{22} U_{zs} = 0$$

The second ocean equation can be regarded as determining u_{zo}. In the first ocean equation we factor out the cylindrical spreading by letting $\widehat{e}_o := \sqrt{r} e_o$ and retaining terms that are $O(1/r)$ or larger. This gives

$$\lambda_o \partial_{rr} \widehat{e}_o + \lambda_o \partial_{zz} \widehat{e}_o + \rho_o \omega^2 \widehat{e}_o = 0$$

or

$$L_o \partial_{rr} \widehat{e}_o + M_o \widehat{e}_o = 0$$

with $L_o = \lambda_o, M_o = \lambda_o \partial_{zz} + \rho_o \omega^2$.

For the seabed equation we solve the fourth equation for U_{zs} and substitute this into the third equation. Upon letting $\widehat{e}_s := \sqrt{r} e_s, \widehat{\epsilon} := \sqrt{r}\epsilon, \widehat{u}_{zs} := \sqrt{r} u_{zs}$ and retaining terms that are $O(1/r)$ and larger we obtain

$$L_s \partial_{rr} \begin{pmatrix} \widehat{e}_s \\ \widehat{\epsilon} \\ \widehat{u}_{zs} \end{pmatrix} + M_s \begin{pmatrix} \widehat{e}_s \\ \widehat{\epsilon} \\ \widehat{u}_{zs} \end{pmatrix} = 0$$

where

$$L_s = \begin{pmatrix} \lambda_s + 2\mu_s & Q & 0 \\ Q & R & 0 \\ 0 & 0 & \mu_s \end{pmatrix}$$

$$M_s = \begin{pmatrix} (\lambda_s + 2\mu_s)\partial_{zz} + p_{11} & Q\partial_{zz} + p_{12} & 0 \\ Q\partial_{zz} + p_{12} & R\partial_{zz} + p_{22} & 0 \\ \left(\lambda_s + \mu_s - \frac{p_{12}}{p_{22}} Q\right)\partial_z & \left(Q - \frac{p_{12}}{p_{22}} R\right)\partial_z & \mu_s \partial_{zz} + \left(p_{11} - \frac{p_{12}^2}{p_{22}}\right) \end{pmatrix}.$$

Both the ocean and seabed equations have the form

$$L \partial_{rr} \widehat{e} + M \widehat{e} = 0$$

Following Collins [6] we write formally

$$\partial_{rr} \widehat{e} + L^{-1} M \widehat{e} = 0$$

and factor to obtain

$$\left(\partial_r \widehat{e} + i\sqrt{L^{-1}M\widehat{e}}\right)\left(\partial_r \widehat{e} - i\sqrt{L^{-1}M\widehat{e}}\right) = 0$$

Solutions to this equation will be superpositions of incoming and outgoing waves. We are interested in the outgoing waves, which are solutions to

$$\partial_r \widehat{e} = i\sqrt{L^{-1}M}\widehat{e} = ik_0 \sqrt{1 + \frac{L^{-1}(M - k_0^2 L)}{k_0^2}}\,\widehat{e}$$

Let $\widetilde{e} = e^{-ik_0 r}\widehat{e}$. Then the differential equation becomes

$$\partial_r \widetilde{e} = ik_0 \left(-1 + \sqrt{1 + \frac{L^{-1}(M - k_0^2 L)}{k_0^2}} \right) \widetilde{e}$$

The square root is eliminated by using the Padé approximation

$$-1 + \sqrt{1+\zeta} \doteq \sum_{j=1}^{n} \frac{a_{jn}\zeta}{1 + b_{jn}\zeta} \tag{21.3}$$

where the Padé coefficients are

$$a_{jn} = \frac{2}{2n+1} \sin^2 \frac{j\pi}{2n+1}, b_{jn} = \cos^2 \frac{j\pi}{2n+1}, j = 1, \ldots, n.$$

Thus the far field solution can be approximated by

$$e(r,z) \doteq \frac{e^{ik_0 r}}{\sqrt{r}} \widetilde{e}(r,z) \tag{21.4}$$

where $\widetilde{e}$ is the solution to

$$\partial_r \widetilde{e} = ik_0 \sum_{j=1}^{n} \frac{a_{jn} L^{-1}(M - k_0^2 L)}{k_0^2 + b_{jn} L^{-1}(M - k_0^2 L)} \widetilde{e}. \tag{21.5}$$

Equation (21.5) is a particular instance of the general problem

$$\partial_r \varphi = A\varphi, A = \sum_{j=1}^{n} A_j, A_j > 0, \varphi(r_0) = \varphi_0$$

treated in Marchuk [9]. We seek approximations to φ at points $r_m = r_0 + m\Delta r$. Each interval $[r_m, r_{m+1}]$ is in turn subdivided in n subintervals and the Crank-Nicolson scheme for discretizing parabolic equations is applied sequentially for each operator A_j

$$\frac{\varphi\left(r_m + \frac{j}{n}\Delta r\right) - \varphi\left(r_m + \frac{j-1}{n}\Delta r\right)}{\Delta r} + A_j \frac{\varphi\left(r_m + \frac{j}{n}\Delta r\right) + \varphi\left(r_m + \frac{j-1}{n}\Delta r\right)}{2} = 0,$$

$j = 1 \ldots, n$. If $\frac{1}{2}\Delta r \|A_j\| < 1, j = 1 \ldots n$ the two-cycle composition $\varphi(r_{m+1}) = T\varphi(r_{m-1})$ where

$$\begin{aligned} T := \quad & \left(1 + \tfrac{1}{2}\Delta r A_1\right)^{-1} \left(1 - \tfrac{1}{2}\Delta r A_1\right) \ldots \quad \left(1 + \frac{1}{2}\Delta r A_n\right)^{-1} \left(1 - \frac{1}{2}\Delta r A_n\right) \\ & \left(1 + \tfrac{1}{2}\Delta r A_n\right)^{-1} \left(1 - \tfrac{1}{2}\Delta r A_n\right) \ldots \quad \left(1 + \frac{1}{2}\Delta r A_1\right)^{-1} \left(1 - \frac{1}{2}\Delta r A_1\right) \end{aligned}$$

can be shown to be the Crank-Nicolson scheme up to order $O((\Delta r)^2)$. For our problem

$$A_j = -\frac{ik_0 a_{jn} L^{-1}(M - k_0^2 L)}{k_0^2 + b_{jn} L^{-1}(M - k_0^2 L)}.$$

Substituting this into the Crank-Nicolson scheme

$$\left(1+\frac{1}{2}\Delta r A_j\right)\varphi\left(r_m+\frac{j}{n}\Delta r\right)=\left(1-\frac{1}{2}\Delta r A_j\right)\varphi\left(r_m+\frac{j-1}{n}\Delta r\right)$$

gives

$$(k_0^2(1-c_{jn}^-)L+c_{jn}^-M)\,\varphi\left(r_m+\frac{j}{n}\Delta r\right)=(k_0^2(1-c_{jn}^+)L+c_{jn}^+M)\,\varphi\left(r_m+\frac{j-1}{n}\Delta r\right)$$

where $c_{jn}^{\pm} := b_{jn} \pm \frac{i}{2}k_0 a_{jn}\Delta r$. Discretization of the depth derivatives in the operator M using finite difference formulas then gives a set of matrix equations

$$B(c_{jn}^-)\varphi\left(r_m+\frac{j}{n}\Delta r\right)=B(c_{jn}^+)\varphi\left(r_m+\frac{j-1}{n}\Delta r\right), j=1,\dots,n$$

where $B(c_{jn}^{\pm})$ are banded diagonal matrices. For computational purposes the seabed must be truncated at some depth. At this point the quantities e_s, ε, and u_{zs} are simply set to zero. Also a pressure distribution at some radial distance r_0 from the source is required to start the scheme. There are various ways to generate such fields. The examples in this article use the Gaussian starter of Greene [7] with $r_0 = 0$.

21.6 A SOURCE OF INSTABILITY FOR PARABOLIC APPROXIMATION

As mentioned in the introduction Wetton and Brooke [10] have pointed out that instability may arise in parabolic approximation when the eigenvalues for the problem have negative imaginary parts. Although the argument only applies to one-term approximations, it is observed empirically that if parabolic approximation is unstable for one-term approximations it is also unstable when more terms are employed. Consider a solution v to a system

$$L\partial_{rr}v + Mv = 0 \tag{21.6}$$

and let $\widehat{v}$ be its parabolic approximation using a one-term Padé approximation. Then $\widehat{v} = we^{ik_0 r}$ where from (21.5) w is the solution to

$$\partial_r w = ik_0\frac{a_{11}L^{-1}(M-k_0^2L)}{k_0^2+b_{11}L^{-1}(M-k_0^2L)}w.$$

If we separate variables in the original system (21.6), we find that $v = \varphi(z)e^{ik_0\sqrt{\kappa}r}$ where φ is a solution to the eigenvalue problem

$$M\varphi = k_0^2\kappa L\varphi. \tag{21.7}$$

The branch cut for $\sqrt{\kappa}$ is chosen to be the positive real axis so that $v(r,z) \to 0$ as $r \to \infty$. Let $\{\kappa_n\}$ and $\{\varphi_n(z)\}$ be the eigenvalues and eigenfunctions for this problem. The solution is then

$$v(r,z) = \sum_n B_n\varphi_n(z)e^{ik_0\sqrt{\kappa_n}r}.$$

On the other hand if we separate variables $w = \psi(z)e^{ik_0\lambda r}$ for the parabolic problem, then we obtain the eigenvalue problem

$$\lambda\psi = \frac{a_{11}L^{-1}(M - k_0^2 L)}{k_0^2 + b_{11}L^{-1}(M - k_0^2 L)}\psi$$

which can be rewritten as

$$M\psi = k_0^2\left(1 + \frac{\lambda}{a_{11} - b_{11}\lambda}\right)L\psi.$$

This is the same eigenvalue problem as for (21.7) and thus the eigenfunctions are the same, and the eigenvalues are related by

$$\kappa_n = 1 + \frac{\lambda_n}{a_{11} - b_{11}\lambda_n} \Rightarrow \lambda_n = \frac{a_{11}(\kappa_n - 1)}{1 + b_{11}(\kappa_n - 1)}.$$

Thus we have

$$\widehat{v} = we^{ik_0 r} = \sum_n B_n\varphi_n(z)e^{ik_0\lambda_n r}e^{ik_0 r} = \sum_n B_n\varphi_n(z)e^{ik_0\left(1+\frac{a_{11}(\kappa_n-1)}{1+b_{11}(\kappa_n-1)}\right)r}$$

Hence the only difference between the solution v to the elliptic problem and its parabolic approximation $\widehat{v}$ is that in the parabolic approximation $\sqrt{\kappa}$ is approximated by

$$P(\kappa) = 1 + \frac{a_{11}(\kappa - 1)}{1 + b_{11}(\kappa - 1)}.$$

The difficulty is that the Padé approximation to $\sqrt{\kappa}$ takes the branch cut along the negative real axis whereas the physically correct branch cut is along the positive real axis. This means that for most eigenvalues κ with negative imaginary parts $\mathrm{Im}P(\kappa)$ will be negative and $\exp(ik_0 P(\kappa)r)$ will grow with r, rather than decay as it should. This can be fixed providing the following conjecture is true:

All eigenvalues of significance in the far field lie on the same side of the real κ-axis as the m_+ branch cut.

As mentioned earlier, this is the case for all sediments surveyed thus far.

To remedy the instability problem note that setting $\kappa = 1 + \zeta$ in (21.3) gives

$$\sqrt{\kappa} \doteq 1 + \sum_{j=1}^{n}\frac{a_{jn}(\kappa - 1)}{1 + b_{jn}(\kappa - 1)}.$$

If $\mathrm{Im}\kappa < 0$ then we want instead

$$\sqrt{\kappa} \doteq -1 - \sum_{j=1}^{n}\frac{a_{jn}(\kappa - 1)}{1 + b_{jn}(\kappa - 1)}.$$

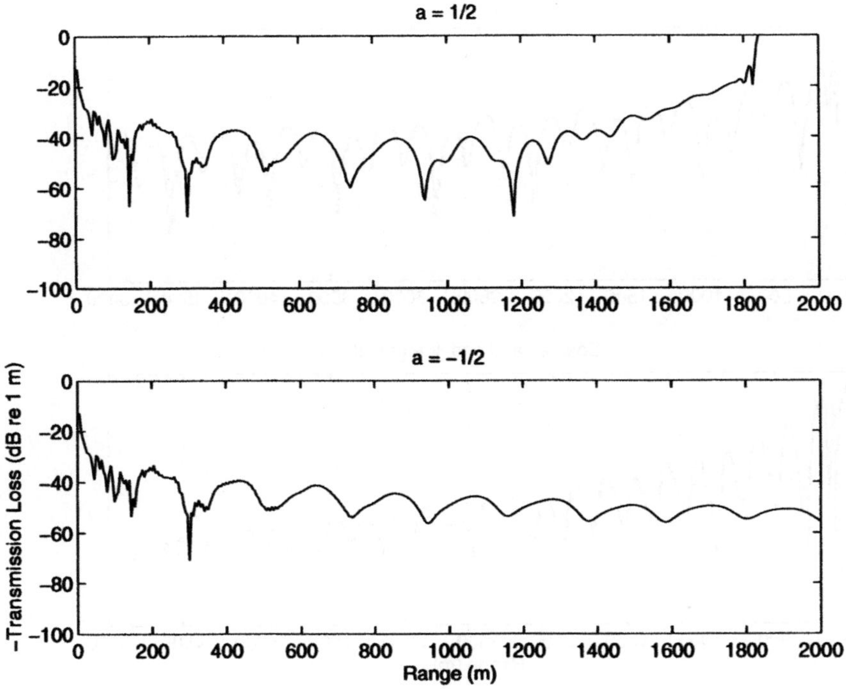

Figure 21.2 An illustration of instability in parabolic approximation.

In a computer code this is easily accomplished by computing m_{0+} and, if $\text{Im}\, m_{0+} < 0$, switching the signs of the a_{jn} and taking the plane wave factor in (21.4) to be $e^{-ik_0 r}$ rather than $e^{ik_0 r}$.

Transmission loss is defined to be $TL = -20 \log(N_0 \,|p_o(r,z)|)$ and is measured in decibels. The normalization factor N_0 is chosen so that $TL = 0$ one meter from the source (hence dB re 1m). For a point source $N_0 = 4\pi$. For parabolic approximation the normalization is incorporated into the starting field whence $N_0 = 1$. Figure 21.2 gives an illustration of the instability for a sediment consisting of coarse sand and fine gravel whose m_+ branch cut lies below the real κ-axis. This is one of the sediments whose eigenvalues map is shown in Figure 21.1. A one-term Padé approximation to the square root operator was used. Parabolic approximation is unstable if $a_{11} = 1/2$, but stable if $a_{11} = -1/2$.

Finally we compare transmission loss curves for the modal and parabolic solutions. The ocean depth is 50 m and the source and receiver depths are both 25 m. The source frequency is 100 Hz. Figure 21.3 shows the comparison for the two sediments of Figure 21.1. Two-term Padé approximations of the square root operator were used for parabolic approximation. As can be seen the two solutions are in reasonably good agreement in the far field. For both sediments the periods of oscillation are slightly

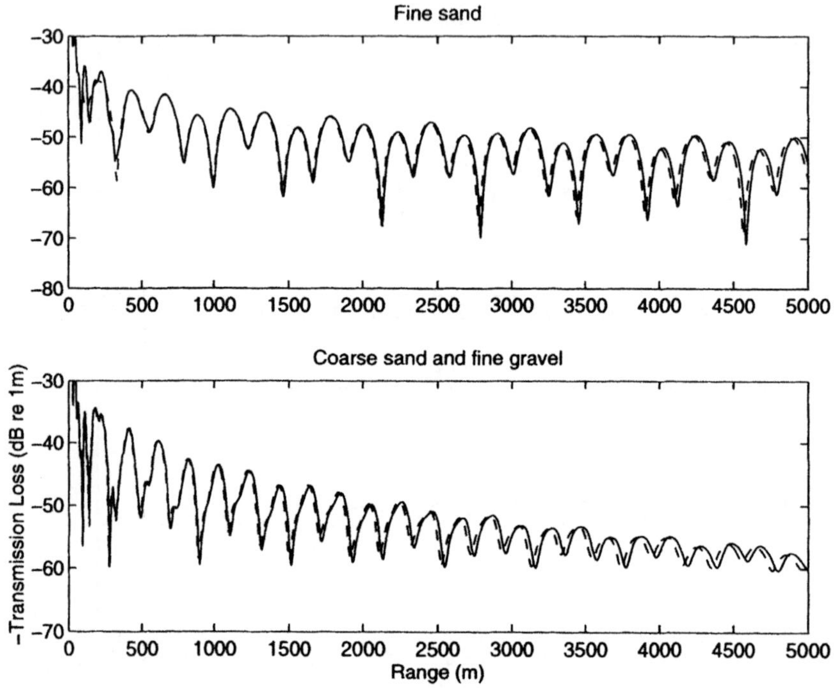

Figure 21.3 Comparison of transmission loss curves generated by parabolic approximation (solid line) with the modal solution (dashed line).

different for the two solutions. The agreement becomes better as the step size for depth discretization in the finite difference formulas used in parabolic approximation is made smaller.

21.7 CONCLUSION

Parabolic approximation, as it is usually implemented, is unstable for coarse-grained sediments because such sediments have eigenvalues with negative imaginary parts. At least in the case of a sediment with constant parameters, this problem can be remedied, providing that it can be ascertained *a priori* which side of the real axis the eigenvalues lie on. In all sediments surveyed thus far the location of m_+ branch cut, which can be calculated directly from the geo-acoustical model, predicts on which side of the real axis the eigenvalues lie, and thus permits the determination of the appropriate Padé coefficients and plane-wave factor for stable parabolic approximation. As can be seen by comparison with the modal solution, the approximations then generated are accurate in the far field.

References

[1] M.A. Biot. Theory of propogation of elastic waves in a fluid-saturated porous solid. i. lower frequency range. *J. Acoust. Soc Am.*, 28(2):168–178, 1956.

[2] M.A. Biot. Theory of propogation of elastic waves in a fluid-saturated porous solid. ii. higher frequency range. *J. Acoust. Soc. Am.*, 28(2):179–191, 1956.

[3] J.L. Buchanan and R.P. Gilbert. Transmission loss in the far field over a seabed with rigid substrate assuming the biot sediment model. *Journal of Computational Acoustics*, 4(1):29–54, 1996.

[4] J.L. Buchanan, R.P. Gilbert, and Y. Xu. Green's function representation for acoustic pressure over a poroelastic seabed. *Applicable Analysis*, 1997. To appear.

[5] M.D. Collins. Higher-order and elastic parabolic equations for wave propagation in the ocean. In D. Lee, A. Cakmak, and R. Vichnevetsky, editors, *Computational Acoustics*, volume 3, pages 167–184. Elsevier Science Publishers, 1990.

[6] M.D. Collins. Higher order pade approximations for accurate and stable elastic parabolic equations with application to interface wave propagation. *J. Acoust. Soc. Am.*, 89(3):1050–1057, 1991.

[7] R.R. Greene. The rational approximation to the acoustic wave equation with bottom interaction. *J. Acoust. Soc. Am.*, 76(6):1764–1773, 1984.

[8] C.W. Holland and B.A. Brunson. The biot-stoll sediment model: An experimental assessment. *J. Acoust. Soc. Am.*, 84(4):1437–1443, 1988.

[9] G.I. Marchuk. Splitting and alternating direction methods. In P.G. Ciarlet and J.L. Lions, editors, *Handbook of Numerical Analysis, Volume I*, pages 197–462. North-Holland, 1990.

[10] B.T.R. Wetton and G.H. Brooke. One-way wave equations for seismo-acoustic propagation in elastic wave guides. *J. Acoust. Soc. Am.*, 87(2):624–632, 1990.

22 SOME INVERSE PROBLEMS IN STRATIFIED MEDIA

Y. Xu†

Dedicated to Professor Robert Gilbert on the occasion of his 65th birthday.

Abstract: We discuss an inverse obstacle scattering problem, namely the problem of shape determination from scattered wave's far-field pattern, for sound-soft obstacle in a stratified medium in this paper. Uniqueness results for given complete far-field data and incomplete far-field data are obtained. An example is constructed to show a special feature of inverse scattering in a stratified medium. [1]

22.1 INTRODUCTION

We consider an inverse obstacle scattering problem for acoustic waves in a stratified medium. By stratified medium here we mean that the refraction index of the medium is a function of one variable. One such medium is an ocean. In a stratified medium, sound waves may be trapped by acoustic ducts and caused to propagate horizontally. Therefore, the scattered energy flux is not spreaded out spherically. Instead there are free-wave far-fields and guided-wave far-fields. Due to the nature of stratified medium, some results which are valid for inverse scattering in a homogeneous

†This research was supported in part by NSF Grant BES-9402539 and the CECA scholarship of University of Tennessee at Chattanooga

H. Florian et al. (eds.), Generalized Analytic Functions, 255–264.

medium may not be true for a stratified medium. For example, a scatterer is determined uniquely by far-field patterns in an open set corresponding to infinitely many linearly independent incident waves. This may not be true for stratified medium, unless the far-field is detected in a window large enough to contain both free-waves and guided-waves. Or two open sets are needed; one for free-wave far-field and one for guided-wave far-field.

We will prove some uniqueness results in this paper. A non-uniqueness example is given to show the difference between inverse scattering problems in a stratified medium and in a homogeneous medium.

Inverse scattering problems in a homogeneous medium have been intensively investigated. There are a large number of publications. We make no attempt in this paper to survey the subject. In classical inverse scattering we mention the recent books of Colton and Kress [3] and papers refereed there. See Cheney and Isaacson [2] for results in a perturbed dissipative half-space, and Gilbert and Xu for results in a finite depth ocean [4].

22.2 FORMULATION OF THE INVERSE PROBLEM AND PRELIMINARY RESULTS

In this section we formulate the inverse problem. First we outline some results for the related direct scattering problem. More discussion may be found in [6] [7].

We assume that the medium is stratified, i.e., the refraction index $n = n(z)$ is a strictly positive function which is continuous except for a finite number of jump discontinuitis. A discontinuity in $n(z)$ represents a change in medium. Moreover, we assume that for some constants h_1, h_2 and positive constants n_-, n_+

$$n(z) = \begin{cases} n_- & \text{for } -\infty < z < h_1 < 0 \\ n_+ & \text{for } 0 < h_2 < z < \infty. \end{cases} \tag{22.1}$$

If $n(z)$ has a finite number of jump discontinities at points $-\infty < z_1 < z_2 < \cdots < z_k < \infty$, we use the following notations:

$$I_l = \{z | z_l < z < z_{l+1}\},\ \Omega_l = R^2 \times I_l,\ l = 0, 1, 2, ..., L,$$

where $z_0 = -\infty$ and $z_{L+1} = \infty$.

$$H(R^3) = C^1(R^3) \cap C^2(\cup_{l=0}^{L} \Omega_l),\ H(R) = C^1(R) \cap C^2(\cup_{l=0}^{L} I_l),$$

$$H(R^3 \setminus \bar{\Omega}) = C^1(R^3 \setminus \bar{\Omega}) \cap C^2(\cup_{l=0}^{L} \Omega_l \setminus \bar{\Omega}).$$

The scattering of acoustic time-harmonic waves by a sound-soft obstacle in such a stratified medium lead to following exterior boundary value problem: given $u^i \in H(R^3)$ satisfying

$$\Delta u^i + k^2 n^2(z) u^i = 0 \text{ in } R^3 \tag{22.2}$$

with positive wave number k, find the scattered field $u^s \in H(R^3 \setminus \overline{\Omega})$ satisfying

$$\Delta u^s + k^2 n^2(z) u^s = 0 \text{ in } R^3 \setminus \overline{\Omega} \tag{22.3}$$

such that the total field $u = u^i + u^s$ satisfies boundary condition

$$u = 0 \text{ on } \partial\Omega. \tag{22.4}$$

Here $\Omega \in \Omega_l$ for some l is a bounded domain with C^2 boundary $\partial\Omega$. To ensure the uniqueness of the exterior problem, we require that the unknown scattered wave u^s satisfies the generalized Sommerfeld radiation condition. Namely, if $\phi_i \in H(R)$, $i \in J_N = \{1, 2, ..., N\}$ are the normalized eigenfunctions of ordinary differential equation

$$p''(z) + k^2[n^2(z) - a^2]p(z) = 0, \quad -\infty < z < \infty, \tag{22.5}$$

and a_i the corresponding eigenvalues, we can decompose the scattered wave u^s into scattered guided-wave u^s_g and scattered free-wave u^s_f as follows:

$$u^s(\mathbf{x}, z) = u^s_f(\mathbf{x}, z) + u^s_g(\mathbf{x}, z), \tag{22.6}$$

$$u^s_g(\mathbf{x}, z) = \sum_{n=1}^{N} \phi_n(z) u_n(\mathbf{x}), \tag{22.7}$$

where

$$u_n(\mathbf{x}) = \int_{-\infty}^{\infty} \phi_n(\zeta) u^s(\mathbf{x}, \zeta) d\zeta. \tag{22.8}$$

$u^s(\mathbf{x}, z)$ satisfies the outgoing generalized Sommerfeld radiation condition [6], [7], if

$$\frac{\partial u^s_f}{\partial R} - i\hat{k}u^s_f = O\left(\frac{1}{R^2}\right), \quad u^s_f = O\left(\frac{1}{R}\right), \tag{22.9}$$

$$\textit{as } R \to \infty, \textit{ uniformly for } (\theta, \phi) \in [0, 2\pi] \times [0, \pi],$$

and

$$\frac{\partial u_n}{\partial r} - ika_n u_n = O\left(\frac{1}{r^{3/2}}\right), \tag{22.10}$$

$$n \in J_N, \textit{ as } r \to \infty, \textit{ uniformly for } \theta \in [0, 2\pi],$$

where $\hat{k} = kn_+$ if $0 \le \phi < \pi/2$, $\hat{k} = kn_-$ if $\pi/2 < \phi \le \pi$.

Here and later in this paper, we use the following conventions. A point in R^3 is described as

$$P = (\mathbf{x}, z) = (x_1, x_2, z) = (r, \theta, z) = (R, \theta, \phi),$$
$$P' = (\xi, \zeta) = (\xi_1, \xi_2, \zeta) = (r', \theta', \zeta) = (\rho, \theta', \phi')$$

in different coordinate systems respectively. There are relations

$$R^2 = r^2 + z^2 = |\mathbf{x}|^2 + z^2 = |P|^2,$$

$$r = R \sin\phi, \; z = R\cos\phi, \; x_1 = r\cos\theta, \; x_2 = r\sin\theta,$$

$$\rho^2 = r'^2 + \zeta^2 = |\xi|^2 + \zeta^2 = |P'|^2,$$
$$r' = \rho \sin\phi', \; \zeta = \rho\cos\phi', \; \xi_1 = r'\cos\theta', \; \xi_2 = r'\sin\theta'.$$

The aforementioned direct scattering problem has been discussed in [6], [7]. We proved that this exterior problem has a unique solution. Moreover, using integral equation methods, we showed that the scattered wave u^s has an asymptotic behavior of the form

$$u^s(P) = \frac{e^{i\hat{k}R}}{R} F_f(\hat{P}) + \frac{1}{\sqrt{r}} \mathbf{A}_g(r) \cdot \mathbf{F}_g(\hat{\mathbf{x}}, z) + R_f(P) + R_g(P), \tag{22.11}$$

where

$$\mathbf{A}_g(r) = (e^{ika_1 r}, e^{ika_2 r}, ..., e^{ika_N r}),$$
$$\mathbf{F}_g(\hat{\mathbf{x}}, z) = (F_g^1(\hat{\mathbf{x}}, z), \; ..., \; F_g^N(\hat{\mathbf{x}}, z)),$$

$\hat{P} = P/|P| \in S_1$, $\hat{\mathbf{x}} = \mathbf{x}/|\mathbf{x}|$, $(\mathbf{x}, z) \in D_1$ and $S_1 = \{P \in R^3 | |P| = 1\}$, $D_1 = \{(\mathbf{x}, z) \in R^3 \mid |\mathbf{x}| = 1\}$. The function $F_f(\hat{P})$ defined on the unit sphere S_1 of R^3 is called *the free-wave far-field pattern*. The function vector $\mathbf{F}_g(\hat{\mathbf{x}}, z)$ defined on the cylinder D_1 in R^3 is called the guided-wave far-field pattern vector. Its jth component $F_g^j(\mathbf{x}, z)$ is called the jth normal mode. The direct sum $F_f(\hat{P}) \oplus \mathbf{F}_g(\hat{\mathbf{x}}, z)$ will be refered to as *far-field pattern*. The vanishing of the far-field pattern $F_f \oplus \mathbf{F}_g$ implies

$$\lim_{R\to\infty} \int_{\Gamma_R} |u^s|^2 ds = 0, \tag{22.12}$$

where $\Gamma_R = \Gamma_R^1 \cup \Gamma_R^2$, $\Gamma_R^1 = \{(\mathbf{x}, z) \in R^3 \mid |\mathbf{x}|^2 + z^2 = R^2, \; |\phi - \pi/2| \geq \pi/4\}$, and $\Gamma_R^2 = \{(\mathbf{x}, z) \in R^3 \mid |\mathbf{x}| = R/\sqrt{2}, \; |z| \leq R/\sqrt{2}\}$. Whence $u^s = 0$ follows by Lemma 3.2 of [6].

As discussed in [6], [7], we are particularly interested in the scattering of incident waves u^i from the set

$$\mathcal{U} := \{u^i(\,\cdot\,; \alpha) | \alpha \in S_1\} \cup \{u_j^i(\,\cdot\,; \hat{\alpha}, \beta) | (\hat{\alpha}, \beta) \in D_1, \;\; 1 \leq j \leq N\}.$$

$u^i(P; \alpha)$ is the distorted plane wave with direction $\alpha = (\sin\gamma\cos\delta, \sin\gamma\sin\delta, \cos\gamma)$ and given by

$$u^i(P; \alpha) = F_f^0(-\alpha; P) = \frac{-ikn_+ \cos\gamma p_1(z, kn_+ \sin\gamma)}{2\pi W(kn_+ \sin\gamma)} e^{ikn_+ r \sin\gamma\cos(\delta - \theta)}, \tag{22.13}$$

$$\text{for } 0 \leq \gamma < \frac{\pi}{2}, \; 0 \leq \delta, \, \theta \leq 2\pi, \; 0 \leq \mathrm{r} < \infty, \; -\infty < \mathrm{z} < +\infty,$$

$$u^i(P; \alpha) = F_f^0(-\alpha; P) = \frac{-ikn_- \cos\gamma p_2(z, kn_- \sin\gamma)}{2\pi W(kn_- \sin\gamma)} e^{ikn_- r \sin\gamma\cos(\delta - \theta)}, \tag{22.14}$$

$$\text{for } \frac{\pi}{2} < \gamma \leq \pi \;\; 0 \leq \delta, \, \theta \leq 2\pi, \; 0 \leq \mathrm{r} < \infty, \; -\infty < \mathrm{z} < +\infty.$$

Here $p_1(z,ka), p_2(z,ka)$ are Jost solutions of the equation (22.5), that is,

$$p_1(z,ka) = e^{-ik\sqrt{n_-^2-a^2}z} + O\left(\frac{1}{|z|}\right), \text{ as } z \to -\infty,$$

$$p_2(z,ka) = e^{ik\sqrt{n_+^2-a^2}z} + O\left(\frac{1}{|z|}\right), \text{ as } z \to \infty.$$

$u_j^i(P;\hat{\alpha},\beta)$ is the jth normal mode wave with direction $(\hat{\alpha},\beta) = (\cos\delta, \sin\delta, \beta)$,

$$u_j^i(P;\hat{\alpha},\beta) = F_g^0(-\hat{\alpha},\beta;P;j) = \sqrt{\frac{ka_j}{2\pi i}}\frac{\phi_j(\beta)\phi_j(z)}{W'(ka_j)}e^{ika_jr\cos(\delta-\theta)} \tag{22.15}$$

$$-\infty < \beta,\zeta < \infty,\ 0 \le \delta,\theta \le 2\pi,\ 0 \le r < \infty,\ j \in J_N.$$

$F_f^0(\alpha;P)$ and $\mathbf{F}_g^0(\hat{\alpha},\beta;P) = (F_g^0(-\hat{\alpha},\beta;P;1),...,F_g^0(-\hat{\alpha},\beta;P;N))$ are from the asymptotic form of the Green's function

$$G(P;P') = \frac{e^{ik R}}{R}F_f^0(\hat{P};P') + \frac{1}{\sqrt{r}}\mathbf{A}_g(r)\cdot\mathbf{F}_g^0(\hat{\mathbf{x}},z;P') + R_f(P) + R_g(P), \tag{22.16}$$

where $R_f(P) = O(1/R^2)$ and $R_g(P) = O(1/r^{\frac{3}{2}})$. We call $F_f^0(\alpha;P)$ the *free-wave far-field pattern from a point source at P in a stratified medium*, and $\mathbf{F}_g^0(\hat{\alpha},\beta;P)$ the *guided-wave far-field pattern vector from a point source at P in a stratified medium.* (c.f. [6], [7]) $F_f^0(\alpha;P)\oplus\mathbf{F}_g^0(\hat{\alpha},\beta;P)$ is refered to as the *radiating far-field pattern from a point source at P.*

Clearly, $u^i(P;\alpha)$ and $u_j^i(P;\hat{\alpha},\beta)$ satisfy (22.2) in the $H(R^3)$ sense for $\alpha \in S_1$ and $(\hat{\alpha},\beta) \in D_1$, $j \in J_N$, respectively.

The scattered wave corresponding to $u^i(P;\alpha)$ and $u_j^i(P;\hat{\alpha},\beta)$ are denoted by $u^s(P;\alpha)$ and $u_j^s(P;\hat{\alpha},\beta)$ respectively. $u^s(P;\alpha)$ and $u_j^s(P;\hat{\alpha},\beta)$ have asymptotic expansions

$$\begin{aligned} u^s(P;\alpha) &= \frac{e^{ikR}}{R}F_f(\hat{P};\alpha) + \frac{1}{\sqrt{r}}\mathbf{A}_g(r)\cdot\mathbf{F}_g(\hat{\mathbf{x}},z;\alpha) \\ &\quad + R_f(P;\alpha) + R_g(P;\alpha), \end{aligned} \tag{22.17}$$

$$\begin{aligned} u_j^s(P;\hat{\alpha},\beta) &= \frac{e^{ikR}}{R}F_f(\hat{P};\hat{\alpha},\beta;j) + \frac{1}{\sqrt{r}}\mathbf{A}_g(r)\cdot\mathbf{F}_g(\hat{\mathbf{x}},z;\hat{\alpha},\beta;j) \\ &\quad + R_f(P;\hat{\alpha},\beta;j) + R_g(P;\hat{\alpha},\beta;j). \end{aligned} \tag{22.18}$$

where $R_f(P,\cdot) = O(1/R^2)$ and $R_g(P,\cdot) = O(1/r^{\frac{3}{2}})$.

The inverse problem we consider in this paper can be generally stated as following: given the combination of the far-field patterns $F_f(\,\cdot\,;\alpha)\oplus\mathbf{F}_g(\,\cdot\,;\alpha)$ and $F_f(\,\cdot\,;\hat{\alpha},\beta;j)\oplus\mathbf{F}_g(\,\cdot\,;\hat{\alpha},\beta;j)$ of the scattered waves $u^s(\,\cdot\,;\alpha)$ and $u_j^s(\,\cdot\,;\hat{\alpha},\beta)$ for several incident waves $u^i(\,\cdot\,;\alpha)$ and $u_j^i(\,\cdot\,;\hat{\alpha},\beta)$ with different incident directions α, $(\hat{\alpha},\beta)$ and different modes $j \in J_N$, determine the shape of the scatterer Ω.

22.3 UNIQUENESS

We want to investigate the conditions under which an obstacle is uniquely determined by knowledge of the far-field patterns for incident waves.

The surface of the obstacle, $\partial\Omega$, may consist of several connected closed components. The given data for the inverse problem are the values of the incident distorted plane waves, the incident mode waves and the corresponding far-field patterns $u^i(\,\cdot\,;\alpha), u_j^i(\,\cdot\,;\hat{\alpha},\beta)$, $F_f(\hat{P};\alpha) \oplus \mathbf{F}_g(\hat{\mathbf{x}},z;\alpha)$ and $F_f(\hat{P};\hat{\alpha},\beta) \oplus \mathbf{F}_g(\hat{\mathbf{x}},z;\hat{\alpha},\beta)$, where $\hat{P},(\hat{\mathbf{x}},z),\alpha$ and $(\hat{\alpha},\beta)$ are from a subset in $S_1 \times D_1 \times S_1 \times D_1$. This subset may be chosen in a number of ways.

Case 1: Complete data of far-field patterns.

Assume $F_f(\hat{P};\alpha) \oplus \mathbf{F}_g(\hat{\mathbf{x}},z;\alpha)$ and $F_f(\hat{P};\hat{\alpha},\beta;j) \oplus \mathbf{F}_g(\hat{\mathbf{x}},z;\hat{\alpha},\beta;j)$ are known for all $\hat{P} \in S_1$, $(\hat{\mathbf{x}},z) \in D_1, j \in J_N$ and $\alpha \in \tilde{S}_1 \subset S_1, (\hat{\alpha},\beta) \in \tilde{D}_1 \subset D_1$, where $\tilde{S}_1, \tilde{D}_1$ are subsets of S_1 and D_1.

Theorem 1 *Assume that Ω_1 and Ω_2 are two sound-soft scatterers. If the far-field patterns $F_f(\hat{P};\alpha) \oplus \mathbf{F}_g(\hat{\mathbf{x}},z;\alpha)$ and $F_f(\hat{P};\hat{\alpha},\beta;j) \oplus \mathbf{F}_g(\hat{\mathbf{x}},z;\hat{\alpha},\beta;j)$, $j \in J_N$ coincide for $\hat{P} \in S_1$ and $(\hat{\mathbf{x}},z) \in D_1$ for incident waves $u^i(\,\cdot\,;\alpha)$ and $u_j^i(\,\cdot\,;\hat{\alpha},\beta)$, $\alpha \in \tilde{S}_1$, $(\hat{\alpha},\beta;j) \in \tilde{D}_1 \times J_N$, then $\Omega_1 = \Omega_2$. Here $\tilde{S}_1$ is a sequence of vectors $\alpha_n \in S_1$ with a limit point $\alpha_0 \in \tilde{S}_1$, $\tilde{D}_1 = \tilde{C}_1 \times R$ and $\tilde{C}_1 \subset C_1 = \{\mathbf{x} \in R^2 |\, |\mathbf{x}| = 1\}$ is a sequence of vectors $\hat{\alpha}_n \in C_1$ with a limit point $\hat{\alpha}_0 \in \tilde{C}_1$.*

Proof: Let v_1^s and v_2^s be the scattered waves with scatterers Ω_1, Ω_2 respectively. Consider the function $v^s = v_1^s - v_2^s$ in $B_R = \{P \in R^3 |\, |P| < R\} \supset \Omega_1 \cup \Omega_2$. Since $v_j^s, (j = 1,2)$ have coincident far-field patterns,

$$\lim_{R\to\infty} \int_{\Gamma_R} |v^s|^2 ds = 0.$$

It implies that $v^s = 0$ in $R^3 \setminus B_R$ [6]. By the unique continuation property of elliptic equations, $v^s = 0$ in the unbounded component D of $R^3 \setminus (\Omega_1 \cup \Omega_2)$. In particular, $v^s = 0$ on ∂D.

Assume first that $\Omega^* = (R^3 \setminus \bar{D}) \setminus \overline{\Omega_2}$ is not empty. Then v_2^s is defined in Ω^* and $v_2 = v_2^s + u^i = 0$ on $\partial\Omega^*$ since $v_2^s = -u^i$ on $\partial\Omega_2$ and $v_2^s = v_1^s = -u^i$ on $\partial\Omega_1 \cap \bar{D}$. Thus v_2 solve the problem

$$\triangle\, v_2 + k^2 n_0^2(z) v_2 = 0 \;\; in\; \Omega^*, \quad v_2 = 0 \;\; on\; \partial\Omega^*. \tag{22.19}$$

That is, v_2 is a solution of the eigenvalue problem

$$\mathcal{A}(u,v) = \lambda\mathcal{B}(u,v), \;\; \text{for any } v \in H_0^1(\Omega^*)$$

with

$$\mathcal{A}(u,v) = \int_{\Omega^*} \text{grad}u \cdot \text{grad}v \; dP$$

and

$$\mathcal{B}(u,v) = \int_{\Omega^*} n_0^2(z)uv\,dP.$$

From the discussion in [5], we know that for such a eigenvalue problem, any eigenvalue $k^2 < \infty$ has finite multiplicity. However, for distinct incident waves from $\mathcal{U}$, the corresponding scattered waves are linearly independent in $L^2(\Omega^*)$. In fact, if v_j are scattered waves corresponding to $u^i(\,\cdot\,;\alpha)$ or $u^i_j(\,\cdot\,;\hat{\alpha},\beta)$ from $\mathcal{U}$, and

$$\sum_{j=1}^{m} c_j v_j = 0 \;\; in \;\; L^2(\Omega^*)$$

where c_j are constants and m is an arbitrary integer, the unique continuation property of elliptic equation follows that the above is valid in D. Rename v_j if necessary, we may assume that the first m_1 of the $\{v_j\}$ are scattered waves corresponding to incident free-waves, and the others are to the incident guided-waves. Thus,

$$\sum_{j=1}^{m_1} c_j[u^i(P;\alpha_j) + O(\frac{1}{|P|})] + \sum_{j=m_1+1}^{m} c_j[u^i_{l_j}(P;\hat{\alpha}_j,\beta_j) + O(\frac{1}{|P|})] = 0, \;\; \text{for } |P| \text{ large}, \tag{22.20}$$

where $l_j \in J_N$. The orthogonality of the eigenfunctions of ordinary differential equation (22.5) follows

$$\sum_{j=1}^{m_1} c_j u^i(P;\alpha_j) + O(\frac{1}{|P|}) = 0, \;\; \text{for } |P| \text{ large}, \tag{22.21}$$

and

$$\sum_{j=m_1+1}^{m} c_j u^i_{l_j}(P;\hat{\alpha}_j,\beta_j) + O(\frac{1}{|P|}) = 0, \;\; \text{for } |P| \text{ large}. \tag{22.22}$$

It implies that $c_j = 0$, for $j = 1, ..., m$. This is a contradiction.

If $\Omega^* = (R^3 \setminus \bar{D}) \setminus \bar{\Omega}_2 = \emptyset$, then $\Omega_1^* = (R^3 \setminus \bar{D}) \setminus \bar{\Omega}_1 \neq \emptyset$. Using the same argument above in Ω_1^* for $v_1 = v_1^s + u^i$, we conclude a contradiction also. Hence, there must be $\Omega_1 = \Omega_2$. □

Corollary 1 *The shape of a scatterer is determined uniquely by* $F_f(\hat{P};\alpha) \oplus \mathbf{F}_g(\hat{\mathbf{x}},z;\alpha)$ *for all* $\hat{P} \in S_1$, $(\hat{\mathbf{x}},z) \in D_1$ *and* $\alpha \in \tilde{S}_1$. □

Corollary 2 *The shape of a scatterer is determined uniquely by* $F_f(\hat{P};\hat{\alpha},\beta;j) \oplus \mathbf{F}_g(\hat{\mathbf{x}},z;\hat{\alpha},\beta;j)$ *for all* $\hat{P} \in S_1$, $(\hat{\mathbf{x}},z) \in D_1$ *and* $(\hat{\alpha},\beta;j) \in \tilde{D}_1 \times J_N$. □

Case 2: Incomplete data of far-field patterns.

2(a). Assume that only the free wave far field patterns $F_f(\hat{P};\alpha)$, $F_f(\hat{P};\hat{\alpha},\beta;j)$ are known for all $\hat{P} \in S_1$ and $\alpha \in \tilde{S}_1 \subset S_1$, $(\hat{\alpha},\beta;j) \in \tilde{D}_1 \times J_N \subset D_1 \times J_N$, where $\tilde{S}_1$ and $\tilde{D}_1$ are defined as in Theorem 2.1.

Theorem 2 *Assume that Ω_1 and Ω_2 are two sound-soft scatterers such that the free-wave far-field patterns coincide for incident free-waves $u^i(\,\cdot\,;\alpha)$ for $\alpha \in \tilde{S}_1$ and incident guided-waves $u_j^i(\,\cdot\,;\hat{\alpha},\beta)$ for $(\hat{\alpha},\beta;j) \in \tilde{D}_1 \times J_N$. Then $\Omega_1 = \Omega_2$.*

Proof: By the reciprocity relation (c.f. [6] Th. 3.3)

$$F_f(\hat{P};\hat{\alpha},\beta;j) = F_g^j(-\hat{\alpha},\beta;-\hat{P}), \quad \hat{P} \in S_1, \ (\hat{\alpha},\beta) \in D_1, \ j \in J_N, \tag{22.23}$$

the condition in the theorem is equivalent to that the free-wave far-field patterns $F_f(\hat{P};\alpha)$ coincide on S_1 for incident free-waves $u^i(\,\cdot\,;\alpha), \alpha \in \tilde{S}_1$, and the guided-wave far-field pattern vectors $\mathbf{F}_g(\hat{\alpha},\beta;\hat{P})$ coincide on $\tilde{D}_1^* = \{(\hat{\alpha},\beta) \in D_1 | (-\hat{\alpha},\beta) \in \tilde{D}_1\}$ for incident free-waves $u^i(\,\cdot\,;\hat{P})$ for all $\hat{P} \in S_1$. The real analyticity of $F_g^j(\hat{\alpha},\beta;\hat{P})$ with respect to δ where $\hat{\alpha} = (\cos\delta, \sin\delta)$ follows that the guided-wave far-field patterns coincide on D_1 for incident free-wave $u^i(\,\cdot\,;\hat{P})$ for all $\hat{P} \in S_1$, in particular for $\hat{P} \in \tilde{S}_1$. Hence, by Corollary 2.1, $\Omega_1 = \Omega_2$.

Remark: Theorem 2 suggests that if the scattered data can be obtained only above a scatterer in a sound channel, one may still determine the shape of the scatterer if one can send in incident waves from sources in the sound channel and off (above) the sound channel. However, that free-wave far-field patterns concide for infinitely many distinct incident waves may not imply that two scatterers are identical. See an example given in next section.

2(b). Assume that only the guided-wave far-field pattern vectors $\mathbf{F}_g(\hat{\mathbf{x}},z;\alpha)$ and $\mathbf{F}_g(\hat{\mathbf{x}},z;\hat{\alpha},\beta;j)$ are known for all $(\hat{\mathbf{x}},z) \in D_1$ and $\alpha \in \tilde{S}_1$, $(\hat{\alpha},\beta;j) \in \tilde{D}_1 \times J_N$.

Theorem 3 *Assume that Ω_1 and Ω_2 are two sound-soft scatterers such that the guided-wave far-field pattern vectors coincide for incident waves $u^i(\,\cdot\,;\alpha)$ for $\alpha \in \tilde{S}_1$ and $u_j^i(\,\cdot\,;\hat{\alpha},\beta)$ for $(\hat{\alpha},\beta;j) \in \tilde{D}_1 \times J_N$. Then $\Omega_1 = \Omega_2$.*

Proof: In view of the reciprocity relation (c.f. [6] Th. 3.3)

$$F_g^j(\hat{\mathbf{x}},z;\alpha) = F_f(-\alpha;-\hat{\mathbf{x}},z;j), \quad \alpha \in S_1, \ (\hat{\mathbf{x}},z) \in D_1, \ j \in J_N, \tag{22.24}$$

and the real analyticity of $F_f(\alpha;\hat{\mathbf{x}},z;j)$ with respect to δ and γ for $\gamma \neq \pi/2$, where

$$\alpha = (\sin\gamma\cos\delta, \sin\gamma\sin\delta, \cos\gamma),$$

the condition in the theorem is equivalent to that the far-field patterns $F_f(\hat{P};\hat{\alpha},\beta;j) \oplus \mathbf{F}_g(\hat{\mathbf{x}},z;\hat{\alpha},\beta;j)$ coincide for incident guided-wave $u_j^i(\,\cdot\,;\hat{\alpha},\beta)$ with $(\hat{\alpha},\beta;j) \in \tilde{D}_1 \times J_N$. The theorem is followed by Corollary 2. □

22.4 AN EXAMPLE FOR NON-UNIQUENESS

In this section we construct an example that for two different obstacles, there are infinitely many distinct incident waves for which the corresponding scattered waves have identical free-wave far-field pattern.

Let $n^2(z)$ be the refraction index such that there exist at least 4 guided modes. That is, the equation (22.5) has at least 4 eigenvalues. The forth eigenvalue and the forth eigenfunction are denoted by a_4 and $\phi_4(z)$, respectively. $\phi_4(z)$ has three zeros, denoted by z_1 z_2 and z_3.

Let Ω_j be the cylinder $\{(\mathbf{x}, z) \in R^3 | \, |\mathbf{x}| < 2, \, z_j < z < z_{j+1}\}$ $j = 1, 2$; and S_1 the surface of the unit ball. Consider a sequence of incident waves $\{u_n^i(P)\}$ defined by

$$u_n^i(P) = \sum_{m=0}^{\infty} \epsilon_m (c_n \cos m\theta + d_n \sin m\theta) \frac{H_m^{(1)}(2ka_4)}{J_m(2ka_4)} J_m(ka_4 r)\phi_4(z), \; P \in R^3, \tag{22.25}$$

where $\epsilon_0 = 1$, $\epsilon_m = 2$ for $m > 0$; and

$$c_n = \int_{S_1} g_n(P')\phi_4(\zeta)J_m(ka_4 r') \cos m\theta' dP',$$

$$d_n = \int_{S_1} g_n(P')\phi_4(\zeta)J_m(ka_4 r') \sin m\theta' dP',$$

and $g_n(P')$, $n = 1, 2, 3, ..$ is a sequence of linearly independent functions on S_1.

$u_n^i(P)$ are solutions of (22.2) and linearly independent. The scattered waves corresponding to $u_n^i(P)$ satisfying (22.3), (22.4) and the radiation condition are

$$u_n^s(P) = -\int_{S_1} g_n(P')H_0^{(1)}(ka_4|\mathbf{x} - \xi|)\phi_4(\zeta)\phi_4(z)dP', \; P \in R^3 \setminus \overline{\Omega}. \tag{22.26}$$

It is obvious that u_n^s satisfy (22.3) and the radiation condition. To see that u_n^s satisfy (22.4), recall that for $r > r'$,

$$H_0^{(1)}(ka_4|\mathbf{x} - \xi|) = \sum_{m=0}^{\infty} \epsilon_m H_m^{(1)}(ka_4 r)J_m(ka_4 r') \cos[m(\theta - \theta')], \tag{22.27}$$

$$u_n^i(P) + u_n^s(P) = \sum_{m=0}^{\infty} \epsilon_m \int_{S_1} g_n(P')\phi_4(\zeta)J_m(ka_4 r') \cos[m(\theta - \theta')]dP' \cdot$$
$$\left\{ \frac{H_m^{(1)}(ka_4 2)}{J_m(ka_4 2)} J_m(ka_4 r)\phi_4(z) - H_m^{(1)}(ka_4 r)\phi_4(z) \right\}.$$

Hence, $u_n^i + u_n^s = 0$ for $r = 2$. In view of

$$\phi_4(z_1) = \phi_4(z_2) = \phi_4(z_3) = 0,$$

we have

$$u_n^i(P) + u_n^s(P) = 0, \text{ for } P \in \partial\Omega_j, \; j = 1, 2.$$

From (22.26), $u_n^s(P)$ for $n = 1, 2, 3, ...$ have vanishing free-wave far-field pattern. This example shows a new feature in inverse scattering problems in a stratified medium; that is, a scatterer may not be determined by free-wave far-field patterns corresponding to infinitely many distinct incident waves.

Notes

1. **Key words.** inverse scattering, uniqueness, acoustic waves, stratified medium. **AMS (MOS) subject classifications.** 35R30, 35J05, 35P25

References

[1] L. Brekhovskikh, *Waves in Layered Media*, Second Edition, Academic Press, New York, (1980).

[2] M. Cheney and D. Isaacson, *Inverse problems for a perturbed dissipative half-space*, Inverse Problems, (to appear).

[3] D. Colton and R. Kress, *Inverse Acoustic and Electromagnetic Scattering Theory*, Springer-Verlag, Berlin Heidelberg New York, (1992).

[4] R. Gilbert and Y. Xu, *Generalized Herglotz functions and inverse scattering problem in a finite depth ocean*, in Invariant imbedding and inverse problems, J. Corones et al. ed. SIAM, Philadelphia, PA, 1992.

[5] Weinberger, H., "Variational Methods for Eigenvalue Approximation," SIAM, Philadelphia, (1974).

[6] Y. Xu, *Scattering of acoustic wave by obstacle in stratified medium*, in Partial Differential Equations with Real Analysis, H. Begehr and A. Jeffray ed., Pitman Research Notes in Mathematics series 263, Longman Sc & Tech., Essex, UK (1992), 147-168.

[7] Y. Xu, *Radiation condition and scattering problem for time-harmonic acoustic waves in a stratified medium with non-stratified inhomogeneity*, *Math. Meth. in Appl. Sci.*, Vol. 18, 51-66(1995).

23 A NOTE ON THE GLOBAL EXISTENCE AND DECAY OF SOLUTIONS OF SYSTEMS OF PARABOLIC CONSERVATION LAWS

A. Jeffrey

Dedicated to my friend and colleague of many years Robert Gilbert on the occasion of his 65th birthday

Abstract: This note summarizes some of the results of a study, the details of which are to be published elsewhere [1, 2], of the global existence and time decay of solutions of a one-dimensional system of parabolic conservation laws of the form

$$u_t + f(u)_x = \varepsilon u_{xx} \ , \ x \in \mathbb{R} \quad , \quad t > 0 \ u(t,x)|_t = 0, \quad u_0(x) \ , \quad x \in \mathbb{R},$$

where u is an $n \times 1$ unknown vector and $f(u)$ is an arbitrary $n \times 1$ smooth vector-valued function defined in a ball of radius r with its center at a fixed point $w_0 \in \mathbb{R}^n, \in > 0$. For an associated system in which $\in = 0$ that is is hyperbolic at a fixed point $w_0 \in \mathbb{R}^n$, conditions are given on $u_0(x)$ that ensure a unique globally smooth solution exists. In addition decay rates for the solution are estimated for

$$\|\partial^k/\partial x^k u(t,x) - w_0\|_{L^2(\mathbb{R},\mathbb{R}^n)} \quad \text{and} \quad \|\partial^k/\partial x^k u(t,x) - w_0\|_{L^\infty(\mathbb{R},\mathbb{R}^n)}.$$

H. Florian et al. (eds.), Generalized Analytic Functions, 265–269.

23.1 BACKGROUND

A Cauchy problem is considered for the following one-dimensional parabolic system of conservation laws that is important in the study of fluid mechanics, and elsewhere,

$$u_t + f(u)_x = \varepsilon u_{xx}, \quad x \in \mathbb{R},\ t > 0 \tag{23.1}$$

subject to the initial data

$$u(t,x)|_{t=0} = u_0(x),\ x \in \mathbb{R}, \tag{23.2}$$

where $u(t,x) = (u_1(t,x), u_2(t,x), \ldots, u_n(t,x))^T$ is an unknown solution vector and $f(u) = (f_1(u), f_2(u), \ldots, f_n(u))^T$ is an arbitrary $n \times 1$ smooth vector-valued function defined in a ball of radius r with its center at a fixed point $w_0 \in \mathbb{R}^n, \varepsilon > 0$.

Associated with the above problem is the one-dimensional system of conservation laws

$$u_t + f(u)_x = 0, \quad x \in \mathbb{R},\ t > 0 \tag{23.3}$$

subject to the initial data given above in (23.2). System (23.3) is **hyperbolic** at a fixed point $w_0 \in \mathbb{R}^n$ if at w_0 the Jacobian ∇f has n real eigenvalues $\lambda_1(w_0), \lambda_2(w_0), \ldots, \lambda_n(w_0)$ and there is a corresponding set of n linearly independent right eigenvectors $r_i(w_0)$, where $i = 1, 2, \ldots, n$. The system is **strictly hyperbolic** if, in addition to the above conditions, the eigenvalues are all *distinct*.

In each of the above cases

$$\nabla f(w_0) r_i(w_0) = \lambda_i(w_0) r_i(w_0), \text{ with } \det(r_1(w_0), r_2(w_0), \ldots, r_n(w_0)) \neq 0, \tag{23.4}$$

for $i = 1, 2, \ldots, n$, so the Jacobian is diagonalizable. A solution corresponding to an eigenvalue $\lambda_i(u)$ of either a hyperbolic or a strictly hyperbolic system is said to be **genuinely nonlinear** in the sense of Lax [3, 4, 5] if

$$r_i(u)\nabla \cdot \lambda_i(u) \neq 0, \tag{23.5}$$

and to be **linearly degenerate** if

$$r_i(u)\nabla \cdot \lambda_i(u) \equiv 0. \tag{23.6}$$

Smooth solutions of a quasilinear hyperbolic system can be studied by diagonalizing the system to obtain transport equations of the form [5, 6]

$$\frac{\partial w_i}{\partial t} + \lambda_i(w)\frac{\partial w_i}{\partial x}, \quad \text{for } 1 \leq i \leq n, \tag{23.7}$$

where the functions $w_i(u)$ are the **Riemann invariants** of the system satisfying

$$\nabla w_i \cdot \nabla f = \lambda_i \nabla w_i, \quad \text{for } i = 1, 2, \cdots, n. \tag{23.8}$$

A necessary and sufficient condition for the existence of the Riemann invariants $w(u) = (w_1(u), w_2(u), \ldots, w_n(u))$ for a strictly hyperbolic system are the Frobenius conditions

$$l_i r_j, r_k = 0, \quad \text{for any } j, k \neq i. \tag{23.9}$$

where l_i is the ith left eigenvector corresponding to $\lambda_i(u)$ and $\cdot, \cdot$ is the Poisson bracket of vector fields in u-space.

A pair of smooth functions $(\eta(u), q(u)) : \mathbf{R} \to \mathbf{R}$ is called an **entropy-entropy flux pair** for (23.3) if all smooth solutions satisfy an additional conservation law

$$\eta(u)_x + q(u)_x = 0, \tag{23.10}$$

so that

$$\nabla q = \nabla \eta \nabla f. \tag{23.11}$$

The function $\eta(x)$ is **convex** if for all $u, w \in \mathbf{R}^n$

$$w^T \eta''(u) w \geq 0, \tag{23.12}$$

and **strictly convex** for all $u, w \in \mathbf{R}^n$ if

$$w^T \eta''(u) w \geq \delta |w|^2, \tag{23.13}$$

for some constant $\delta > 0$.

23.2 MAIN RESULTS

For general strictly hyperbolic conservation laws (23.3) with either genuinely nonlinear or linearly degenerate characteristic fields, Glimm [7] established a general global existence theorem when the initial data is sufficiently small and, as a consequence, it is natural to wish to identify and construct admissible discontinuous solutions $u(t, x)$ of the Cauchy problem (23.3) subject to the initial data (23.2) as limits of the solutions $u^\varepsilon(t, x)$ of (23.1) and (23.2) as $\varepsilon \to 0$. This problem is related to the so-called entropy conditions and to the method of vanishing viscosity used to exclude non-physical solutions from weak solutions of hyperbolic systems [6], and to the study of viscous shock profile solutions.

The following questions arise, and for their resolution rely on the use of entropy-entropy flux pairs, Gronwall's lemma and various results established elsewhere:

(i) For each fixed $\in > 0$ and arbitrary smooth flux functions $f(u)$ does the Cauchy problem (23.1), (23.2) always have a unique globally smooth solution?

(ii) If the answer to the first question is in the affirmative, how can the temporal behavior of globally smooth solutions be obtained?

(iii) If the answer to the first question is in the affirmative, does the solution decay to the temporal asymptotic solution determined by (ii) at an optimal rate, in

the sense that its decay rate in some norm coincides with the decay rate in the same norm for the solution of the corresponding linear part of the Cauchy problem?

The answers to the above problems are provided in the following theorems, the detailed proofs of which are to be published elsewhere [1, 2].

Theorem 1 (Global esistence) *Let the corresponding inviscid system associated with system (23.1) be hyperbolic at some fixed point* $w_0 \in \mathbf{R}^n$, *then for each fixed constant* $r > 0$, *if* $u_0(x) - w_0 \in L^\infty \cap L^1(\mathbf{R}, \mathbf{R}^n)$ *with* $\|u_0(x) - w_0\|_{L^1(\mathbf{R},\mathbf{R}^n)}$ *sufficiently small and* $\|u_0(x) - w_0\|_{L^\infty(\mathbf{R},\mathbf{R}^n)} < r$, *the Cauchy problem (23.1), (23.2) admits a unique globally smooth solution* $u(t, x)$ *satisfying*

$$\|u(t,x) - w_0\|_{L^\infty(\mathbf{R}^+ \times \mathbf{R},\mathbf{R}^n)} \leq r.$$

Theorem 2 (Temporal asymptotic decay) *Let the conditions of Theorem 1 be satisfied, then the globally smooth solution* $u(t, x)$ *that exists by virtue of the theorem satisfies the following temporal decay estimates. For each* $k = 0, 1, \ldots,$

$$\left\| \frac{\partial^k}{\partial x^k} [u(x,t) - w_0] \right\|_{L^2(\mathbf{R},\mathbf{R}^n)} \leq C(1+t)^{-(2k+1)/4}$$

and

$$\left\| \frac{\partial^k}{\partial x^k} [u(x,t) - w_0] \right\|_{L^\infty(\mathbf{R},\mathbf{R}^n)} \leq C(1+t)^{-(k+1)/2}$$

23.3 GENERAL REMARKS

Establishing global existence involves two main steps, the first of which is the study of local existence, and the second is the determination of a *priori* estimates for the solution. These steps were followed when deriving the above results and involve ideas related to the paper by Hoff and Smoller [8, 9] and its extension by Zhao [10]. The results reported in Theorems 1 and 2 are somewhat more general than those in [8, 9], because although they require the inviscid system to be hyperbolic at some fixed point $w_0 \in \mathbf{R}^n$, they do not require the systems to satisfy the additional conditions specified in [8, 9] of the form

$$c|u|^p \leq \eta(u) \leq d|u|^p, p \geq 1, c > 0, d > 0,$$

where $u_0(x) - w_0 \in L^\infty \cap L^p(\mathbf{R}, \mathbf{R}^n)$ with $\|u_0(x) - w_0\|_{L^p(\mathbf{R},\mathbf{R}^n)}$ sufficiently small.

Related results are to be found in Schonbek [11], and for the case of global weak solutions for isentropic equations of a polytropic gas in Jeffrey and Lu [12].

References

[1] Zhao, H., Jeffrey, *A Global existence and optimal temporal decay estimates for systems of parabolic conservation laws I: The one-dimensional case* (to appear).

[2] Jeffrey. A and Zhao.*H Global existence and optimal temporal decay estimates for systems of parabolic conservation laws II:Multi-dimensional case* (to appear).

[3] Lax, P.D., *Hyperbolic systems of conservation laws*, Comm. Pure Appl.Math. 10(1957), 537-566.

[4] Smoller, J.A., *Shock Waves and Reaction Diffusion Equations*, Springer Verlag, New York, 1983.

[5] Jeffrey, A., *Quasilinear Hyperbolic Systems and Waves*, Research Note in Mathematics No.5, Pitman, London 1976.

[6] LeVeque, R. J.,*Numerical Methods for Conservation Laws*, Birkhauser Verlag, Boston, 1990.

[7] Glimm, J., *Solutions in the large for nonlinear hyperbolic systems of equations*, Comm. Pure Appl. Math. 18(1965),95-105.

[8] Hoff, D., Smoller, J.,*Solutions in the large for certain nonlinear parabolic systems*, Analyse Nonlineaire, 2(1985), 213-235.

[9] Hoff, D., Smoller, J., *Global existence for systems of parabolic conservation laws*, Journal of Differential Equations, 68(1987), 210-220.

[10] Zhao, H.J., *Solutions in the large for certain nonlinear parabolic systems in arbitrary spatial dimensions*, Applicable Analysis, 59(1995), 349-376.

[11] Shoenbek, M., *Uniform decay rate for papabolic conservation laws*, Nonlinear Analysis, 10(9)(1986),943-956.

[12] Jeffrey, A, Lu, Y.G., *A study of the global weak solution for the isentropic equations of a polytropic gas*, Amer. Math. Soc. Studies in Advanced Mathematics, 3(1997), 261-270.

24 ANALYTICAL AND NUMERICAL TREATMENT OF HELE-SHAW MODELS WITH AND WITHOUT REGULARIZATION

M. Reissig
and F. Hübner

Dedicated to Prof. R.P. Gilbert on the occasion of his 65th birthday.

24.1 TECHNICAL BACKGROUND - HISTORICAL REMARKS

As an introduction we want to make some historical remarks about Hele-Shaw problems and we want to give a practical motivation, simultaneously.
Hele-Shaw, an English physicist, did the first experiments in 1898. He investigated the streamlines of a viscous fluid between two parallel, tilted, less distant plates.
This experimental arrangement, the two plates, large enough, with a small distance and an approximately incompressible fluid between them is called the Hele-Shaw cell (see figure 1).

H. Florian et al. (eds.), Generalized Analytic Functions, 271–287.

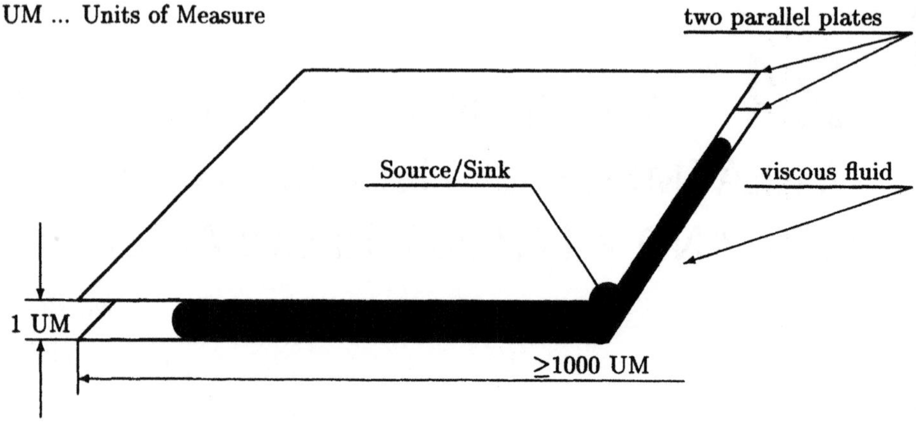

figure 1 (Hele-Shaw cell)

Hele Shaw derived his model under the following principle physical assumptions:

- The fluid we will consider here is a Newtonian and incompressible one. Incompressibility allows us to handle the fluid as a potential flow.
- Thermal effects will be neglected.
- Furthermore the flow is assumed as stationary.
- The velocity perpendicular to the bordering plates is equal to zero. That is a reasonable assumption because the plates are very small separated (exactly speaking, the ratio of the horizontal expansion l to the vertical distance h is of order 10^{-3}). This also causes that the problem can be reduced to a two-dimensional one.
- To neglect the presence of an outer (second) fluid and its influence to our problem it is assumed that the viscosity of the outer fluid μ'' is small compared with the inner one μ' ($\mu'' << \mu'$).
- An other essential simplification is that the pressure on the boundary of the fluid is supposed to be constant ($p = const.$). That means, surface tension and time-dependence (and other possible influences respectively) are neglected.

Further investigations, for instance by Stokes, offered that a lot of problems in fluid mechanics can be transformed into the Hele-Shaw problem.
This model can also be used for several problems in electrochemistry, poroelasticity, underground flows and other problems being equivalent to 2-dimensional potential flows.
We obtain the Hele-Shaw problem as a special case of the Stefan problem if we assume stationarity.
In 1934 Leibenzon, an Russian oil industrialist, found independently of Hele-Shaw an

other very important application. He obtained the same results like Hele-Shaw by inquiring into oil fields. This is one of the typical natural examples for a Hele-Shaw cell: the ratio of the vertical expansion by the horizontal one is nearby 10^{-3}, the viscosity of the groundwater can be neglected with respect to the viscosity of oil, the exterior pressure is supposed to be constant because of the approximately infinite reservoir of groundwater.

The first exact solutions of the Hele-Shaw problem driven by a single sink at the origin were given by Polubarinova-Kochina [17] and Galin [5].

It turned out that, in general, the Hele-Shaw moving boundary problem with suction is ill-posed in a variety of respects, mainly involving irregular behaviour of the free boundary. The solution of Polubarinova-Kochina [17] has a finite-time blow-up assume constant (atmospheric) pressure at the free boundary and cusp formation in the free boundary before it reaches the sink.

Using complex-analytic methods, many explicit solutions to the Hele-Shaw problem can be found by mapping the flow region conformally onto the unit disk in an auxiliary plane (see the complex model (M2) in section 24.2).

When the map is a polynomial of order ≥ 2 , finite time blow-up always occurs through the development of a cusp before the moving boundary reaches the sink, but for non polynomial maps other forms of blow-up are also possible ([9], [12]).

24.2 DERIVATION OF THE HELE-SHAW MODELS

Consider a two-dimensional potential flow of incompressible fluid which is modeled by the velocity vector field $\vec{v} = (v_1, v_2)$ in the plane. The assumption about potentiality of the flow states that the fluid velocity $\vec{v}$ is linearly proportional to the gradient of the fluid pressure, which can be written as

$$\vec{v} = -grad\, p. \tag{24.1}$$

For the incompressible flow the divergence-free property

$$div\ \vec{v} = 0 \tag{24.2}$$

must be also satisfied. Thus the potential function of the vector velocity field, i.e. the pressure p, is a harmonic function

$$div(grad\, p) = \Delta p = 0. \tag{24.3}$$

We shall study the motion of fluid within a simply connected bounded region. For this we suppose that

$$\partial\Omega(0) \tag{24.4}$$

is given and has some regularity properties which are stated in the next section. Hereinafter only one sink in the fluid region is supposed. This means that without loss of generality we can suppose that for all times t the origin belongs to the fluid region $\Omega(t)$ where the potential function has a prescribed logarithmic singularity of the form

$$p(w_1, w_2, t) \sim -\frac{q(t)}{2\pi} log \mid \vec{w} \text{ as } \vec{w} \mid \to 0, \vec{w} = (w_1, w_2). \tag{24.5}$$

The real function $q(t)$ models injection ($q(t) > 0$) and suction ($q(t) < 0$) respectively. To find a mathematical model one has to take into consideration the following two conditions:

a) the **dynamic boundary condition**

$$p = 0, \tag{24.6}$$

together with (24.5) it means, that p is simply the Green's function for an aposteriori known region $\Omega(t)$;

b) the **kinematic boundary condition**

$$\vec{v} = -grad\, p = \frac{d\Gamma(t)}{dt}, \tag{24.7}$$

where $\Gamma(t) = \partial\Omega(t)$, this relation follows from the mass conservation, it means, that the velocity of the fluid at the boundary must coincide with the velocity of the point of the boundary as a geometric object.

The problem (24.3), (24.4), (24.6) and (24.7) is called the **classical Hele-Shaw problem.**

There are different approaches to derive a mathematical model for this problem. Let us sketch a first one (see [8]).

Let us introduce for all $t \in I = (-\epsilon, \epsilon)$ an one-parametric family of C^2-diffeomorphisms

$$\vec{u} = (u_1(\vec{x}, t), u_2(\vec{x}, t)) \tag{24.8}$$

of the unit circle $\partial U = \{(x_1, x_2) \in \mathbb{R}^2 : |(x_1, x_2)| = 1\}$ onto $\Gamma(t)$.

It gives the velocity of the boundary as $\frac{d\Gamma}{dt} = \partial_t \vec{u}(\vec{x}, t)$, so that (24.7) can be written as

$$\partial_t \vec{u}(\vec{x}, t) = -grad\, p(\vec{u}(\vec{x}, t), t) \text{ for all } (\vec{x}, t) \in \partial U \times I. \tag{24.9}$$

The initial condition (24.4) takes the form

$$\vec{u}(\vec{x},0) = \vec{u}_0(\vec{x}). \tag{24.10}$$

Let us note that condition (24.6), together with (24.3) and (24.5), means that p is, up to a constant factor, the Green's function $\mathcal{G}_{\Omega(t)}$ for the homogeneous Dirichlet problem for Laplace's equation in $\Omega(t)$, that is,

$$p = \frac{q}{2\pi}\mathcal{G}_{\Omega(t)}. \tag{24.11}$$

Thus the motion of the boundary of the fluid region for the Hele-Shaw flows with condition (24.6) is described by

$$\partial_t \vec{u}(\vec{x},t) = -\frac{q}{2\pi} grad\, \mathcal{G}(\vec{u}(\vec{x},t),t) \tag{24.12}$$

for all $(\vec{x},t) \in \partial U \times I$. Pay attention that $\mathcal{G}_{\Omega(t)}$ is a **nonlocal operator** in the physical plane.

Real mathematical model **(M1)** (see[8]):
Find a C^2-mapping $\vec{u} : \partial U \times I \to \mathbb{R}^2$ satisfying

(i) $\vec{u}(\vec{x},t) \in \Gamma(t)$ for all $(\vec{x},t) \in \partial U \times I$;

(ii) $\vec{u}(.,t) : \partial U \in \Gamma(t)$ is a C^2-diffeomorphism for each fixed t (bijective and whose derivative never vanishes);

(iii) $\partial_t \vec{u}(\vec{x},t) = -\frac{q}{2\pi} grad\, \mathcal{G}_{\Omega(t)}(\vec{u}(\vec{x},t),t)$ for all $(\vec{x},t) \in \partial U \times I$;

(iv) the boundary $\Gamma(0)$ of $\Omega(0)$ is described by $\vec{u}(\vec{x},0) = \vec{u}_0(\vec{x})$, which is a diffeomorphism of the unit circle.

There is a second mathematical model which describes the classical Hele-Shaw problem as a nonlinear boundary value problem for the **Riemann mapping function** from the unit disk onto the fluid region (for the derivation see [22]).

Complex mathematical model **(M2)** (see [5], [17], [22]):
Find a mapping $f = f(\rho,t) : \overline{U} \times (-\epsilon,\epsilon) \to \mathbb{C}$, analytic and univalent as a function of ρ in a neighbourhood of $\overline{U}$, continuously differentiable with respect to t in $(-\epsilon,\epsilon)$ satisfying

(i) $\Re\left(\frac{1}{\rho}\frac{\partial f}{\partial t}\overline{\frac{\partial f}{\partial \rho}}\right) = -1$ on $|\rho| = 1$;

(ii) $f(\rho,0) = f_0(\rho)$, $f_0(0) = 0$, analytic and univalent in a neighbourhood of $|\rho| = 1$, f_0 describes $\Omega(0)$;

(iii) $f(0,t) = 0$.

Remark 1: Due to [8] a necessary condition for a solution of model (M1) to exist is that $\Gamma(0)$ is an analytic curve. But the existence of classical solutions remained open.

Remark 2: From the results of [25] one gets the existence (locally in time) and uniqueness of solutions of model (M2) which depend analytically on the time variable, too. The fairly complicated proof was based on the standard method of successive approximations, but the uniform estimates in this nonlocal nonlinear case (condition *(i)*) are obtained by numerous nontrivial estimates of singular integrals.

Remark 3: The idea of Polubarinova-Kochina to find exact solutions of (M2) was to choose the ansatz $f(\rho, t) = a_1(t)\rho + a_2(t)\rho^2$. Then an easy calculation yields that $a_1(t)$ and $a_2(t)$ have to satisfy the system of algebraic equations

$$\begin{aligned} a_1^2(t) + 2a_2^2(t) &= a_1^2(0) + 2a_2(0) - 2t \\ a_1^2(t)a_2(t) &= a_1^2(0)a_2(0). \end{aligned}$$

If $a_1(0) = 1$ and $a_2(0) = \frac{1}{4}$, then blow-up occurs at $t = T_* \approx 0.566$ (see figure 2).

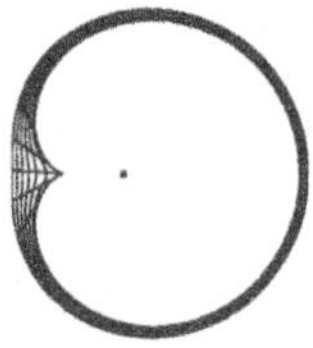

figure 2
The classical example of Polubarinova-Kochina

24.3 MATHEMATICAL TREATMENT - APPLICATION OF THE METHOD OF SCALES OF BANACH SPACES

24.3.1 The abstract Cauchy-Kovalevsky Theorem of Nishida

The next theorem is an effective tool for the consideration of the models (M1) and (M2). It generalizes the classical Cauchy-Kovalevsky theorem for nonlinear systems of partial differential equations to abstract operator equations in **scales of Banach spaces**. We cite here the version from [16].

Theorem 1 *Let us consider the abstract Cauchy problem*

$$\frac{dw}{dt} = \mathcal{L}(t, w), \ w(0) = 0 \tag{24.13}$$

satisfying the following conditions in a scale of Banach space $\{\mathbf{B}_s, \|.\|_s\}_{0<s\leq 1}$ *(a family of continuously embedded Banach spaces* $\{\mathbf{B}_s, \|.\|_s\}_{0<s\leq 1}$ *is called a scale of Banach spaces if for all* $0 < s' \leq s \leq 1$ *the norm of the canonical embedding operator* $\|\mathbf{I}_{s\to s'}\| \leq 1$*).*
(C, K, R and T are certain positive constants independent of s'*, s and t):*

(i) the right-hand side $\mathcal{L}(t,w)$ *is a continuous in t mapping of* $[-T,T] \times \{w \in \mathbf{B}_s : \|w\|_s \leq R\}$ *into* $\mathbf{B}_{s'}$ *for all* $0 < s' < s \leq 1$*;*

(ii) the continuous function $\mathcal{L}(t,0)$ *satisfies* $\|\mathcal{L}(t,0)\|_s \leq \frac{K}{1-s}$ *for all* $s \in (0,1)$*;*

(iii) it holds for all $0 < s' < s \leq 1$, $t \in [-T,T]$ *and* w_1, w_2 *belonging to* $\{w : \|w\|_s < R\}$ *the "Lipschitz condition"*
$\|\mathcal{L}(t,w_1) - \mathcal{L}(t,w_2)\|_{s'} \leq \frac{C}{s-s'}\|w_1 - w_2\|_s.$

Under these assumptions there exists one and only one solution
$w \in C^1\left((-a_0(1-s), a_0(1-s)), \mathbf{B}_s\right)_{0<s<1} : \|w(t)\|_s < R,$
where a_0 *is a suitable positive constant.*

Remark 4: There exist a lot of applications of this theorem to applied problems. In general one can apply it to prove properties of solutions which are near to analyticity. Such properties are of interest for Hele-Shaw flows. The theorem gives the existence of solutions as well as for the backward and forward Cauchy problem, the suction and injection problem, respectively.

24.3.2 The application of the abstract Cauchy-Kovalevsky theorem

Let the initial fluid region be a simply connected domain with analytic boundary. We suppose that there exists a point sink (suction problem) or a point source (injection problem).

Main question: Can we prove that for a small period of time there exists a unique solution of the classical Hele-Shaw problem for both suction and injection as well and the fluid region remains simply connected and has an analytic boundary?

Main result: Yes, we can! On the one hand one expects this result. On the other hand it is clear that one cannot expect a global-in-time solution. For the suction problem see figure 2 (example of Polubarinova-Kochina), in the injection problem the fluid region can lose the property to be simply connected.

If we want to apply the abstract Cauchy-Kovalevsky theorem to the models (M1) and (M2), then there arise some preparatory questions:

- Can we reduce the model to an abstract Cauchy problem of type (24.13)?
- How can we construct an appropriate scale of Banach spaces for this problem?

- Is it possible to show that the operator $\mathcal{L} = \mathcal{L}(t, w)$ is a **singular operator of first order** in this scale (this means, that the coefficient $\frac{C}{s-s'}$ appears in condition (*iii*) of the theorem)? If we replace this coefficient by $\frac{C}{(s-s')^\alpha}$, $\alpha > 1$, then the theorem is false.

We want to sketch the answers to the questions in the case of model (M2). The nonlinear boundary condition $\Re\left(\frac{\partial_t f \overline{\partial_\rho f}}{\rho}\right) = -1$ on $|\rho| = 1$ can be written as $\Re\left(\frac{\partial_t f}{\rho\, \partial_\rho f}\right) = -|\partial_\rho f|^{-2}$ on $|\rho| = 1$. Using $f(0,t) = 0$ and the univalence of f the function $\frac{\partial_t f}{\rho\, \partial_\rho f}$ is holomorphic in $\{|\rho| < 1\}$. With the additional assumption $\Im\left(\frac{\partial_t f}{\rho\, \partial_\rho f}\right)(0,t) = 0$ the Schwarz formula yields the abstract Cauchy problem

$$\frac{\partial f}{\partial t} + \rho \frac{\partial f}{\partial \rho} \frac{1}{2\pi i} \int_{|z|=1} \left|\frac{\partial f}{\partial z}\right|^{-2} \frac{z+\rho}{z-\rho} \frac{dz}{z} = 0, \; f(\rho, 0) = f_0(\rho). \tag{24.14}$$

Substituting $v = (\partial f/\partial z)^{-1}$ and differentiating the above equation by $\partial/\partial\rho$ gives finally the equivalent problem

$$\frac{\partial v}{\partial t} - \rho \mathbf{T}_t(v) \frac{\partial v}{\partial \rho} + v \frac{\partial}{\partial \rho} (\rho \mathbf{T}_t(v)) = 0, \; v(\rho, 0) = (\partial f_0/\partial \rho)^{-1}, \tag{24.15}$$

where $\mathbf{T}_t(v) := \frac{1}{2\pi i} \int_{|z|=1} |v(z)|^2 \frac{z+\rho}{z-\rho} \frac{dz}{z}$.
The Cauchy problem (24.15) is of type (24.13) with the Schwarz operator $\mathbf{T}_t(v)$ as a non-local operator.

As an appropriate scale of Banach spaces we choose ($G_a = \{z \in \mathbb{C} : |z| < a\}$)

$$\{\mathbf{B}_s, \|.\|\}_{0<s\leq 1} = \left\{H(G_{r_1+s(r_2-r_1)}) \cap C(\overline{G}_{r_1+s(r_2-r_1)}),\; sup\, |.|\right\}, \tag{24.16}$$

that means, that $\mathbf{B}_s$ is a space of functions which are holomorphic in $G_{r_1+s(r_2-r_1)}$ and continuous in its closure. The choice of $1 < r_1 < r_2$ corresponds to the demand that the solutions should be holomorphic, even univalent, in a neighbourhood of the unit disk.

Using the theory of Cauchy-type integrals one can prove (see [21]):

- there exists an analytic continuation of $\mathbf{T}_t(v)$ to $\tilde{\mathbf{T}}_t(v)$ mapping continuously each space $\mathbf{B}_s$ into itself;
- $\partial/\partial\rho$ is a singular operator of first order in the scale $\{\mathbf{B}_s, \|.\|_s\}_{0<s\leq 1}$.

These both results are essential for the application of Theorem 3.1. If one understands the abstract Cauchy-Kovalevsky theory, then this approach represents a simplified proof of the results from [25].

Remark 5: The condition *(iii)* of model (M1) has the form of (24.13), consequently, a reduction is not necessary. But in opposite to the above consideration the study of the nonlocal operator $(\nabla\mathcal{G}_{\Omega(t)})(\vec{u}(\vec{x},t),t)$ makes difficulties. One has to define scales of Hölder-Gevrey spaces of order 1 on the unit circle. Moreover, one has to prove analytic type estimates for the gradient of the Dirichlet problem for harmonic functions up to the boundary in this scale (see [19]). The results of [19] imply the existence of classical solutions for the model (M1) (compare with Remark 1).

24.4 PROPOSALS FOR REGULARIZATIONS

What do we know up to now about the classical Hele-Shaw suction problem?
On the one hand we have the local existence in time of analytic solutions. On the other hand there appears a finite-time blow up, a cusp formation, respectively, before the moving boundary reaches the sink. This means that

- the curvature of the free boundary becomes infinite at the cusp,
- the velocity of points of the moving boundary tends to infinity at the cusp.

For this reason two regularizations of the suction problem are proposed by incorporating extra terms at the free boundary conditions to penalize large curvatures or large normal velocities.

Regularized dynamic boundary conditions:

(a) include a Gibbs-Thomson term proportional to the curvature
$p + \sigma\kappa = 0$ on $\Gamma(t)$, κ is the curvature of $\Gamma(t)$, $\sigma > 0$,
large curvatures are penalized ([10], [24]);

(b) kinetic undercooling condition
$p + \alpha\frac{\partial p}{\partial n} = 0$ on $\Gamma(t)$, $\alpha > 0$,
large normal velocities are penalized ([23]).

The authors have studied the **kinetic undercooling regularization**. Which influence has the kinetic undercooling condition on the mathematical models?

Real mathematical model **(M3)** (see [11]):
Find an analytic mapping $\vec{u} : \partial U \times I \to \mathbb{R}^2$, $I = (-\epsilon, \epsilon)$, satisfying

(i) $\vec{u}(\vec{x},t) \in \Gamma(t)$ for all $(\vec{x},t) \in \partial U \times I$;

(ii) $\vec{u}(.,t) : \partial U \to \Gamma(t)$ is an analytic diffeomorphism for each fixed t;

(iii) $\partial_t\vec{u}(\vec{x},t) = -\frac{q}{2\pi} grad\, \mathcal{G}_{\Omega(t)}(\vec{u}(\vec{x},t),t)$ for all $(\vec{x},t) \in \partial U \times I$, where $\mathcal{G}_{\Omega(t)}$ is the Green's function for the problem:
$-\Delta\mathcal{G} = \delta_0$ in $\Omega(t)$, $\frac{\partial\mathcal{G}}{\partial n} + \frac{1}{\alpha}\mathcal{G} = 0$ on $\Gamma(t)$;

(iv) the boundary $\Gamma(0)$ of $\Omega(0)$ is described by $\vec{u}(\vec{x},0) = \vec{u}_0(\vec{x})$ which is an analytic diffeomorphism of the unit circle.

In [11] the local existence in time of solutions is proved by applying the abstract Cauchy-Kovalevsky theory. In opposite to (M1) we have to suppose the property of analyticity because a result mentioned in Remark 1 is not known in the case of undercooling regularization.

Complex mathematical model (M4):
Find a mapping $f = f(\rho, t)$ with the same properties as those one of model (M2) and a holomorphic function $w_{reg} = w_{reg}(\rho, t) : \overline{U} \times I \to \mathbb{C}$ satisfying *(ii)*, *(iii)* of model (M2) and

(i) $\Re\left(\frac{1}{\rho}\frac{\partial f}{\partial t}\overline{\frac{\partial f}{\partial \rho}}\right) = -1 + \Re(\rho w_{reg})$ on $|\rho| = 1$;

(iv) $\Im(\rho w_{reg}) = \alpha \partial_\theta \left(|\frac{\partial f}{\partial \rho}|^{-1} \Re\left(\frac{1}{\rho}\frac{\partial f}{\partial t}\overline{\frac{\partial f}{\partial \rho}}\right)\right)$ on $|\rho| = 1$.

If $\alpha = 0$, then $\Im(\rho w_{reg}) = 0$. Consequently, $w_{reg} \equiv 0$ and we obtain the classical Hele-Shaw model. In an unpublished paper [20] the local existence in time of solutions $(f,\ w_{reg})$ is proved. Therefore one has to study a nonlinear abstract Cauchy-Kovalevsky problem and a Riemann-Hilbert-Poincare problem to determine f and w_{reg}.

Both results give a local existence result in t. But the global-in-time question whether it is possible to remove all of the initial fluid region in the Hele-Shaw cell with suction remains an intriguing open problem. Nevertheless the local result motivates numerical calculations (nontrivial exact solutions in the case of kinetic undercooling regularization are unknown). In the next chapter we represent some of these from [14].

24.5 NUMERICAL RESULTS

To handle model (M4) numerically it is useful to rewrite condition *(iv)* by applying the Hilbert-operator. Thus we obtain

(iv')

$$\Re[\rho w_{reg}] = \alpha \frac{\partial}{\partial \theta} \mathcal{H} \left[\left|\frac{\partial f}{\partial \rho}\right|^{-1} (-1 + \Re[\rho w_{reg}]) \right] \tag{24.17}$$

by using $\Im(\rho w_{reg})(0) = 0$.
Dividing condition *(i)* by $\left|\frac{\partial f}{\partial \rho}\right|^2$ ($\frac{\partial f}{\partial \rho} \neq 0$; $\forall \rho \in U$) gives the Schwarz problem for the holomorphic function $\frac{\partial f}{\partial t} / \rho \frac{\partial f}{\partial \rho}$ on the unit circle:

$$\frac{\partial f}{\partial t}\frac{1}{\rho}\left(\frac{\partial f}{\partial \rho}\right)^{-1} = \mathcal{S}\left[\left|\frac{\partial f}{\partial \rho}\right|^{-2}(-1 + \Re\,[\rho w_{reg}])\right],$$

which consequently ends in

$$\begin{aligned} \frac{\partial f}{\partial t}(\rho, t) &= \rho \frac{\partial f}{\partial \rho}(\rho, t)\, \mathcal{S}\left[\left|\frac{\partial f}{\partial \rho}\right|^{-2}(-1 + \Re[\rho w_{reg}])\right] \\ &= -\rho \frac{\partial f}{\partial \rho}(\rho, t)\, \mathcal{S}\left[\left|\frac{\partial f}{\partial \rho}\right|^{-2}\right] + \rho \frac{\partial f}{\partial \rho}(\rho, t)\, \mathcal{S}\left[\left|\frac{\partial f}{\partial \rho}\right|^{-2} \Re[\rho w_{reg}]\right]. \end{aligned} \tag{24.18}$$

If we note that the Schwarz operator $\mathcal{S}$ and the Hilbert operator $\mathcal{H}$ are defined by

$$[\mathcal{S}u](\rho) := \frac{1}{2\pi i}\int_{\partial U} u(\eta)\,\frac{\eta-\rho}{\eta+\rho}\frac{d\eta}{\eta}$$

and

$$[\mathcal{H}u]\left(e^{i\theta}\right) := \frac{1}{2\pi}\int_0^{2\pi} u\left(e^{i\tau}\right)\cot\frac{\tau-\theta}{2}d\tau$$

respectively, then we can use the identity

$$\mathcal{S}[u] = u - i\mathcal{H}[u]$$

and the notation $f(\rho,t) = u(\rho,t) + iv((\rho,t)$ to rewrite (24.18) into

$$\begin{aligned}
u_t &= u_\theta\mathcal{H}\left[\tfrac{1}{u_\theta^2+v_\theta^2}\right] + v_\theta\left(\tfrac{1}{u_\theta^2+v_\theta^2}\right)\\
&- u_\theta\mathcal{H}\left[\left(\tfrac{1}{u_\theta^2+v_\theta^2}\right)\Re[\rho w_{reg}]\right] + v_\theta\left(\tfrac{1}{u_\theta^2+v_\theta^2}\right)\Re[\rho w_{reg}],\\
v_t &= v_\theta\mathcal{H}\left[\tfrac{1}{u_\theta^2+v_\theta^2}\right] + u_\theta\left(\tfrac{1}{u_\theta^2+v_\theta^2}\right)\\
&- v_\theta\mathcal{H}\left[\left(\tfrac{1}{u_\theta^2+v_\theta^2}\right)\Re[\rho w_{reg}]\right] - u_\theta\left(\tfrac{1}{u_\theta^2+v_\theta^2}\right)\Re[\rho w_{reg}].
\end{aligned}$$

For a discretization we share the unit circle ($|\rho| = 1$) into a equidistant grid with $n = 2N$ ($N = 1,2,3,\ldots$) grid points $\rho_j = e^{i\theta_j}$, where $\theta_j = \frac{i-1}{n}2\pi$, $j = 1,2,3,\ldots,n$. Likewise, we introduce a time discretization $t_k = k\Delta t$, Δt is suitable chosen.

With the following discretizations

$$(\partial_\theta u)_{j,k} = \frac{u_{j+1,k}-u_{j-1,k}}{2\Delta\theta} \text{ (3-points difference scheme)}$$

or

$$(\partial_\theta u)_{j,k} = \frac{8\left(u_{j+1,k}-u_{j-1,k}\right)-\left(u_{j+2,k}-u_{j-2,k}\right)}{12\Delta\theta} \text{ (5-points difference scheme).}$$

and

$$\partial_t u_{j,k} = \frac{u_{j,k+1}-u_{j,k}}{\Delta t} \text{ (2-layers time discretization)}$$

or

$$\partial_t u_{j,k} = \frac{u_{j,k+1}-u_{j,k-1}}{2\Delta t} \text{ (3-layers time discretization)}$$

respectively, we arrive at

$$
\begin{aligned}
u_{j,k+1} &= u_{j,k} \\
&+ \Delta t\left\{-\frac{(\partial_\theta v)_{j,k}}{\left(\partial_\theta u_{j,k}\right)^2+\left(\partial_\theta v_{j,k}\right)^2} + (\partial_\theta u)_{j,k}\,\mathcal{H}\left[\frac{1}{(\partial_\theta u)^2+(\partial_\theta v)^2}\right]_{j,k}\right. \\
&+ \left.\frac{(\partial_\theta v)_{j,k}\Re[\rho w_{reg}]_{j,k}}{\left(\partial_\theta u_{j,k}\right)^2+\left(\partial_\theta v_{j,k}\right)^2} - (\partial_\theta u)_{j,k}\,\mathcal{H}\left[\frac{\Re[\rho w_{reg}]}{(\partial_\theta u)^2+(\partial_\theta v)^2}\right]_{j,k}\right\}, \\
v_{j,k+1} &= v_{j,k} \qquad\qquad (24.19) \\
&+ \Delta t\left\{\frac{(\partial_\theta u)_{j,k}}{\left(\partial_\theta u_{j,k}\right)^2+\left(\partial_\theta v_{j,k}\right)^2} + (\partial_\theta v)_{j,k}\,\mathcal{H}\left[\frac{1}{(\partial_\theta u)^2+(\partial_\theta v)^2}\right]_{j,k}\right. \\
&- \left.\frac{(\partial_\theta u)_{j,k}\Re[\rho w_{reg}]_{j,k}}{\left(\partial_\theta u_{j,k}\right)^2+\left(\partial_\theta v_{j,k}\right)^2} - (\partial_\theta v)_{j,k}\,\mathcal{H}\left[\frac{\Re[\rho w_{reg}]}{(\partial_\theta u)^2+(\partial_\theta v)^2}\right]_{j,k}\right\}.
\end{aligned}
$$

Consequently we obtain the following algorithm:

1. computation of $f(\rho_j, t_k)$ for $t_k = 0$ and all $j = 1, 2, ..., n$,
2. stop computation if $\left|\frac{\partial f}{\partial \rho}(\rho, t_k)\right|^2$ is less than a given precision,
3. plot $f(\rho, t) = u(\rho, t) + iv(\rho, t)$ at chosen values for t,
4. solve (24.17) for $\Re[\rho w_{reg}]$ at $t = t_k$,
5. insert $\Re[\rho w_{reg}]$ in (24.19) and determine $f(\rho, t_{k+1})$,
6. set $f(\rho, t_k) = f(\rho, t_{k+1})$, $t_k = t_{k+1}$ and continue with 2.

Now some numerical results will be presented to illustrate the influence of undercooling effects and different discretization parameters. The arising variable λ is the ratio of the remaining domain by the initial one. The following table gives a comparison of blow-up times calculated by several parameters of the discretization for the Polubarinova-Kochina equation (T_* for the unregularized example is ≈ 0.566).

grid points N	time step Δt	discretization scheme	T_* $\alpha = 0.01$	$\alpha = 1$	$\alpha = 10^6$
32	0.05	3-point	0.65	2.00	3.00
32	0.05	5-point	0.65	1.95	2.95
32	0.01	3-point	0.65	1.98	2.93
32	0.01	5-point	0.63	1.89	2.76
32	0.005	3-point	0.650	1.985	2.930
32	0.005	5-point	0.625	1.890	2.760
48	0.05	3-point	0.65	1.95	3.00
48	0.05	5-point	0.65	1.95	2.85
48	0.01	3-point	0.66	1.98	2.96
48	0.01	5-point	0.65	1.90	2.77
48	0.005	3-point	0.655	1.980	2.955
48	0.005	5-point	0.655	1.895	2.760
64	0.05	3-point	0.65	1.95	3.00
64	0.05	5-point	0.65	1.95	2.85
64	0.01	3-point	0.66	1.98	2.96
64	0.01	5-point	0.65	1.90	2.77
64	0.005	3-point	0.655	1.980	2.960
64	0.005	5-point	0.655	1.895	2.760

The next figures show the influence of different values of α on the classical Polubarinova-Kochina example.

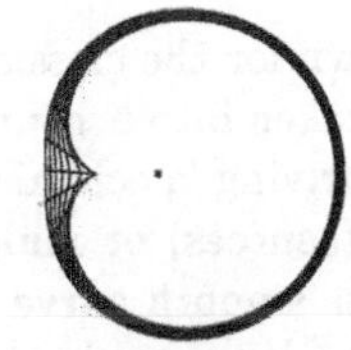

figure 3
$\alpha = 0$
$T_* = 0.566$
$\lambda = 0.866$

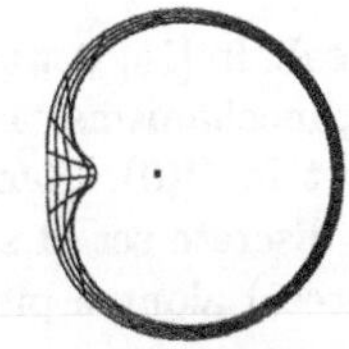

figure 4
$\alpha = 0.01$
$T_* = 0.65$
$\lambda = 0.81$

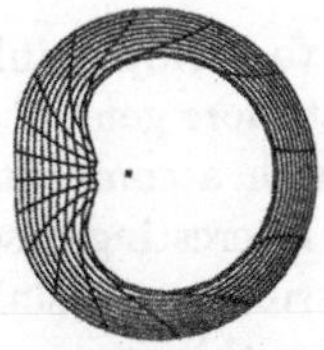

figure 5
$\alpha = 1$
$T_* = 1.93$
$\lambda = 0.44$

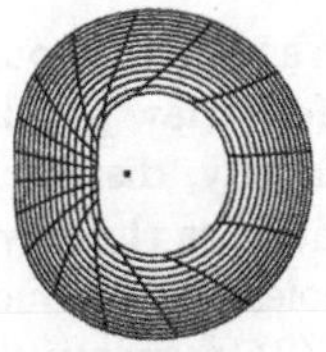

figure 6
$\alpha = 10^6$
$T_* = 2.74$
$\lambda = 0.22$

Here are two further examples of different types of initial functions, a polynomial and a rational one respectively.

polynomial:

$$f_0(\rho) = \rho - \tfrac{1}{72}\rho^6$$

figure 7
$\alpha = 0$
$T_* = 1.23$
$\lambda = 0.618$

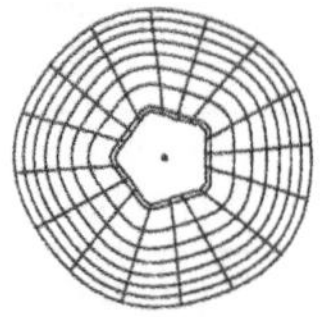

figure 8
$\alpha = 1$
$T_* = 2.87$
$\lambda = 0.086$

rational function:

$$f_0(\rho) = \rho \frac{4-\sqrt{5/2}\rho}{2(\rho-\sqrt{5/2})}$$

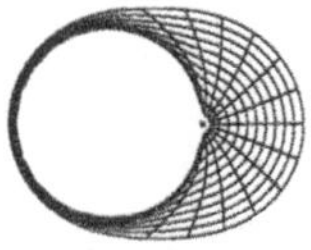

figure 9
$\alpha = 0$
$T_* = \pi$
$\lambda = 0.494$

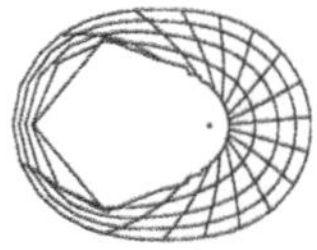

figure 10
$\alpha = 1$
$T_* = 3.75$
$\lambda = 0.26$

24.6 CONCLUDING REMARKS

We restricted ourselves to a simple sink (source). In [15] it was shown for the classical Hele-Shaw problem that more general driving mechanisms can be taken into account, namely, distributions with a compact support in $\Omega(0)$. Such a driving mechanism includes the physically interesting cases of a discrete set of sinks (sources) or multipoles, a continuous distribution of sinks (sources) along a piecewise smooth curve in $\Omega(0)$, respectively. Under this assumption the same main result can be proved. One expects the same generalization for models with kinetic undercooling regularization.

It is a pleasure for us to mention contributions of Prof.R.P.Gilbert to Hele-Shaw flows. Among other things he is interested in generalizations of the classical Hele-Shaw model to higher-dimensional cases. In [1] a generalization of model (M1) was proposed and following the approach of [8] the existence of global weak solutions was proved. The methods of [1] were applied to nonNewtonian Hele-Shaw flows in [2], in the model of type (M1) the Laplace operator is replaced by a self adjoint elliptic operator of the form $\sum_{i,k=1}^{n} \partial_{x_i}(a_{ik}\partial_{x_k}u) + cu$. But the existence of classical solutions remained open. As far as the authors know all attempts to apply the abstract Cauchy-Kovalevsky theory to higher-dimensional Hele-Shaw models failed up to now. New results of R.P. Gilbert are published in [6] and [7].

Finally, let us mention some papers about Hele-Shaw flows with surface tension regularization. In opposite to the models (M1) to (M4) these models cannot be interpreted as abstract Cauchy-Kovalevsky problems, the corresponding evolution problems are not type-independent. One has to use parabolic theory, see for example [4], [18] and the references therein. In [3] a model which includes both regularizations is studied. One has local existence and for special geometries of $\Omega(0)$ even global-in-time existence results. One can find some very interesting hypotheses for Hele-Shaw flows with surface tension and the role of this regularization in [13].

References

[1] Begehr, H. and Gilbert, R.P., Hele-Shaw type flows in R^n, Nonlinear Analysis **10** (1986), 65-85.

[2] Begehr, H. and Gilbert, R.P., Non-Newtonian Hele-Shaw flows in $n \geq 2$ dimensions, Nonlinear Analysis **11** (1987), 17-47.

[3] Deckelnick, K. and Elliott, Ch.M., Local and global existence results for anisotropic Hele-Shaw flows, Centre for Math. Anal. and Appl., Research Report 96/06, University of Sussex, Brighton.

[4] Escher, J. and Simonett, G., Classical solutions for Hele-Shaw models with surface tension, Adv. in Diff. Equ. **2** (1997) 4, 619-642.

[5] Galin, L.A., Unsteady filtration with a free surface (in Russian), Dokl. Akad. Nauk SSSR **47** (1945), 250-253.

[6] Gilbert, R.P. and Shi, P., Nonisothermal, nonNewtonian Hele-Shaw flows, Part I: Mathematical formulation, in: Proceedings on Transport Phenomenon, 1067-1088, Technomic, Lancaster, 1992.

[7] Gilbert, R.P. and Shi, P., Nonisothermal, nonNewtonian Hele-Shaw flows, Part II: Asymptotics and existence of weak solutions, Nonlinear Analysis **27**(1996), 539-559.

[8] Gustafsson, B., Applications of variational inequalities to a moving boundary problem for Hele-Shaw flows, SIAM J. Math. Analysis **16** (1985), 279-300.

[9] Hohlov, Yu.E. and Howison, S.D., Ill-posed moving boundary problems (in preparation).

[10] Hohlov, Yu.E., Howison, S.D., Huntingford, C., Ockendon, J.R. and Lacey, A.A., A model for non-smooth free boundaries in Hele-Shaw flows, Quart. J. Mech. Appl. Math. **47** (1994), 107-128.

[11] Hohlov, Yu.E. and Reissig, M., On classical solvability for the Hele-Shaw moving boundary problems with kinetic undercooling regularization, Euro. J. of Appl. Math. **6** (1995), 421-439.

[12] Howison, S.D. and Hohlov, Yu.E., About the classification of solutions for Hele-Shaw moving boundary problems (in Russian), Dokl. Rossi. Akad. Nauk **325** (1992), 1161-1166.

[13] Howison, S.D. and Richardson, S., Cusp development in free boundaries, and two-dimensional slow viscous flows, Euro. J. of Appl. Math. **6** (1995), 441-454.

[14] Hübner, F., Numerical treatment of Hele-Shaw models with and without regularization, Research Report, Technische Universität Bergakademie Freiberg, 1997.

[15] May, I., Untersuchungen zur Existenz und Eindeutigkeit von analytischen Lösungen für Hele-Shaw Strömungen mit kompakten Quellen und Senken, Diplomarbeit, Technische Universität Bergakademie Freiberg, 1994.

[16] Nishida, T., A note on a theorem of Nirenberg, J. Diff. Geom. **12** (1977), 629-633.

[17] Polubarinova-Kochina, P.Ya., On the motion of the oil contour (in Russian), Dokl. Akad. Nauk SSSR **47** (1945), 254-257.

[18] Prockert, G., Parabolic evolution equations for quasistationary free boundary problems in capillary fluid mechanics, PhD thesis, University of Eindhoven, 1996.

[19] Reissig, M., About the existence and uniqueness of analytic solutions for a moving boundary problem for Hele-Shaw flows in the plane, Nonlinear Analysis **23** (1994), 565-576.

[20] Reissig, M., About a complex model for Hele-Shaw flows with kinetic undercooling regularization (in preparation).

[21] Reissig, M. and v.Wolfersdorf, L., A simplified proof for a moving boundary problem for Hele-Shaw flows in the plane, Arkiv för matematik **31** (1993), 101-116.

[22] Richardson, S., Hele-Shaw flows with a free boundary produced by the injection of fluid into a narrow channel, J. Fluid Mech. **56** (1972), 609-618.

[23] Romero, L.A., The fingering problem in a Hele-Shaw cell, PhD thesis, California Institute of Technology, 1981.

[24] Tanveer, S., Evolution of Hele-Shaw interface for small surface tension, Phil. Trans. R. Soc. Lond. A **343** (1993), 155-204.

[25] Vinogradov, Yu.P. and Kufarev, P.P., On a filtration problem (in Russian), Prikl. Mat. Mech. **12** (1948) (English translation as University of Delaware Applied Mathematics Institute Technical Report **182A**, 1984).

25 GENERALIZED ANALYTIC FUNCTIONS IN MECHANICS

E. Obolashvili

Introduction

The famous mathematicians Leonard Euler, Pier Laplace noted that the nature is created according to the mathematical plan and its creater is the Lord-God. Classical partial differential equations of mathematical physics, electromagnetic fields, relativity quantum mechanics, heat processes and their some natural generalizations can be obtained using Dirac operator without applying any physical laws. This makes us think that Dirac operator with Clifford analysis is one of those mathematical tools with help of which the nature is designed.

The well-known Cauchy-Riemann operator really is Dirac operator in the space of one complex variable. Therefore generalized Cauchy-Riemann equations (GCRE) can be written using this operator. Remarkable in physics Maxwell and Dirac equations can be simply formulated in the language of the Clifford algebra [11], [7],. We will see that from these equations considered in the stationary case GCRE can be obtained [9].

In general the theory of generalized analytic functions (GAF), as the solutions of GCRE, has quite a wide range of applications. First note that it is closely related with quaziconformal mappings, the surface theory, the theory of elasticity, the shell theory and so on. It should be noted that I. Vekua considerably broadened and deepened the study of equilibrium of stressed membrane shells by the methods of the theory of GAF, besides their relation with the problem of infinitesimal surface bendings was

H. Florian et al. (eds.), Generalized Analytic Functions, 289–297.

given [14], [15]. In [1] the theory of GAF is used for the solutions of axially symmetric problems of elasticity.

The equilibrium equations of a membrane shells can be derived from the general equilibrium equations of an elastic body in any curvilinear coordinate system and then for the components of the stress tensor and the displacement vector of the membrane shells of positive curvature general complex representations [14] can be obtained [10]. Thus in this work we will consider only few applications of GAF in the membrane shell theory and in the electromagnetic fields and in the relativity quantum mechanics.

As below only elliptic and hyperbolic equations will be considered the Clifford algebras $R_{(n)}$ and $R_{(n,n-1)}(n \geq 1)$ will be used. They are associative real 2^n - dimensional noncommutative ($n \geq 2$) with the basis e_A, $A(\alpha_1 \dots \alpha_k)$, $0 \leq \alpha_1 < \dots < \alpha_k \leq n$ and with the multiplication rules.

$$e_0^2 = e_0, \quad e_j^2 = -e_0, \quad j = 1, 2, \dots, n-1$$

$$e_j e_k + e_k e_j = 0, \quad k \neq j, = 1, 2, \dots n, \tag{25.1}$$

$$e_n^2 = -e_0 \quad \text{in the case} \quad R_{(n)} \tag{25.2}$$

$$e_n^2 = e_0 \quad \text{in the case} \quad R_{(n,n-1)}, \tag{25.3}$$

where e_0 is identity element. For any element of these spaces two conjugations are defined:

$$\overline{u} = \sum_A u_A \overline{e}_A, \quad \widetilde{u} = \sum_A u_A \widetilde{e}_A \tag{25.4}$$

$$\overline{e}_0 = e_0, \quad \overline{e}_j = \widetilde{e}_j = -e_j, \quad j = 1, \dots n, \tag{25.5}$$

$$\overline{e}_A = \overline{e}_{\alpha_k} \dots \overline{e}_{\alpha_1}, \quad \widetilde{e}_A = \overline{e}_{\alpha_1} \dots \overline{e}_{\alpha_k}.$$

Consider the modification of the Dirac operator [4]

$$\overline{\partial} = \sum_0^n \frac{\partial}{\partial x_k} e_k = \frac{\partial}{\partial x_0} e_0 + D, \tag{25.6}$$

$$\partial = \frac{\partial}{\partial x_0} e_0 - D,$$

where D is Dirac operator. In any dimensional space the equation will be regarded:

$$\overline{\partial} u + \widetilde{u} h = 0 \tag{25.7}$$

It is obvious this equation in $R_{(n)}$ will be elliptic and in $R_{(n,n-1)}$ will be hyperbolic.The equation (25.7) is remarkable with that in one Clifford algebra it can be generalized Moisil-Theodorescu system, in the second algebra it can be Dirac or Maxwell equations, and as their partial cases metaparabolic equations or GCRE.

25.1 APPLICATIONS GAF IN THE MEMBRANE SHELL THEORY

In the classical membrane shell theory the stresses are completly defined by the condition of the shell equilibrium without taking into account the interdependence of stress and strains [6]. Membrane theory is frequently used for the approximate calculation of shells. By analogy with the case of a plane stressed state of the elastic body [8] the surface stressed state is defined and from the equilibrium equations of a continuous medium in a curvelinear coordinate system [12] the equations of equilibrium of membrane shells can be obtained [10].

Let S be a smooth surface, (x_1, x_2) the curvelinear coordinates of a point of S. It is assumed that each point x of the given elastic body can be uniquely charaterized by (x_1, x_2, x_3), where $|x_3|$ is the distance from x to S. The contravariant components of the stress tensor will be denoted by $F^{ik}(i, k = 1, 2, 3)$.

The stressed state of an elastic body will be called the surface parallel to S if at every point of the body one has the conditions.

$$F^{3k} = 0, \quad k = 1, 2, 3, \tag{25.8}$$

This assumption is invariant with respect to the transformation of the coordinates of the type

$$\overline{x}_\alpha = \overline{x}_\alpha(x_1, x_2), \quad \alpha = 1, 2,$$
$$\overline{x}_3 = x_3.$$

Then taking into account (25.8) one can obtain the equilibrium equations of an elastic body when it is in the surface stressed state. From there the equations of the classical theory of membrane shells can be obtained [10].

Let the curvature of the middle surface S of a shell be positive. Then on S there exists the conjugate isometric coordinate system (x_1, x_2), i.e. the second basic quadratic form of the surface will be represented as

$$d\sigma^2 = D(dx_1^2 + dx_2^2), \quad D > 0.$$

Without loss of generality the volume forces can be supposed zero. Then from the equilibrium equations for membrane shells for stress tensor general complex Vekua representations can be obtained:

$$F^{11} = -F^{22} = P\,\mathrm{Re}[w], \quad F^{12} = -P\,\mathrm{Im}\,w, \tag{25.9}$$

where w, called the stress complex function, is any solution of GCRE

$$\frac{\partial w}{\partial \overline{z}} + A\overline{w} = 0, \quad P = [a^{3/2}D]^{-\frac{1}{2}}, \quad A = \frac{1}{2}(a_1 - ia_2), \tag{25.10}$$

$$a_1 = \frac{1}{2}\frac{\partial ln\sqrt{a}D}{\partial x_1} - 2\Gamma_{12}^2, \quad a_2 = -\frac{1}{2}\frac{\partial ln\sqrt{a}D}{\partial x_2} + 2\Gamma_{12}^1, \tag{25.11}$$

a is the discriminant of the principal metric tensor of the surface S; Γ_{ik}^j-are Christoffel symbols of the second kind on S calculated by the metric tensor.

It is known, that $A \equiv 0$ is the necessary and sufficient condition for the middle surface of the shell to be the second order surface of positive curvature. The sufficiency of this condition was shown by I.N. Vekua (cf. [14, 15]) and necessity in [6]. It is interesting to define a class of surfaces of positive curvature for which $A = Const \neq 0$.

Defining stress tensor the strain tensor can be expressed by Hook's law for isotropic homogeneous elastic body. After that we will have the equations for displacement components u_k $(k = 1, 2)$ the soliutions of which will be represented in the form:

$$u_1 = q\,\mathrm{Re}[V], \quad u_2 = q\,\mathrm{Im}[V] \tag{25.12}$$

where $q = (\sqrt{a}D)^{1/2}$, V is the solution of the equation [14].

$$\frac{\partial V}{\partial \bar{z}} - \overline{AV} = 0 \tag{25.13}$$

V is called the complex displacement function. The equations (25.10) and (25.14) are adjoint.

In the classical membrane theory of shells of positive curvature two boundary conditions are given, one condition is static and the other one geometric. The static condition is used for defining the stress tensor and the geometric one for defining the displacement vector [14]. By force of representations (25.9), (25.12), the both cases can be reduced to the problems of defining GAF with some given boundary conditions. Therefore in mathematical point of view the problem to define stress tensor and the problem to define displacement for membrane shell are one and the same. The various boundary conditions are cinsidered in [10]. For the second order membrane shells with positive curvature many boundary value problems can be solved in quadratures while for any membrane shells though positive curvature they can not be solved explicitly, since the corresponding problems for GAF, even GCRE with constant coefficients,can not be solved but for holomorphic functions are solved. One of the main reasons is that the Riemann-Schwartz symmetric principle for holomorphic functions is just for many domains but for GHF is just only a few domains. These questions will be discussed below.

Let G be a domain in the plane $z = x + iy$, its boundary consist the segment L of the straight line $y = 0$ or the arc Γ of the circle $|z| = 1$ with the centre inside G.

Riemann-Schwartz principle: Let $\Phi(z) = u - iv$ be holomorphic function in G and

$$u(x, 0) = 0 \quad \text{or} \quad v(x, 0) = 0, \quad x \in L \tag{25.14}$$

then the function

$$\Psi(z) = \begin{cases} \Phi(z), & z \in G, \\ \mp\overline{\Phi(\bar{z})}, & z \in G^*, \end{cases} \tag{25.15}$$

where G^* is symmetric domain of G with respect to L, will be holomorphic in $G + G^* + L$, the sign $(-)$ is for the first condition in (1.7) and $(+)$ for the second one. But if we have the conditions

$$u(x, y) = 0, \quad \text{or} \quad v(x, y) = 0, \quad x + iy = t \in \Gamma \tag{25.16}$$

then

$$\Psi(z) = \begin{cases} \Phi(z), & z \in G, \\ \mp\overline{\Phi(z^*)}, & z \in G^*, \end{cases} \tag{25.17}$$

where in this case G^* is the symmetric domain of G with respect to Γ, i.e. it is mirror image to Γ, will be holomorphic in $G + G^* + \Gamma$. Let now L be a straight line $\alpha x + \beta y = 0$, besides without loss of generality one can assume $\alpha^2 + \beta^2 = 1$. The symmetric points with respect to L will be z and

$$z^* = -(\alpha + i\beta)^2 \bar{z} \tag{25.18}$$

It is easy to concluse that if $\Psi(z)$ is defined in the symmetric points correspondly as (25.17) with the conditions on L like (25.16), then for holomorphic functions the symmetric principle of reflection will be again just.

For GHF, i.e. for the solution of (25.10) with $A = a + ib$ when the boundary of G contains the arc of circle or the segment of straight line L where (α, β) and (a, b) are independent, the symmetric principle of reflection is not just. But it is possible to define (a, b) by (α, β) or vice versa in such a way that the symmetric principle will be just. Really, by the transformation (25.18) the equation (25.10) will be transformed into

$$\overline{\frac{\partial w(z^*)}{\partial \overline{z^*}}} - \overline{A}(\alpha + i\beta)^2 w(z^*) = 0 \tag{25.19}$$

If

$$\overline{A(z^*)}(\alpha + i\beta)^2 + A(z) = 0, \tag{25.20}$$

then for the function $\Psi(z) = \overline{w(z^*)}$ one will have the same equation (25.10) and with the conditions like (25.16) on L the above formulated principle will be right again. For instance, if L is the segment of the axis $y = 0$, then $z^* = \bar{z}$ and $A(z) - \overline{A(\bar{z})} = 0$, i. e. if A is constant then it must be real one. If L is the segment of $x = 0$, then $z^* = -\bar{z}$ and $A(z) + \overline{A(-\bar{z})} = 0$, i. e. if A is constant its real part must be zero. Thus for holomorphic functions the class of domains for which symmetric principle of reflection is just, is more wide then for GHF. That is why there exists many boundary value problems, which can be solved in quadratures for holomorphic functions, but for GHF it is impossible [9].

Note that for harmonic and metaharmonic functions one has the same situation, besides for harmonic and biharmonic functions too. For instanse in the quarter of the plane $(x > 0, y > 0)$ harmonic function with Dirichlet boundary condition vanishing at infinity can be defined in quadrature, but biharmonic function with condition when on the boundary it and its normal derivative are given, can not be defined in quadrature. The reason is that by these homogeneous conditions the symmetric principle is not right. For biharmonic function $u(x, y)$ this principle is right when, for instance, on $y = 0$ one has the conditions

$$u(x, 0) = 0, \quad \frac{\partial^2 u}{\partial y^2} = 0$$

Thus the symmetric principle of reflection has the important applications to solve boundary value problems.

25.2 MAXWELL AND DIRAC EQUATIONS AND THEIR RELATIONS TO GCRE

The Clifford algebra $R_{(3,2)}$ is especially interesting as it is defined over fourdimensional Minkovski space which is particularly important in the applications. First consider hyperbolic regular equation for $u(x) \in R_{(3,2)}$:

$$\overline{\partial} u = 0 \tag{25.21}$$

$$u(x) = u_0 e_0 - u_1 e_1 - u_2 e_2 - \varphi e_3 - \psi e_1 e_2 - u_3 e_3 e_1 - u_4 e_2 e_3 - u_5 e_{123} \tag{25.22}$$

Then by force of (25.1),(25.3),(25.6) one can obtain from (25.21):

$$\operatorname{div} E - \frac{\partial \varphi}{\partial x_3} = 0, \quad \operatorname{div} H + \frac{\partial \psi}{\partial x_3} = 0,$$
$$\operatorname{grad} \psi + rotE + \frac{\partial H}{\partial x_3} = 0, \quad \operatorname{grad} \phi + \operatorname{rot} H - \frac{\partial E}{\partial x_3} = 0, \tag{25.23}$$

where $E = (u_0, u_1, u_2)$, $H = (u_5, u_4, u_3)$, the operators div, grad, rot are taken with respect to x_0, x_1, x_2. These equations are invariants with respect to the Lorentz transformations. It does not mean that each equation in them is individually invariant, but after transformation of (25.21) as some basis will be transformed too, one will have the equations as (25.23).

If $\psi = \varphi = 0$, (25.23) will be well-known Maxwell equations of electromagnetic fields. Thus (25.23) can be considered as generalization of Maxwell equations. Now we will see that (25.23) are connected with Dirac equations of relativity quantum mechanics of the electron. Consider only the principal part of Dirac equations in the form as it is given in the famous monograph by R.Courant and D.Hilbert [5]

$$[\frac{\partial}{\partial t} - \sum_{0}^{2} \alpha_k \frac{\partial}{\partial x_k}]u = 0, \quad t \equiv x_3 \tag{25.24}$$

where $u(u_0^1, u_1^1, u_2^1, u_3^1)$ is four-component complex vector of the variables x_0, x_1, x_2, t, α_k are 4×4 matrices defined as

$$\alpha_0 = \begin{pmatrix} 0 & 0 & 0 & 1 \\ 0 & 0 & 1 & 0 \\ 0 & 1 & 0 & 0 \\ 1 & 0 & 0 & 0 \end{pmatrix}, \quad \alpha_1 = \begin{pmatrix} 0 & 0 & 0 & -i \\ 0 & 0 & i & 0 \\ 0 & -i & 0 & 0 \\ i & 0 & 0 & 0 \end{pmatrix}, \quad \alpha_2 = \begin{pmatrix} 0 & 0 & 1 & 0 \\ 0 & 0 & 0 & -1 \\ 1 & 0 & 0 & 0 \\ 0 & -1 & 0 & 0 \end{pmatrix}.$$

It is easy to see that $\alpha_k^2 = e_0$, $\alpha_k \alpha_j + \alpha_j \alpha_k = 0$, $j \neq k = 0, 1, 2$. and are related to the known Dirac's Matrices. Representing $u_k^1 = u_k + i v_k (k = 0, 1, 2, 3)$ and multiplying matrices on vectors, the equation (25.24) can be written as real eight equations. They will not be brought here for their long expression. One can directly check that there exists four possibilities to write these eight real equations in the form (25.23):

1. $E = (u_0, -v_0, -u_1)$, $\quad H = (-v_2, -u_2, v_3)$, $\quad v_1 = \psi$, $\quad u_3 = \varphi$

2. $E = (v_0, u_0, -v_1), \quad H = (u_2, -v_2, -u_3), \quad u_1 = -\psi, \quad v_3 = \varphi$
3. $E = (v_1, -u_1, v_0), \quad H = (u_3, v_3, u_2), \quad v_2 = \varphi, \quad u_0 = -\psi$
4. $E = (u_1, v_1, u_0), \quad H = (-v_3, u_3, -v_2), \quad v_0 = \psi, \quad u_2 = \varphi$

Thus Dirac equations of quantum mechanics can be considered as generalization of Maxwell equations (25.23) written in vectorial form.

Some scientist physics are suprised that Dirac equations (25.24) can be written in the vectorial form (25.23) and have the relation to the Maxwell equations.

Particular cases. 1. Let all quantities in (25.23) depend only on x_0, x_1, x_2, i.e. the stationary process will be considered. Then it will be two separated Moisil-Theodorescu systems. Probably for this reason last time this system is called by some mathematicians as Maxwell equations. 2. Let all quantities in (25.23) depend only on two space coordinates, for instance on x_0, x_1 and on the time variable $t \equiv x_3$. Then (25.23) will represent two separated systems for the quantities (u_0, u_1, u_3, φ) and $(u_5, u_4, -u_2, -\psi)$ respectively, besides, both systems have the same form. Thus it is sufficient to consider only one of them:

$$\begin{aligned} &\frac{\partial u_0}{\partial x_0} + \frac{\partial u_1}{\partial x_1} - \frac{\partial \varphi}{\partial t} = 0, \quad \frac{\partial \varphi}{\partial x_0} + \frac{\partial u_3}{\partial x_1} - \frac{\partial u_0}{\partial t} = 0, \\ &\frac{\partial u_0}{\partial x_1} - \frac{\partial u_1}{\partial x_0} - \frac{\partial u_3}{\partial t} = 0, \quad \frac{\partial \varphi}{\partial x_1} - \frac{\partial u_3}{\partial x_0} - \frac{\partial u_1}{\partial t} = 0. \end{aligned} \tag{25.25}$$

Consider the complex functions $P = u_0 + \varphi - i(u_1 + u_3)$, $Q = u_0 - \varphi - i(u_1 - u_3)$, and the complex variable $z = x_0 + ix_1$. Then (25.25) will be equivalent to two separated complex equations:

$$\begin{aligned} &2\frac{\partial P}{\partial \bar{z}} - \frac{\partial \overline{P}}{\partial t} = 0, \\ &2\frac{\partial Q}{\partial \bar{z}} + \frac{\partial \overline{Q}}{\partial t} = 0. \end{aligned}$$

These equations are particular case of the metaparabolic equations general form of which was considered in [3]. Thus Maxwell or Dirac equations with two space variables and time variable are related to metaparabolic equations.

Now consider the equation (25.7) with $u(x) \in R_{3.2)}$ defined by (25.22) and $h = h_0 e_0 - \sum_1^3 h_k e_k$. Then by force of (25.1), (25.3),(25.6) it can be written in the form:

$$\begin{aligned} &\operatorname{div} E - \frac{\partial \varphi}{\partial t} + (E \cdot A) - \varphi h_3 = 0 \\ &\operatorname{grad} \psi + \operatorname{rot} E + \frac{\partial H}{\partial t} + [E \times A] + \psi A + H h_3 = 0 \\ &\operatorname{div} H + \frac{\partial \psi}{\partial t} - (H \cdot A) - \psi h_3 = 0 \\ &\operatorname{grad} \varphi + \operatorname{rot} H - \frac{\partial E}{\partial t} - [H \times A] - \varphi A + E h_3 = 0 \end{aligned} \tag{25.26}$$

where $A = (h_0, h_1, h_2)$, E and H are defined as in (25.23). These equations can be considered as generalized Maxwell or Dirac equations. If all equations in (25.26) do not depend on t and $h_3 = 0$, then it will be two separated generalized Moisil-Theodorescu systems; And besides if all quantities do not depend on x_2 and $h_2 = 0$, then (25.26) will be four separated GCRE. May be it means that the processes characterized by the equations (25.26) considering only with two space variables (x_0, x_1) can be represented by four independent stationary processes in the plane (x_0, x_1) defined by GCRE. That is why to study the properties of GAF and to solve correctly posed boundary value problems for them are very applicable.

I would like to express my gratitude to professor Dr. Wolfgang Tutschke for offering me the idea to write this paper about applications of GAF in mechanics. I do not think that all possible applications are pointed out above. The properties of holomorphic and generalized holomorphic functions are so remarkable and visible that their applications can not be exhausted.

References

[1] Aleksandrov A. and Solov'ev J., Space problems of the theory of elasticity (Applications of the theory of complex variable functions), M. "Nauka", 1978 (russian).

[2] Begehr H.,Complex Analytic Methods for Partial Differential Equations. World scientific, 1994.

[3] Begehr H. and Gilbert R. Transformations,Transmutations and kernel functions. Vol. 1, Longman, scientific and Technical, 1992.

[4] Brackx F., Delanghe R.,Sommen F., Cliffrd Analysis. Pitman, 1982.

[5] Courant R., Hilbert D., Methods of Mathematical physics, II., New-York, 1962.

[6] Flugge, W. Static and Dynamic der Schalen. Springer, Berlin, 1957.

[7] Hestenes D., Space-Time Algebra. Gordon and Breach, New-York, 1966.

[8] Muskhelishvili N.,Singular Integral Equations. Groningen, Noordhoff 1953.

[9] Obolashvili E., Some *PDE* in Clifford Analysis. Banach Center Publications, v.37, 1996, 173-176.

[10] Obolashvili E.,Some Applications of *GAF* in the shell theory. Complex Analysis, Trends and Applications. (ed. E. Lanckau, W.Tutschke) Akademie-Verlag, Berlin, 1983.

[11] Riesz H., Clifford numbers and spinors. Lecture series 38, Maryland, 1958.

[12] Sneddon J. and Berry D., The classical theory of elasticity. Springer, Berlin, 1958.

[13] Tutschke W., Solutions of Initial Value problems in classes of Generalized Analytic functions. Leipzig, Teubner-Text zur Math. 1989.

[14] Vekua I., Generalized Analytic functions. Pergamon Press, Oxford, 1962.

[15] Vekua I., Shell theory, General methods of construction. Pitman, Boston..., 1985.

26 INTEGRABILITY AND THE SCHRÖDINGER EQUATION

R. Carroll

Herrn R. P. Gilbert zu seinem 65. Geburtstag gewidmet

26.1 INTRODUCTION

We sketch here the author's embellishments to a paper [11] by Faraggi and Matone with title "Duality of X and ψ and a statistical interpretation of space in quantum mechanics" and add a number of new observations. In particular ideas from dispersionless KP and KdV are used (essentially a WKB device here) which allow one to relate a "canonical" symplectic form $dX \wedge dP$ to $[-i\epsilon/2(\Re\mathcal{F})^2]d\Im\mathcal{F} \wedge d\Re\mathcal{F}$, where $\mathcal{F}$ is a "prepotential" introduced in [11], motivated by work on Seiberg-Witten (SW) duality. In fact the main feature of the background mathematics associated with KdV or KP is to provide additional structure allowing the introduction of X into the theory in a manner commensurate with its role in [11]. Further, we can relate the framework here to work of Olave [17] which develops a context connecting classical and quantum mechanics (QM). The presentation here is based on [2] where more details can be found. Some of the connections are incomplete in the last step $QM \to WKB \to KdV$ but are complete going the other way. For references to duality ideas stemming from SW theory see also to [1, 5, 9, 10, 12, 14, 15] (with apologies for omissions).

H. Florian et al. (eds.), Generalized Analytic Functions, 299–311.

26.2 BACKGROUND

The point of departure is the Schrödinger equation

$$\mathcal{H}\psi_E = -\frac{\hbar^2}{2m}\psi_E'' + V(X)\psi_E = E\psi_E \tag{26.1}$$

where X is the quantum mechanical (QM) space variable with $\psi_E' = \partial\psi_E/\partial X$ and we write $\epsilon = \hbar/\sqrt{2m}$ (E is assumed real). In [2] we discussed the possible origin of this from a Kadomtsev-Petviashvili (KP) situation $L_+^2\psi = \partial\psi/\partial t_2$ where $L_+^2 = \partial_x^2 - v(x,t_i)$ and e.g. $\tau_2 = -i\sqrt{2m}T_2$ so $\partial_{t_2} = \epsilon\partial_{T_2} = -i\hbar\partial_{\tau_2}$ (one writes $X = \epsilon x$ and $T_i = \epsilon t_i$ in the dispersionless theory). As seen below the format of dispersionless theory is related to WKB methods and in fact we will expand the standard dispersionless theory in the WKB direction. This leads to an approximation

$$\epsilon^2\psi_E'' - V(X,T_i)\psi_E \sim \epsilon\frac{\partial\psi_E}{\partial T_2} = -i\hbar\frac{\psi_E}{\partial\tau_2} \tag{26.2}$$

corresponding to the Schrödinger equation. This is also related to the Korteweg-deVries (KdV) equation and it's dispersionless form dKdV as indicated below. For the approximation one assumes e.g. $v = v(x,t_i) \to v(X/\epsilon, T_i/\epsilon) = V(X,T_i) + O(\epsilon)$. This is standard in dispersionless KP = dKP and certainly realizable by quotients of homogeneous polynomials for example. In fact it is hardly a restriction since given e.g. $F(X) = \sum_0^\infty a_n X^n$ consider $\tilde{f}(x,t_i) = a_0 + \sum_1^\infty (x^n/\prod_2^{n+1} t_i)$. Then $\tilde{f}(X/\epsilon, T_i/\epsilon) = a_0 + \sum_1^\infty (X^n/\prod_2^{n+1} T_i)$ and one can choose the T_i recursively so that $1/T_1 = a_1$, $1/T_1T_2 = a_2, \cdots$, leading to $F(X) = \tilde{F}(X,T_i)$. Further, when $\psi_E = exp(S/\epsilon)$ for example, one has $\epsilon\psi_E' = S_X\psi_E$ with $\epsilon^2\psi_E'' = \epsilon S_{XX}\psi_E + (S_X)^2\psi_E$ so in (26.2) we are neglecting an $O(\epsilon)\psi_E$ term from v, and for $\psi_E = exp(S/\epsilon)$ another $\epsilon S_{XX}\psi_E$ term is normally removed in dispersionless theory. Then for $\mathcal{H}$ independent of τ_2 for example one could assume V is independent of T_2 and write formally in (26.2), $\hat{\psi}_E = exp(E\tau_2/i\hbar)\cdot\psi_E$, with $\mathcal{H}\psi_E = E\psi_E$, which is (26.1). Since in the QM problem one does not however run $\hbar \to 0$ (hence $\epsilon \not\to 0$) one should argue that these $O(\epsilon)$ terms should be retained, and we will develop this approach, which essentially corresponds to WKB (with some background structure). In particular one could ask for $v(X/\epsilon, T_i/\epsilon) = V(X,T_i) + \epsilon\hat{V}(X,T_i) + O(\epsilon^2)$ and retain the $\epsilon\hat{V}$ term along with ϵS_{XX}, in requiring e.g. $S_{XX} = \hat{V}$ (this is covered below - an additional term also arises). In fact the passage from $v \to V$ or $V + \epsilon\hat{V}$ is the only "assumption" in our development and this admits various realizations; the impact here only involves some possible minor restrictions on the class of quantum potentials to which the theory applies. The background mathematics behind V determined by KP or KdV essentially generates some additional structure which allows us to insert X into the theory in a manner commensurate with its role in [11]. The formulation of [11] then entails some constraints on the background objects as indicated in the text. We emphasize that inserting S is familiar from WKB; we are introducing in an ad hoc manner additional variables T_i or T_i, λ or k, etc. to spawn a KP or KdV theory. We do not assume or even suggest that this is in any way connected a priori with the physics of the quantum mechanical problem (although of course it conceivably could be since integrability ideas are important in quantum mechanics). This procedure generates a nice Hamilton-Jacobi (HJ) theory which guides one to insert X into the machinery, but the insertion itself is at "ground level" and simply reflects a WKB formulation; neither the underlying KdV or KP dynamics nor the HJ theory is directly used here. Once X is involved connections to [11] are immediate. Actually the procedure could be reversed as a way of introducing duality ideas into the ϵ-dispersionless theory of the paper (or [2]) and this should probably be related to the duality already studied in Whitham theory (cf. [5, 9, 12, 14, 15]). Thus start with KdV or KP, go to the Schrödinger equation and $dKdV_\epsilon$ or dKP_ϵ, develop the HJ theory, and then use [11] to create duality; more generally start from a finite zone KdV situation with associated Whitham dynamics on a Riemann surface and compare dualities.

We list first a few of the equations from [11], as written in [2], without a discussion of philosophy (some of which will be mentioned later). Thus $\mathcal{F}$ is a prepotential and, since E is real, ψ_E and $\bar{\psi}_E = \psi_E^D$ both satisfy (26.1) with $\psi_E^D = \partial\mathcal{F}/\partial\psi_E$. The Wronskian in (26.1) is taken to be $W = \psi'\bar{\psi} - \psi\bar{\psi}' = 2\sqrt{2m}/i\hbar = 2/i\epsilon$ and one has ($\psi = \psi(X)$ and $X = X(\psi)$ with $X_\psi = \partial X/\partial\psi = 1/\psi'$)

$$\mathcal{F}' = \psi'\bar{\psi};\ \mathcal{F} = \frac{1}{2}\psi\bar{\psi} + \frac{X}{i\epsilon};\ \frac{\partial\bar{\psi}}{\partial\psi} = \frac{1}{\psi}\left[\bar{\psi} - \frac{2}{i\epsilon}X_\psi\right] \tag{26.3}$$

(ψ always means ψ_E but we omit the subscript occasionally for brevity). Setting $\phi = \partial\mathcal{F}/\partial(\psi^2) = \bar{\psi}/2\psi$ with $\partial_\psi = 2\psi\partial/\partial(\psi^2)$ and evidently $\partial\phi/\partial\psi = -(\bar{\psi}/2\psi^2) + (1/2\psi)(\partial\bar{\psi}/\partial\psi)$ one has a Legendre transform pair

$$-\frac{X}{i\epsilon} = \psi^2\frac{\partial\mathcal{F}}{\partial(\psi^2)} - \mathcal{F};\ -\mathcal{F} = \phi\frac{1}{i\epsilon}X_\phi - \frac{X}{i\epsilon} \tag{26.4}$$

One obtains also (♠ • ♠) $|\psi|^2 = 2\mathcal{F} - (2X/i\epsilon)$ $(\mathcal{F}_\psi = \bar{\psi})$; $-(1/i\epsilon)X_\phi = \psi^2$; $\mathcal{F}_{\psi\psi} = \partial\bar{\psi}/\partial\psi$. Further from $X_\psi\psi' = 1$ one has $X_{\psi\psi}\psi' + X_\psi^2\psi'' = 0$ which implies

$$X_{\psi\psi} = -\frac{\psi''}{(\psi')^3} = \frac{1}{\epsilon^2}\frac{(E-V)\psi}{(\psi')^3} \tag{26.5}$$

$$\mathcal{F}_{\psi\psi\psi} = \frac{E-V}{4}(\mathcal{F}_\psi - \psi\partial_\psi^2\mathcal{F})^3 = \frac{E-V}{4}\left(\frac{2X_\psi}{i\epsilon}\right)^3 \tag{26.6}$$

Although a direct comparison of (26.6) to the Gelfand-Dickey resolvant equation ((♣♣) below) is not evident (V' is lacking) a result of T. Montroy which expands $\mathcal{F}_{\psi\psi\psi}$ shows that in fact (26.6) corresponds exactly to

$$\epsilon^2\mathcal{F}''' + 4(E-V)\left(\mathcal{F}' - \frac{1}{i\epsilon}\right) - 2V'\left(\mathcal{F} - \frac{X}{i\epsilon}\right) = 0 \tag{26.7}$$

which is (♣♣) since $\Xi = |\psi|^2 = 2\mathcal{F} - (2X/i\epsilon)$.

Next there is a so-called eikonal transformation (cf. [16]) which can be related to

[11] as in [2]. We consider real A and S with $\psi = Ae^{(i/\hbar)S}$; $p = ASin[(1/\hbar)S]$; $q = ACos[(1/\hbar)S]$. Then introducing new variables $\chi = A^2 = |\psi|^2$; $\xi = (1/2\hbar)S$ it follows that there will be a Hamiltonian format with symplectic form (♠) $\delta p \wedge \delta q = \delta\xi \wedge \delta\chi = \tilde{\omega}$. It is interesting to write down the connection between the (S, A) or (χ, ξ) type variables and the variables from [11] and it will be useful to take now $\psi = Aexp(iS/\epsilon)$ ($\epsilon = \hbar/\sqrt{2m}$) with $\xi \sim S/2\epsilon$. Then

$$\mathcal{F} = \frac{1}{2}\chi + \frac{X}{i\epsilon};\ \mathcal{F}' = \psi'\bar{\psi} = \frac{1}{2}\chi' + \frac{i}{\epsilon}P\chi \tag{26.8}$$

for $S' = S_X = P$ and there is an interesting relation (♣) $P\chi = -1 \Rightarrow \delta\chi = -(\chi/P)\delta P$. Further from $\phi = (1/2)exp[-(2i/\epsilon)S]$ and $\psi^2 = \chi exp(4i\xi)$ we have

$$\psi^2\phi = \frac{1}{2}\chi = -\frac{1}{\epsilon}\phi X_\phi;\ \xi = \frac{S}{2\epsilon} = \frac{i}{4}log(2\phi) \tag{26.9}$$

Now the theory of the Seiberg-Witten (SW) differential λ_{SW} following [2, 5, 9, 10, 12, 14, 15] for example involves finding a differential λ_{SW} of the form QdE or $td\omega_0$ (in the spirit of [14] or [10, 12] respectively) such that $d\lambda_{SW} = \omega$ is a symplectic form. In the present context one can ask now whether the form $\tilde{\omega}$ of (♠) makes any sense in such a context. Evidently this is jumping the gun since there is no Riemann surface in sight (see however [2] for a Riemann surface with some justification as in [9]); the motivation to consider the matter here comes from the following formulas which express $\tilde{\omega}$ nicely in terms of the duality variables of [11] (another version of a "canonical" symplectic form in terms of $\mathcal{F}$ alone is given below). Thus a priori $\psi = \Re\psi + i\Im\psi$ has two components which are also visible in $\psi = Aexp(iS/\epsilon)$ as A and S. The relation $P\chi = \chi(\partial S/\partial X) = -1$ indicates a dependence between A and S' (but not A and S) which is a consequence of the duality between ψ and X. Then $2AS'\delta A + A^2\delta S' = 0$ or $\delta S' = -(2S'/A)\delta A \equiv (\delta S'/S') = -2(\delta A/A)$, whereas $\delta\psi/\psi \sim 2(\delta A/A) + (i/\epsilon)\delta S$. It follows that $\Re(\delta\psi/\psi) = -(\delta S'/S')$ and $\Im(\delta\psi/\psi) = (\delta S/\epsilon)$. The sensible thing seems to be to look at the complex dependence of $X(\psi)$ and $\psi(X)$ in terms of two real variables and $\delta\xi \wedge \delta\chi$ will have a nice form in transforming to the variables of [11]. In particular from $\psi^2\phi = (1/2)\chi$ with $\delta\chi = 4\phi\psi\delta\psi + 2\psi^2\delta\phi$ we obtain $(\delta\psi/\psi) = 2(\delta\chi/\chi) - (\delta\phi/\phi)$. Hence one can write

$$\delta\xi \wedge \delta\chi = \frac{i}{4}\frac{\delta\phi}{\phi} \wedge \chi\frac{\delta\chi}{\chi} = \frac{i}{2}\delta\phi \wedge \delta\psi^2 \sim \frac{i}{2}\delta\bar{\psi} \wedge \delta\psi \tag{26.10}$$

(note $\delta\phi = (1/2\phi)\delta\bar{\psi} - (\bar{\psi}/2\psi^2)\delta\psi$) and in an exploratory spirit the differentials $\lambda = (i/2)\phi\delta\psi^2$ or $\lambda = (i/2)\psi^2\delta\phi$, along with $\lambda = (i/2)\bar{\psi}\delta\psi$ or $\lambda = (i/2)\psi\delta\bar{\psi}$, might merit further consideration.

We refer now to [6, 7, 8, 18] for dispersionless KP (= dKP) and consider here $\psi = exp[(1/\epsilon)S(X, T, \lambda)]$ instead of $\psi = Aexp(S/\epsilon)$ (more details are given in Section 3). Thus $P = S' = S_X$ and $P^2 = V - E$ but $E \neq \pm\lambda^2$ (unless otherwise stated) and this does not define S via $P = S_X$ unless we have a KdV situation (which does not seem a priori desirable but will be developed later with modifications of the dispersionless theory); thus generally λ is the λ of $S(T_n,\lambda)$ from KP theory and we recall that ψ always means ψ_E as in [11]. Some routine calculation yields (recall $X_\psi = 1/\psi'$ and $\psi' = (P/\epsilon)\psi$)

$$\phi = \frac{1}{2}e^{-(2i/\epsilon)\Im S};\ \frac{1}{\epsilon}X_\phi = -ie^{(2/\epsilon)S};\ X_\psi = \frac{\epsilon}{P}e^{-S/\epsilon} \tag{26.11}$$

$$\frac{1}{\epsilon}X_{\psi\psi} = \frac{E-V}{P^3}e^{-S/\epsilon};\ \mathcal{F}_\psi = \bar{\psi} = e^{\bar{S}/\epsilon};\ \mathcal{F}_{\psi\psi} = e^{-(2i/\epsilon)\Im S} - \frac{2}{iP}e^{-2S/\epsilon} \tag{26.12}$$

$$|\psi|^2 = e^{(2/\epsilon)\Re S};\ \frac{S}{\epsilon} = \frac{1}{2}log|\psi|^2 - \frac{1}{2}log(2\phi);\ \bar{P} = \bar{S}_X = P - \frac{2}{i\psi\bar{\psi}} \tag{26.13}$$

Summarizing one has

$$\Im\mathcal{F} = -\frac{X}{\epsilon};\ \Re\mathcal{F} = \frac{1}{2}|\psi|^2 = \frac{1}{2}e^{\frac{2}{\epsilon}\Re S} = -\frac{1}{2\Im P} \tag{26.14}$$

In the present situation $|\psi|^2 = exp[(2/\epsilon)\Re S]$ and $2\phi = exp[-(2i/\epsilon)\Im S]$ can play the roles of independent variables (cf. (26.13). The version here of $P\chi = -1$ is $\chi\Im P = -1$, while $\psi^2\phi = (1/2)|\psi|^2 = (1/2)\chi$ again, and we obtain as above the formula (26.10). Now note that for $L = \partial + \sum_1^\infty u_i\partial^{-i}$, $L_+^2 = \partial^2 + 2u_1$, and $u_1 = \partial^2 log(\tau)$ where τ is the famous tau function (cf. Section 3). This implies $v = -2\partial^2 log(\tau)$ here, from which $V = -2F_{XX}$ for $\tau = exp[(1/\epsilon^2)F]$ in the dispersionless theory. We recall also the Gelfand- Dickey resolvant equation (cf. [3]) for $\Xi = \psi\bar{\psi}$, namely, in the present notation (♣♣) $\epsilon^2\Xi''' - 4V\Xi' - 2V'\Xi + 4E\Xi' = 0$ (direct calculation). Using $\Xi = 2\mathcal{F} - (2X/i\epsilon)$, $\Xi' = 2\mathcal{F}' - (2/i\epsilon)$, $\Xi'' = 2\mathcal{F}''$, and $\Xi''' = 2\mathcal{F}'''$, we obtain then from (♣♣) (cf. also (26.7))

$$\epsilon^2\mathcal{F}''' + \left(\mathcal{F}' - \frac{1}{i\epsilon}\right)(8F'' + 4E) + 4F'''\left(\mathcal{F} - \frac{X}{i\epsilon}\right) = 0 \tag{26.15}$$

which provides a relation between F and $\mathcal{F}$. We will see in Section 3 how to embellish all this with a new modification of the dKP and dKdV theory.

26.3 DISPERSIONLESS THEORY

We give next a brief sketch of some ideas regarding dispersionless KP (dKP) following mainly [6, 7, 8, 13, 18] to which we refer for philosophy. We will make various notational adjustments as we go along and subsequently will modify some of the theory. One can think of fast and slow variables with $\epsilon x = X$ and $\epsilon t_n = T_n$ so that $\partial_n \to \epsilon\partial/\partial T_n$ and $u(x,t_n) \to \tilde{u}(X,T_n)$ to obtain from the KP equation $(1/4)u_{xxx} + 3uu_x + (3/4)\partial^{-1}\partial_2^2 u = 0$ the equation $\partial_T\tilde{u} = 3\tilde{u}\partial_X\tilde{u} + (3/4)\partial^{-1}(\partial^2\tilde{u}/\partial T_2^2)$ when $\epsilon \to 0$ $(\partial^{-1} \to (1/\epsilon)\partial^{-1})$. For the underlying mathematical theory write (t_n) for

(x, t_n) (i.e. $x \sim t_1$ here) and consider $L_\epsilon = \epsilon\partial + \sum_1^\infty u_{n+1}(\epsilon, T)(\epsilon\partial)^{-n}$. Here L is the Lax operator and one takes now

$$\psi = exp\left(\frac{1}{\epsilon}S(T,\lambda) + O(1)\right); \; \tau = exp\left(\frac{1}{\epsilon^2}F(T) + O\left(\frac{1}{\epsilon}\right)\right) \tag{26.16}$$

We recall that $\partial_n L = [B_n, L]$, $B_n = L_+^n$, $L\psi = \lambda\psi$, and $\psi = \tau(T - (1/n\lambda^n))\cdot exp[\sum_1^\infty T_n\lambda^n]/\tau(T)$. Putting in the ϵ and using ∂_n for $\partial/\partial T_n$ now, with $P = S_X$, one obtains (•) $\lambda = P + \sum_1^\infty U_{n+1}P^{-n}$; $P = \lambda - \sum_1^\infty P_i\lambda^{-1}$; $\partial_n S = \mathcal{B}_n(P)$ which implies $\partial_n P = \hat{\partial}\mathcal{B}_n(P)$. Further $\hat{\partial} \sim \partial_X + (\partial P/\partial X)\partial_P$ and $B_n = \sum_0^n b_{nm}\partial^m$ so one has $\mathcal{B}_n = \sum_0^n b_{nm}P^m$. We list a few additional formulas which are easily obtained (cf. [6]); thus, writing $\{A, B\} = \partial_P A\partial A - \partial A\partial_P B$ one has $\partial_n\lambda = \{\mathcal{B}_n, \lambda\}$ and we can write $S = \sum_1^\infty T_n\lambda^n + \sum_1^\infty S_{j+1}\lambda^{-j}$ with $S_{n+1} = -(\partial_n F/n)$, $\partial_m S_{j+1} = -(F_{mn}/n)$, $V_{n+1} = -nS_{n+1}$, with

$$\mathcal{B}_n = \lambda^n + \sum_1^\infty \partial_n S_{j+1}\lambda^{-j}; \; \partial S_{n+1} \sim -P_n \sim -\frac{\partial V_{n+1}}{n} \sim -\frac{\partial\partial_n F}{n} \tag{26.17}$$

We sketch next a few formulas from [5, 6, 13]. First it will be convenient to rescale the T_n variables and write $t' = nt_n$, $T'_n = nT_n$, $\partial_n = n\partial'_n = n(\partial/\partial T'_n)$. Then $\partial'_n S = \lambda_+^n/n$; $\partial'_n\lambda = \{\mathcal{Q}_n, \lambda\}$ ($\mathcal{Q}_n = \mathcal{B}_n/n$); etc. Now think of (P, X, T'_n), $n \geq 2$, as basic Hamiltonian variables with $P = P(X, T'_n)$. Then $-\mathcal{Q}_n(P, X, T'_n)$ will serve as a Hamiltonian via (♠♠) $\dot{P}'_n = dP'/dT'_n = \partial\mathcal{Q}_n$; $\dot{X}'_n = dX/dT'_n = -\partial_P\mathcal{Q}_n$. The function $S(\lambda, X, T_n)$ plays the role of part of a generating function $\hat{S}$ for the Hamilton-Jacobi (HJ) theory with action angle variables $(\lambda, -\xi)$ where $d\lambda/dT'_n = \dot{\lambda}'_n = \partial_\xi R_n = 0$; $d\xi/dT'_n = \dot{\xi}'_n = -\partial_\lambda R_n = -\lambda^{n-1}$ (note that $\dot{\lambda}'_n = 0 \sim \partial'_n\lambda = \{\mathcal{Q}_n, \lambda\}$). The motivation here for HJ theory is to insert X from the dispersionless context into the framework in fundamental manner, commensurate with its role in [11] (cf. also [4, 6]). For KdV one looks at the dispersionless theory based on k where $\lambda^2 \sim (ik)^2 = -k^2$. There results, for $P = S_X$, the formula $P^2 + q = -k^2$, and we write $\mathcal{P} = (1/2)P^2 + p = (1/2)(ik)^2$ with $q \sim 2p \sim 2u_2$. One has $\partial k/\partial T_{2n} = \{(ik)^{2n}, k\} = 0$ and from $ik = P(1 + qP^{-2})^{1/2}$ we obtain

$$ik = P\left(1 + \sum_1^\infty \binom{\frac{1}{2}}{m} q^m P^{-2m}\right); \; P = ik - \sum_1^\infty P_j(ik)^{-j} \tag{26.18}$$

(cf. (•) with $u_2 = q/2$). The flow equations become then $\partial'_{2n+1}P = \hat{\partial}\mathcal{Q}_{2n+1}$; $\partial'_{2n+1}(ik) = \{\mathcal{Q}_{2n+1}, ik\}$. Note here some rescaling is needed since we want $(\partial^2 + q)_+^{3/2} = \partial^3 + (3/2)q\partial + (3/4)q_x = B_3$ instead of our previous $B_3 \sim 4\partial^3 + 6q\partial + 3q_x$. Thus we want $\mathcal{Q}_3 = (1/3)P^3 + (1/2)qP$ to fit the notation above.

Now the dKP theory as in [6, 7, 13, 18] involves a parameter $\epsilon \to 0$ and we recall $L = \partial + \sum_1^\infty u_{n+1}(t)\partial^{-n} \to L_\epsilon = \epsilon\partial + \sum_1^\infty u_{n+1}(\epsilon, T)(\epsilon\partial)^{-n}$ where $t \sim (t_k)$, $T \sim (T_k)$, and $X = T_1$ with $u_{n+1}(\epsilon, T) = U_{n+1}(T) + O(\epsilon)$ as in Section 2. Then for

$\psi = exp(S/\epsilon)$ one has $L\psi = \lambda\psi \to \lambda = P + \sum_1^\infty U_{n+1}P^{-n}$ where $P = S_X$ with $S = S(X, T_k, \lambda)$ ($k \geq 2$). Here all the terms which are $O(\epsilon)$ are passed to zero and in view of $\epsilon \not\to 0$ in the QM situation where $\epsilon = \hbar/\sqrt{2m}$ one thinks of rewriting some of the dKP theory in order to retain $O(\epsilon)$ terms at least (and dropping $O(\epsilon^2)$ terms). We will call this dKP_ϵ theory and it essentially corresponds to WKB with the proviso that there is a background mathematics providing some additional structure. We recall now $S = \sum_1^\infty T_n\lambda^n + \sum_1^\infty S_{j+1}\lambda^{-j}$; $\mathcal{B}_n = \partial_n S = \lambda^n + \sum_1^\infty \partial_n S_{j+1}\lambda^{-j}$ and via $log\psi = (S/\epsilon) + O(1) \sim log\tau[\epsilon, T_n - (\epsilon/n\lambda^n)] - log\tau + \sum_1^\infty T_n\lambda^n/\epsilon$ with $log\tau = (F/\epsilon^2) + O(1/\epsilon)$ and $F[T_n - (\epsilon/n\lambda^n)] - F(T_n) \sim -\epsilon\sum_1^\infty(\partial_n F/n\lambda^n) + O(\epsilon^2)$ one obtains $S_{n+1} = -(\partial_n F/n)$. Consider now the next order terms via F, i.e.

$$F\left(T_n - \frac{\epsilon}{n\lambda^n}\right) - F(T_n) = -\epsilon\sum_1^\infty\left(\frac{\partial_n F}{n\lambda^n}\right) + \frac{\epsilon^2}{2}\sum\left(\frac{F_{mn}}{mn}\right)\lambda^{-m-n} + O(\epsilon^3) \tag{26.19}$$

Thus $\Delta log\tau = (1/\epsilon^2)\Delta F$ has $O(1)$ terms $(1/2)\sum(F_{mn}/mn)\lambda^{-m-n}$ which correspond to the $O(1)$ terms in $log\psi$. Hence we have a natural way of writing $\tilde{S} = S^0 + \epsilon S^1$ with $S^0 = S$ and

$$S^1 = \frac{1}{2}\sum\left(\frac{F_{mn}}{nm}\right)\lambda^{-m-n};\ \tilde{S}_X \sim P + \frac{\epsilon}{2}\sum\left(\frac{F_{1mn}}{nm}\right)\lambda^{-m-n} \tag{26.20}$$

One could also envision $F = F^0 + \epsilon F^1$, etc. with e.g. $\hat{V} = P_X + 2PP^1 = -2F^1_{XX}$ but we restrict matters here to $F = F^0$ and $\hat{V} = 0$ (see below).

We will see that $dKdV_\epsilon$ has some attractive features whereas dKdV is untenable. Thus note first that the equation $\mathcal{F} = (1/2)\psi\bar{\psi} + (X/i\epsilon) \sim (1/2)exp[(2/\epsilon)\Re S] + (X/i\epsilon)$ has ϵ at various levels which is confusing. Moreover $|\psi|^2 = exp[(2/\epsilon)\Re S]$ should be bounded by 1 which suggests a dKP_ϵ format with $S \to \tilde{S} = S^0 + \epsilon S^1$ ($S^0 \sim S$), $\Re S^0 = 0$, and

$$|\psi|^2 = e^{2\Re S^1} = exp\left[\Re\sum\left(\frac{F^0_{mn}}{mn}\right)\lambda^{-m-n}\right] \tag{26.21}$$

For this to occur we need ($\bullet\bullet\bullet$) $\Re S^0 = \Re\sum_1^\infty T_n\lambda^n + \Re\sum_1^\infty S_{j+1}\lambda^{-j} = 0$ where one expects $S_{j+1} = -(\partial_j F^0/j)$ to be real. This suggests that it would be productive to think of KdV after all with $\lambda = ik$ imaginary, $T_{2n} = 0$, and $\partial_{2n}F^0 = 0$ as indicated below (so $S_{2n+1} = 0$ and only λ^{-j} terms occur in ($\bullet\bullet\bullet$) for j odd). Arguments against dKdV are indicated below but $dKdV_\epsilon$ seems appropriate as subsequently indicated. For $dKdV_\epsilon$ one establishes $F^0_{m,2n} = 0$ as in [7] (cf. below) so in (26.21) one only has terms ($\bullet\heartsuit\bullet$) $(F^0_{(2m+1)(2n+1)}/(2n+1)(2m+1)) \cdot \lambda^{-2(m+n)-2}$ which would be real for $\lambda = ik$. Thus S^0 and $P = S^0_X$ are imaginary while S^1 and $P^1 = \partial_X S^1$ are real. In [2] one digresses here to a general discussion of HJ theory for dKP and shows why dKP_ϵ (actually $dKdV_\epsilon$) is needed (we omit much of this). The main points are sketched. First the conditions under which the formulas of [11] are valid with $E = \pm\lambda^2$ real involve λ either real or pure imaginary. A little thought shows that a KdV situation here with $\lambda = ik$, $\lambda^2 = -k^2 = -E$ would seem to work and we try this here to see what a KdV situation (first without $dKdV_\epsilon$) will involve. We will have then P purely imaginary with U_j and P_j real and note that only odd powers of P or k appear in (26.18). Look now at $ik = P(1 + \sum_1^\infty U_m P^{-2m})$ and for $P = iQ$ we see that $(ik)_+^{2n+1} = \mathcal{B}_{2n+1}$ will be purely imaginary. Further $\partial_P \mathcal{B}_{2n+1}$ will involve only even powers of P and hence will be real. Thus write now

$$\mathcal{B}_{2n+1} = \sum_0^n b_{nj} P^{2j+1};\ \partial_P \mathcal{B}_{2n+1} = \sum_0^n (2j+1) b_{nj} P^{2j} \tag{26.22}$$

and we have (♣ • ♣) $(d/dT_n)\Im\mathcal{F} = -(1/\epsilon)\dot{X}_n = (1/\epsilon)\sum_0^n (2j+1) b_{nj} P^{2j}$. Then the condition $P = iQ$ leads to a compatible KdV situation (♣ • ♣) and further $\dot{P}_n = dP/dT_n = \partial \mathcal{B}_n = \sum_0^n \partial(b_{nj}) P^{2j+1}$ which is realistic (and imaginary). Now we note that there is danger here of a situation where $\Re P = 0$ implies $\Re S = 0$ which in turn would imply $|\psi|^2 = 1$ (going against the philosophy of keeping $|\psi|^2$ as a fundamental variable) and this is one reason we will need $dKdV_\epsilon$ with (26.21) - (•♡•), etc. Thus in general (••) $S = \sum_1^\infty T_n \lambda^n - \sum_1^\infty (\partial_n F/n)\lambda^{-n}$ and $P = \lambda - \sum_1^\infty (F_{1n}/n)\lambda^{-n}$; $\mathcal{B}_m = \lambda^n - \sum_1^\infty (F_{mn}/n)\lambda^{-n}$ and for KdV (with $\lambda = ik$) it follows from the residue formula (cf. [7]) that $F_{nm} = F_{mn} = Res_P\,(\lambda^m d\lambda_+^n)$ that $F_{m,2n} = 0$ and from a $\bar{\partial}$ analysis (cf. [6, 7]) $\partial_j F = (j/2i\pi)\int\int \zeta^{j-1}\bar{\partial}_\zeta S d\zeta \wedge d\bar{\zeta}$. The $\partial_j F$ and F_{1j} can be computed explicitly as in [7] and in particular $F_{1,2n} = 0$ with $(P^2 - U = -k^2)$ $F_{1,2n-1} = (-1)^n\,(U/2)^n \prod_1^n [(2j-1)/j]$. A further calculation along the same lines also shows that $F_{2n} = \partial_{2n} F = 0$ for KdV. Generally F will be real along with the F_{mn} and we recall that the expression for $\mathcal{B}_{2m+1}$ arising from (••) is an alternate way of writing (26.22). For $\lambda = ik$, P and $\mathcal{B}_{2m+1}$ will be purely imaginary but S could be complex via $\sum_1^\infty T_n \lambda^n$ since all powers $\lambda^n = (ik)^n$ will occur in (••). Thus $\Re S \neq 0$ and we have a perfectly respectable situation, provided the T_{2n} are real. However T_{2n} imaginary as in KP1 (cf. [2]), or as in (26.2), would imply $\Re S = 0$ and $|\psi|^2 = 1$ which is not desirable. Another problem is that if $\Re S \neq 0$ is achieved via the times then $|\psi|^2 \sim exp[(1/\epsilon)\sum T_{2n}\lambda^{2n}]$ will not necessarily be ≤ 1. This and other arguments rejecting dKP_ϵ lead one now to $dKdV_\epsilon$ as the natural framework (cf. [2] for details).

Now for $dKdV_\epsilon$, in view of (26.21) - (•••), etc., there is no problem with $\Re S^0 = 0$ while happily $\Re S^1 \neq 0$ and $|\psi|^2 \leq 1$ is realistic. The equation (26.18) applies now but we cannot write $ik \sim \tilde{P}(1 + q\tilde{P}^{-2})^{\frac{1}{2}}$. Indeed other terms will arise involving P_X for example since, for $\tilde{S} = S^0 + \epsilon S^1$ with $\psi = exp(\tilde{S}/\epsilon) = exp[(S^0/\epsilon) + S^1]$, we have $\tilde{P} = \partial\tilde{S} = S_X^0 + \epsilon S_X^1 = P + \epsilon P^1$ so that $\epsilon\partial\psi = \tilde{P}\psi = (P + \epsilon P^1)\psi$, $\epsilon^2\partial^2\psi = (\epsilon P_X + \epsilon^2 P_X^1)\psi + 2\epsilon P^1 P\psi + P^2\psi + \epsilon^2 (P^1)^2\psi$, etc. along with $\epsilon\partial(\psi/\tilde{P}) = -\epsilon(\tilde{P}_X/\tilde{P}^2)\psi + \psi = \psi - \epsilon((P_X/P^2)\psi + O(\epsilon^2)$ from which $(\epsilon\partial)\psi \to \psi$ or $(\epsilon\partial)^{-1}\psi \to \psi/\tilde{P}$ in some sense. Continuing such calculations we obtain terms of $O(\epsilon)$ in $(1/\psi)(\epsilon\partial)^{-n}\psi$ of the form $(1/\psi)\binom{n+1}{2}(\epsilon\partial)^{-n}\,(P_X\psi/P^2)$ Hence from $L\psi = \lambda\psi$ we get to first order

$$\lambda = P + \sum_1^\infty U_{n+1} P^{-n} + \epsilon\left(P^1 + \sum_1^\infty \hat{U}_{n+1} P^{-n} + \right.$$

$$+\sum_1^\infty \left\{\frac{n(n+1)}{2\psi}\right\}(\epsilon\partial)^{-n}\left[\frac{P_X\psi}{P^2}\right] - P^1\sum_1^\infty nU_{n+1}P^{-n-1}\Bigg) \tag{26.23}$$

Note here also that $P = ik - \sum_1^\infty P_n(ik)^{-n}$ inverts (26.18) with $P_n = 0$ for n even ($P_n = F_{1n}/n$ here - cf. [7] where there is an index shift in the P_n); this shows that $P = iQ$. Note here that the constraint $|\psi|^2\Im\tilde{P} = -1 \equiv |\psi|^2\Im P = -1$ and this can be written $exp(2S^1)\Im P = -1$. In any event this leads to expressions for ik, $(ik)_+^{2n+1}$, etc. and in particular for $P = iQ$ imaginary and S^1, P^1 real we obtain $2S^1 + log(\Im P) = i\pi \Rightarrow 2P^1 = -(\Im P_X/\Im P) \Rightarrow P_X = -2PP^1$. We do not pursue this however since in fact the HJ theory is not crucial here as far as $\tilde{P} = P + \epsilon P^1$ is concerned. Given $S = S^0 + \epsilon S^1$ and $F = F^0$ we know $\tilde{P} = P + \epsilon P^1$ is correct and that is all that is needed for the formulas of [11]. Further calculations suggest that one can obtain exact balances for the HJ theory (perhaps with constraints) but higher powers of ϵ should be included.

26.4 FORMULAS FOR FUNDAMENTAL VARIABLES

Thus we take $\lambda^2 = -E$ and specify $dKdV_\epsilon$. We can still label ψ as ψ_E but now one can imagine a $T_2 \sim \tau$ variable inserted e.g. via $\psi = \psi(X, T_{2n+1})exp(E\tau/i\hbar)$ $(n \geq 0)$ with $i\hbar\psi_\tau = E\psi$ and $\epsilon^2\psi'' - V\psi = -E\psi = \lambda^2\psi$ where $V = V(X, T_{2n+1})$ etc. Consider $\mathcal{F} = (1/2)\psi\bar{\psi} + (X/i\epsilon)$ with $\psi = exp[(1/\epsilon)S^0 + S^1]$, $\Re S^0 = 0$ as in $(\bullet\bullet\bullet)$, and $|\psi|^2 = exp(2\Re S^1)$ as in (26.21). Here

$$S^0 = i\left[\sum_1^\infty T_{2n+1}(-1)^n k^{2n+1} + \sum_1^\infty S_{2n}(-1)^n k^{-2n+1}\right] \tag{26.24}$$

and explicitly

$$\mathcal{F} = \frac{1}{2}exp\left[\sum_1^\infty\left(\frac{F^0_{(2m+1)(2n+1)}}{(2m+1)(2n+1)}\right)(-1)^{m+n+1}k^{-2(m+n+1)}\right] + \frac{X}{i\epsilon} \tag{26.25}$$

Thus the ϵ "problem" has been removed from the $|\psi|^2$ term but ϵ still occurs as a scale factor with X. Look now at (26.13) with P replaced by $\tilde{P}$ to obtain $|\psi|^2\Im\tilde{P} = -1$ which in view of the ϵ independence of $|\psi|^2$ suggests that $\Im P^1 = 0$ which in fact is true from $(\bullet\heartsuit\bullet)$. Thus $|\psi|^2\Im P = -1$ as before but $P = S^0_X$ now. Next for $\phi = \bar{\psi}/2\psi$ we have $\phi = (1/2)exp[-(2i/\epsilon)\Im S]$ and S^0 is imaginary as in (26.24) with S^1 real as indicated in $(\bullet\heartsuit\bullet)$. Consequently

$$\phi = \frac{1}{2}exp\left[-\frac{2i}{\epsilon}\Im S^0\right] = \tag{26.26}$$

$$= \frac{1}{2}exp\left[-\frac{2i}{\epsilon}\left(\sum_0^\infty(-1)^n T_{2n+1}k^{2n+1} + \sum_1^\infty(-1)^n S_{2n}k^{-2n+1}\right)\right]$$

One can also return to the discussion at the end of Section 4.1 and suggest again that $X = -\epsilon\Im\mathcal{F}$ and (for $P = iQ$)

$$P = i\Im\tilde{P} = i\Im P^0 = i\Im P = iQ^0 = -\frac{i}{|\psi|^2} = -\frac{i}{2\Re\mathcal{F}} \tag{26.27}$$

are fundamental variables. Note also from (26.24), $log(2\phi) = -(2/\epsilon)S^0$, so

$$log(2\phi) = -4i\xi = -\frac{2i}{\epsilon}\left(\sum(-1)^n T_{2n+1}k^{2n+1} + \sum(-1)^n S_{2n}k^{-2n+1}\right) \tag{26.28}$$

¿From $dX \wedge dP$ we obtain now as a possibly fundamental symplectic form

$$dX \wedge dP = -\epsilon d(\Im\mathcal{F}) \wedge \frac{i}{2(\Re\mathcal{F})^2}d(\Re\mathcal{F}) = -\frac{i\epsilon}{2(\Re\mathcal{F})^2}d(\Im\mathcal{F}) \wedge d(\Re\mathcal{F}) \tag{26.29}$$

which has a certain charm and seems intrinsically related to the duality idea based on $\mathcal{F}$ (note this not $dX \wedge d\tilde{P}$, which would involve an additional term $dX \wedge dP^1$, where a relation to $dX \wedge dP$ could be envisioned via $P^1 = -(1/2)\partial_X log\,P$ - hence (26.29) is based only on WKB ideas). The constraint $|\psi|^2\Im P = -1$ becomes $exp[2\Re S^1]\Im S^0_X = -1$ which can be written out in terms of $F^0 = F$ and $\partial_X S_{2n}$ (cf. [2]). Let us also compute the form $\omega = \delta\xi \wedge \delta\chi$ from (26.10) in one of its many forms. First recall S^0 is imaginary and S^1 is real with $log(2\phi) = -(2/\epsilon)S^0 = -4i\xi$ and $\chi = |\psi|^2 = exp(2S^1)$. Therefore formally, via $\xi = -(i/2\epsilon)S^0$, we have $\omega = \delta\xi \wedge \delta\chi = -(i\chi/\epsilon)\,\delta S^0 \wedge \delta S^1$. The difference here from (26.29) for example is that the term $X = -\epsilon\mathcal{F}$ has no relation to S^0 or S^1 a priori.

26.5 QUANTUM AND CLASSICAL MECHANICS

We go now to a fascinating series of papers by Olave (cf. [17]) which develop quantum mechanics (QM) via the density matrix and classical structures. A synopsis of sections in [17] is given in [2] but here we only mention a few key ideas. There are a number of possible connections with the theory of [2, 11] which we hope to explore further. One starts from three axioms: **(A)** Newtonian mechanics is valid for all particles which constitute the systems in the "ensemble". **(B)** For an isolated system the joint probability density function is conserved, i.e. $(d/dt)\mathcal{P}(x,p,t) = 0$. **(C)** The Wigner-Moyal infinitesimal transformation defined by (5.1) below is adequate for the description of any non-relativistic quantum system (note the emphasis here on the infinitesimal aspect).

$$\rho\left(x + \frac{\delta x}{2}, x - \frac{\delta x}{2}, t\right) = \int \mathcal{P}(x,p,t)exp\left(i\frac{p\delta x}{\hbar}\right)dp \tag{26.30}$$

(we use small x, p here to follow [17] and for comparison with [2, 11] or the rest of this paper one should convert this to large X, P). Using these axioms one produces a believable version of nonrelativistic quantum mechanics. First from (**B**) one obtains (•◇•) $-(\partial\rho/\partial t)+(i\hbar/m)(\partial^2\rho/\partial x\partial(\delta x))-(i/\hbar)\delta V\rho=0$ leading to

$$\left\{\frac{\hbar^2}{2m}\left[\frac{\partial^2}{\partial y^2}-\frac{\partial^2}{\partial(y')^2}\right]-[V(y)-V(y')]\right\}\rho(y,y',t)=-i\hbar\frac{\partial}{\partial t}\rho(y,y',t) \tag{26.31}$$

which is called Schrödinger's first equation for the density function. Now one assumes

$$(A)\ \rho(y,y',t)=\psi^*(y',t)\psi(y,t)\quad (B)\ \psi(y,t)=R(y,t)exp\left(\frac{iS(y,t)}{\hbar}\right) \tag{26.32}$$

where ψ will be called a probability amplitude (perhaps not the best terminology). Next one expands ρ in (26.32A) in terms of δx to get

$$(A)\ \frac{\partial\chi}{\partial t}+\frac{\partial}{\partial x}\left(\chi\frac{S_x}{m}\right)=0\quad (B)\ \frac{\partial}{\partial x}\left[\frac{1}{2m}\left(\frac{\partial S}{\partial x}\right)^2+V+\frac{\partial S}{\partial t}-\frac{\hbar^2}{2mR}\frac{\partial^2 R}{\partial x^2}\right]=0 \tag{26.33}$$

where $\chi=R^2=lim\,\rho[x+(\delta x/2),x-(\delta x/2)]$ as $\delta x\to 0$. Then (26.33B) is equivalent to (•♣•) $(\hbar^2/2m(\partial^2\psi/\partial x^2)-V(x)\psi=-i\hbar(\partial\psi/\partial t)$ which will be called Schrödinger's second equation. In summary: If one can write (26.32A) then ψ given by (26.32B) satisfies (•♣•) along with the equation of continuity (26.33A). One then defines operators (•♠•) $\hat{p}'=-i\hbar\partial/\partial(\delta x)$; $\hat{x}'=x$ where the apostrophe indicates operators acting on the density function and without the apostrophe operators act on the "probability amplitude".

In [17] one argues by showing that the result of the operation upon the density function ρ of the momentum and position operators defined by (•♠•) represents, respectively, the mean values for momentum and position for the ensemble components. The mean values are calculated within the limit $\delta x\to 0$ so that calculation of the density function for infinitesimally close points can be done without loss of generality. This does not imply that only the element for which $\delta x=0$ contributes. The kinematic evolution of the density function is governed by (•◇•) which mixes all contributions. The limits for determining χ or mean values must be taken after (•◇•) is solved. Then Schrödinger's second equation has the operator form $\hat{H}\psi=i\hbar(\partial/\partial t)\psi$. Note that one has now $[\hat{x},\hat{p}]=i\hbar$ and it can be shown that in fact $\Delta x\Delta p\geq\hbar/2$ (Heisenberg uncertainty). However one has $[\hat{x}',\hat{p}']=0$ so $\Delta x\Delta p\geq 0$ in this context. This is natural since no hypotheses about the mean square deviations associated to the classical function $\mathcal{P}$ were made. The uncertainty principle results from writing the density function as the product (26.32A). Thus instead of representing a fundamental property of nature the uncertainty principle simply represents a limitation of the description based on Schrödinger's second equation (•♣•). Consequently quantum mechanics as developed above is only applicable to problems where the density function can be decomposed as in (26.32A). Further one remarks that the dispersion relations do not impose any constraint upon the behavior of nature but only upon our capacity to describe nature by means of quantum theory. If e.g. $\Delta q\Delta p<\hbar/2$ in some situation then quantum theory does not apply or will not give good results.

We note that for a stationary problem $\partial_t\chi = 0$ the condition (26.33A) becomes $\chi P = c$ which is compatible with (♣) where $\chi P = -1$. The equation (•♣•) on the other hand corresponds to $(P^2/2m) + V - (\hbar^2/2m)(\sqrt{\chi}''/\sqrt{\chi}) = 0$ which in turn corresponds to $\dot{\xi} = 0$ for ξ as in (♠). In fact returning to (26.33B) or (•♣•) one can consider this as a Hamilton Jacobi (HJ) equation for one particle subject to an effective potential $V_{eff} = V(x) - (\hbar^2/2mR)(\partial^2 R/\partial x^2)$ and this can be related to the HJ theory sketched earlier for dKP and/or dKdV.

The Wigner-Moyal transformation is developed and operator ordering problems are discussed at length in [17]. One obtains an unambiguous procedure giving the same result as Weyl ordering. Further, thermodynamic behavior is integrated into the picture with an extensive discussion and in the direction of using this framework we suggest here a formula

$$\tilde{\mathcal{F}} = \frac{1}{2}\psi\left(x + \frac{\delta x}{2}\right)\bar{\psi}\left(x - \frac{\delta x}{2}\right) + \frac{X}{i\epsilon} = \frac{1}{2}\rho\left(x + \frac{\delta x}{2}, x - \frac{\delta x}{2}\right) + \frac{X}{i\epsilon} \tag{26.34}$$

where all x variables here should be capitalized and $t \sim T_n$ $(n \geq 2)$. We recall that in [11] the equation $|\psi|^2 = 2\mathcal{F} - (2X/i\epsilon)$ is interpreted as describing the space variable as a macroscopic thermodynamic quantity with the microscopic information encoded in the prepotential. Then QM can be reformulated in terms of (♠ • ♠) with the Schrödinger equation replaced by the third order equation (26.6). Here $\hbar$ can be considered as the scale of the statistical system (cf. [1, 11]). These comments from [11] seem completely adaptable to a connection such as (26.34) with the theory of [17]. Now suppose we define $\tilde{\mathcal{F}}$ as in (26.34) and use (26.32B) so that $\chi = R^2 = lim \rho[x + (\delta x/2), x - (\delta x/2)]$ as $\delta x \to 0$, leading to (•♣•). Then $\tilde{\mathcal{F}} \to \mathcal{F} = (1/2)\chi + (X/i\epsilon)$ with the mixing equation (•◇•) in the background. Further if the t dependence is restricted to $t = t_2$ of the form $exp(-iEt/\hbar)$ for suitable t then the Schrödinger equation (•♣•) has the form (26.1). Some thermodynamic analogies are suggested in [2] and further developments in connection with the renormalization group as in [1] are anticipated.

References

[1] G. Bonelli and M. Matone, hep-th 9602174, Phys. Rev. Lett., 76 (1996), 4107-4110

[2] R. Carroll, hep-th 9607219, 9610216, 9702138, 9705229; Nucl. Phys. B, 502 (1997), to appear

[3] R. Carroll, Topics in soliton theory, North-Holland, 1991

[4] R. Carroll, Applicable Analysis, 49 (1993), 1-31; 56 (1995), 147-164

[5] R. Carroll, solv-int 9606005, Proc. Second World Congress Nonlinear Analysts, North-Holland, to appear

[6] R. Carroll, Jour. Nonlin. Sci., 4 (1994), 519-544; Teor. Mat. Fizika, 99 (1994), 220-225

[7] R. Carroll and Y. Kodama, Jour. Phys. A, 28 (1995), 6373-6387

[8] R. Carroll, Proc. NEEDS'94, World Scientific, 1995, pp. 24-33

[9] R. Carroll and J. Chang, solv-int 9612010, Applicable Anal., 64 (1997), 343-378

[10] R. Donagi and E. Witten, Nucl. Phys. B, 460 (1996), 299-334

[11] A. Faraggi and M. Matone, hep-th 9606063, Phys. Rev. Lett., 78 (1997), 163-166

[12] H. Itoyama and A. Morozov, hep-th 9511126, 9512161, 9601168

[13] Y. Kodama and J. Gibbons, Fourth Workshop on Nonlinear and Turbulent Processes in Physics, World Scientific, 1990, pp. 166-180

[14] I. Krichever and D. Phong, hep-th 9604199

[15] M. Matone, Phys. Lett. B, 357 (1995), 342-348

[16] G. Marmo and G. Vilasi, hep-th 9605191

[17] L. Olave, quant-ph 9503020, 9503021, 9503022, 9503024, 9503025, 9509012, 9509013, 9511028, 9511039, 9601002, 9607002, 9607003, 9609003, 9609023

[18] K. Takasaki and T. Takebe, Inter. Jour. Mod. Phys. A, Supp. 1992, pp. 889-922; Rev. Math. Phys., 7 (1995), 743-808

GPSR Compliance

The European Union's (EU) General Product Safety Regulation (GPSR) is a set of rules that requires consumer products to be safe and our obligations to ensure this.

If you have any concerns about our products, you can contact us on ProductSafety@springernature.com

In case Publisher is established outside the EU, the EU authorized representative is:

Springer Nature Customer Service Center GmbH
Europaplatz 3
69115 Heidelberg, Germany

Batch number: 10755284

Printed by Printforce, the Netherlands